Clinical Biochemistry

Clinical Biochemistry

Lajja Das

RANDOM

RANDOM PUBLICATIONS

NEW DELHI (INDIA)

Clinical Biochemistry

ISBN 978-93-51113-86-7

Published in 2014 in India by

RANDOM PUBLICATIONS

4376-A/4B, Gali Murari Lal, Ansari Road
New Delhi-110 002
Phone : +91-11-43580356, +91-11-23289044
e-mail: randomexports@gmail.com, sales@randompublications.com, info@randompublications.com

Reprint 2021

Type Setting by : Keystoneprintads, Delhi-110051
Digitally Printed at : Replika Press Pvt. Ltd.

Preface

Clinical biochemistry is the area of clinical pathology that is generally concerned with analysis of bodily fluids. Clinical biochemistry is an analytical and interpretative science. The analytical part involves the determination of the level of chemical components in body fluids and tissues.

Clinical Biochemistry is one of the most rapidly advancing areas of laboratory and clinical medicine. The marked increase in the number and availability of laboratory diagnostic procedures has helped in the solution of clinical problems. Individual laboratory tests are rarely ordered and reported singly; usually combinations of lab tests are used. The physician should however be judicious in selecting the tests that really give a clue to the diagnosis of a disease. Some biochemical investigations however are done routinely for most patients e.g. qualitative tests with urine.

Clinical Biochemistry tests comprise over one third of all hospital laboratory investigation. clinical biochemistry is that branch of laboratory medicine in which chemical and biochemical methods are applied to the study of disease while in theory this embraces all non-morphological studies, in practice it is usually, though not exclusively, confined to studies on blood and urine because of the relative ease in obtaining such specimens although analysis are made on other body fluids such as gastric aspirate and cerebrospinal fluid.

Biochemistry analysis is one of the most important areas within clinical laboratory analysis. The term biochemistry usually refers to determine the concentration or activity of a protein, carbohydrate, lipid, electrolyte, enzyme, or small molecule in easily-collected bodily fluids such as blood, serum, plasma, or urine. However, it is not necessarily limited to these determinations. The analysis of virtually any biologically-active substance can be generalised and defined as biochemistry.

Automated analyzers usually include the mechanized versions of basic manual laboratory techniques and procedures, and several ways have been developed for automating them. When initially introduced, automation mimicked manual test procedures and was applied to those tests requested most often. All the individual steps in the procedure are duplicated. Analytical methods, which are quicker and with fewer steps as well as modification of existing protocols are being developed as the manufactures have integrated computer hardware and software into analyzers to provide automatic process control and data processing capabilities.

Activity is a measure of the rate at which a process takes place. Enzymatic activity is usually estimated by measuring the rate at which a substrate is converted to a product. This activity is affected by many things e.g. temperature, time over which the activity is measured, incubation conditions etc. All of these must be defined when reporting the activity and so units or activity include fixed values for each variant. As the combination of the variations as large so many different such units are in use. International standardization of all these variables is almost impossible and as such the international unit of enzyme activity has not been widely accepted the units for each such test in the laboratory should be clear to all concerned.

This book is essential reading for all students studying clinical biochemistry, especially those on biomedical science students.

I thank all members of my team who have helped in the preparation of the book. My special thanks go to "Random Publication" who have published the book.

– Lajja Das

Contents

1

Introduction

Clinical chemistry (also known as chemical pathology, clinical biochemistry or medical biochemistry) is the area of clinical pathology that is generally concerned with analysis of bodily fluids. The discipline originated in the late 19th century with the use of simple chemical tests for various components of blood and urine. Subsequent to this, other techniques were applied including the use and measurement of enzyme activities, spectrophotometry, electrophoresis, and immunoassay. Most current laboratories are now highly automated to accommodate the high workload typical of a hospital laboratory. Tests performed are closely monitored and quality controlled.

All biochemical tests come under chemical pathology. These are performed on any kind of body fluid, but mostly on serum or plasma. Serum is the yellow watery part of blood that is left after blood has been allowed to clot and all blood cells have been removed. This is most easily done by centrifugation, which packs the denser blood cells and platelets to the bottom of the centrifuge tube, leaving the liquid serum fraction resting above the packed cells. This initial step before analysis has recently been included in instruments that operate on the "integrated system" principle. Plasma is in essence the same as serum, but is obtained by centrifuging the blood *without* clotting. Plasma is obtained by centrifugation *before* clotting occurs. The type of test required dictates what type of sample is used. A large medical laboratory will accept samples for up to about 700 different kinds of tests. Even the largest of laboratories rarely do all these tests themselves, and some must be referred to other labs.

This large array of tests can be further sub-categorised into sub-specialities of:

- General or routine chemistry - commonly ordered blood chemistries (e.g., liver and kidney function tests).
- Special chemistry - elaborate techniques such as electrophoresis, and manual testing methods.
- Clinical endocrinology - the study of hormones, and diagnosis of endocrine disorders.

- Toxicology - the study of drugs of abuse and other chemicals.
- Therapeutic Drug Monitoring - measurement of therapeutic medications blood levels to optimize dosage.
- Urinalysis - chemical analysis of urine for a wide array of diseases, along with other fluids such as CSF and effusions
- Fecal analysis - mostly for detection of gastrointestinal disorders.

Clinical Biochemistry is one of the most rapidly advancing areas of laboratory and clinical medicine. The marked increase in the number and availability of laboratory diagnostic procedures has helped in the solution of clinical problems. Individual laboratory tests are rarely ordered and reported singly; usually combinations of lab tests are used. The physician should however be judicious in selecting the tests that really give a clue to the diagnosis of a disease. Some biochemical investigations however are done routinely for most patients e.g. qualitative tests with urine.

A trend is also emerging to conduct certain biochemical investigations, which could reveal predisposition to specific disease processes in healthy individuals. The physician can then suggest preventive measures to the person e.g. elevated levels of plasma cholesterol persisting for a long time contribute to the development of coronary artery disease and increased blood pressure. These individuals should be advised to avoid cholesterol-containing food so that the development of a serious disease could be retarded. However, one must be careful in interpreting these investigations and bear in mind the concept of normal variation from individual to individual.

Metabolic changes associated with specific disorders may give rise to a change in the biochemical profile of a particular body fluid, e.g. blood glucose in diabetes mellitus; glucose levels in the cerebrospinal fluid in bacterial meningitis (which are greatly reduced). Hence, specific parameters are looked for in a specific body fluid when a disease is suspected. From a clinical point of view, one purpose of performing a test could be corroborating a particular diagnosis or ruling it out. Other tests may be done to assess the severity of a disease process or monitor its progress. Still others may evaluate or monitor the effectiveness or potential side effects of a particular therapeutic regimen; and certain tests can give a clue about the prognosis or probable outcome of a disease.

TYPICAL ROLES

A clinical chemistry department within a hospital provides a link between front line clinical staff and the basic sciences employing analytical and interpretative skills to aid the clinician in the prevention, diagnosis and treatment of disease. Diseases such as heart attacks, kidney failure, viral and bacterial infection, infertility, diabetics, high cholesterol, thyroid problems or measuring drug levels to make sure people are on the best dose are some of the many areas where the Clinical Chemistry Laboratory becomes involved in a persons. Some regional laboratories are involved in screening services such as phenylketonuria and cystic fibrosis in newborn babies, genetic screening and screening for illegal drug use.

TESTS

Common clinical chemistry tests include:

Electrolytes

- Sodium
- Potassium
- Chloride
- Bicarbonate

Renal (Kidney) Function Tests

- Creatinine
- Blood urea nitrogen

Liver Function Tests

- Total protein (serum)
- Albumin
- Globulins
- A/G ratio (albumin-globulin)
- Protein electrophoresis
- Urine protein
- Bilirubin; direct; indirect; total
- Aspartate transaminase (AST)
- Alanine transaminase (ALT)
- Gamma-glutamyl transpeptidase (GGT)
- Alkaline phosphatase (ALP)

Cardiac Markers

- Troponin
- Myoglobin
- CK-MB
- B-type natriuretic peptide (BNP)

Minerals

- Calcium
- Magnesium
- Phosphate
- Potassium

Blood Disorders

- Iron
- Transferrin

- TIBC
- Vitamin B12
- Folic acid

Miscellaneous

- Glucose
- C-reactive protein
- Glycated hemoglobin (HbA1c)
- Uric acid
- Arterial blood gases ($[H^+]$, P_{CO2}, P_{O2})
- Adrenocorticotropic hormone (ACTH)
- Toxicological screening and forensic toxicology (drugs and toxins)
- Neuron-specific enolase (NSE)
- fecal occult blood test (FOBT)

PANEL TESTS

A set of commonly ordered tests are combined into a panel:

- Basic metabolic panel (BMP) - 8 tests - sodium, potassium, chloride, bicarbonate, blood urea nitrogen (BUN), creatinine, glucose, calcium
- Comprehensive metabolic panel (CMP) - 14 tests - above BMP plus total protein, albumin, alkaline phosphatase (ALP), alanine amino transferase (ALT), aspartate amino transferase (AST), bilirubin.

MANUAL VS AUTOMATION IN CLINICAL LABORATORY

Automation in clinical laboratory is a process by which analytical instruments perform many tests with the least involvement of an analyst. The International Union of Pure and Applied Chemistry (IUPAC) define automation as "the replacement of human manipulative effort and facilities in the performance of a given process by mechanical and instrumental devices that are regulated by feedback of information so that an apparatus is self-monitoring or self-adjusting". Presently no currently available clinical instrument fully meets this definition, however the term 'automation' is applicable to the individual steps in many analytical processes and modern instrumentation is improvising with more and more intelligence built into new generations of laboratory analyzers to soon come up to the IUPAC definition.

Automated instruments enable laboratories to process a much larger workload without a relative increase in manpower. Automation in clinical laboratories has evolved from fixed automation whereby an instrument performs a repetitive task by itself, and has progressed to programmable automation, which permits it to perform a variety of different tasks. Intelligent automation has recently been introduced into a few individual instruments or systems to enable them to self-monitor and respond appropriately to changing conditions. Instead of resorting to manual means automation leads to reduction in variability of results and error of analysis by doing away with jobs that are repetitive and monotonous for an individual and that can lead to boredom or casual attitude. However, the improved reproducibility attained by automation is not necessarily associated with improved accuracy of test results since accuracy is mainly influenced by the analytical methods used. The significant improvement in quality of laboratory tests in recent years is due the combination of well-designed automated instrumentation with good analytical methods and effective quality assurance programs. Automation may initially incur high costs for procurement of the equipments but is economical in the long run due to the reduction in the manpower required to perform the tasks.

Automated analyzers usually include the mechanized versions of basic manual laboratory techniques and procedures, and several ways have been developed for automating them. When initially introduced, automation mimicked manual test procedures and was applied to those tests requested most often. All the individual steps in the procedure are duplicated. Analytical methods, which are quicker and with fewer steps as well as modification of existing protocols are being developed as the manufactures have integrated computer hardware and software into analyzers to provide automatic process control and data processing capabilities.

2

Clinical Significances of Sphingolipids

One of the most clinically important classes of sphingolipids are those that confer antigenic determinants on the surfaces of cells, particularly the erythrocytes. The ABO blood group antigens are the carbohydrate moieties of glycolipids on the surface of cells as well as the carbohydrate portion of serum glycoproteins.

When present on the surface of cells the ABO carbohydrates are linked to sphingolipid and are therefore of the glycosphingolipid class. When the ABO carbohydrates are associated with protein in the form of glycoproteins they are found in the serum and are referred to as the secreted forms. Some individuals produce the glycoprotein forms of the ABO antigens while others do not. This property distinguishes secretors from non-secretors, a property that has forensic importance such as in cases of rape.

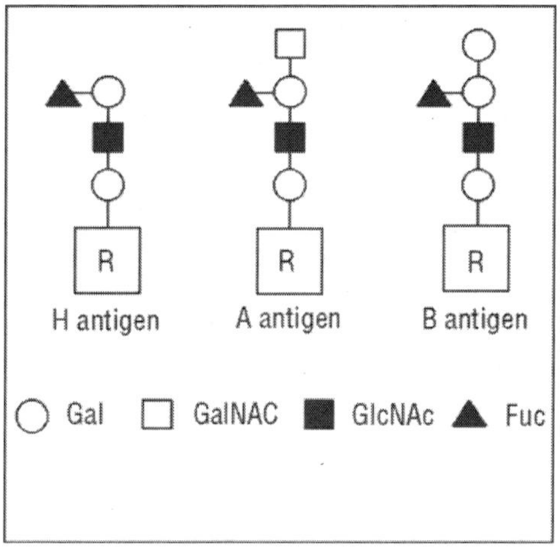

Fig. Strucures of the ABO Blood Group Carbohydrates

A significant cause of death in premature infants and, on occasion, in full term infants is respiratory distress syndrome (RDS) or hyaline membrane disease. This condition is caused by an insufficient amount of pulmonary surfactant. Under normal conditions the surfactant is synthesized by type II endothelial cells and is secreted into the alveolar spaces to prevent atelectasis following expiration during breathing. Surfactant is comprised primarily of dipalmitoyllecithin; additional lipid components include phosphatidylglycerol and phosphatidylinositol along with proteins of 18 and 36 kDa (termed surfactant proteins).

During the third trimester the fetal lung synthesizes primarily sphingomyelin, and type II endothelial cells convert the majority of their stored glycogen to fatty acids and then to dipalmitoyllecithin. Fetal lung maturity can be determined by measuring the ratio of lecithin to sphingomyelin (L/S ratio) in the amniotic fluid. An L/S ratio less than 2.0 indicates a potential risk of RDS. The risk is nearly 75-80% when the L/S ratio is 1.5. The carbohydrate portion of the ganglioside, GM_1, present on the surface of intestinal epithelial cells, is the site of attachment of cholera toxin, the protein secreted by *Vibrio cholerae*. These are just a few examples of how sphingolipids and glycosphingolipids are involved in various recognition functions at the surface of cells. As with the complex glycoproteins, an understanding of all of the functions of the glycolipids is far from complete.

DISORDERS ASSOCIATED WITH ABNORMAL SPHINGOLIPID METABOLISM

Disorder	Enzyme Deficiency	Accumulating Substance	Symptoms
Tay-Sachs disease	Hexosaminidase A	GM_2 ganglioside	mental retardation, blindness, early mortality
Gaucher's disease	Glucocerebrosidase	Glucocerebroside	hepatosplenomegaly, mental retardation in infantile form, long bone degeneration
Fabry's disease	Galactosidase A	Globotriaosylceramide; also called ceramide trihexoside (CTH)	kidney failure, skin rashes
Niemann-Pick disease, more info below Types A and B Type C1 Type C2 Type D	Sphingomyelinase see info below see info below	Sphingomyelin LDL-derived cholesterol LDL-derived cholesterol	all types lead to mental retardation, hepatosplenomegaly, early fatality potential
Krabbe's disease; globoid leukodystrophy	Galactocerebrosidase	Galactocerebroside	mental retardation, myelin deficiency
Sandhoff-Jatzkewitz disease	Hexosaminidase A and B	Globoside, GM_2 ganglioside	same symptoms as Tay-Sachs, progresses more rapidly
GM_1 gangliosidosis	GM_1 ganglioside:-galactosidase	GM_1 ganglioside	mental retardation, skeletal abnormalities, hepatomegaly
Sulfatide lipodosis; metachromatic leukodystrophy	Arylsulfatase A	Sulfatide	mental retardation, metachromasia of nerves
Fucosidosis	L-Fucosidase	Pentahexosylfucoglycolipid	cerebral degeneration, thickened skin, muscle spasticity
Farber's lipogranulomatosis	Acid ceramidase	Ceramide	hepatosplenomegaly, painful swollen joints

Based upon genetic linkage analyses as well as enzyme studies and the characterization of accumulating lysosomal substances, Niemann Pick disease should be divided into type I and type II; type I has 2 subtypes, A and B (257200), which show deficiency of *acid sphingomyelinase*. Niemann Pick disease type II likewise has 2 subtypes, type C1 and C2 (NPC) and type D (NPD). It is obviously confusing to use the abbreviation NPD for Niemann Pick disease in some cases and for subtype D of Niemann Pick disease in other cases.

Recent studies identified the gene for NPC1. This gene contains regions of homology to mediators of cholesterol homeostasis suggesting why LDL-cholesterol accumulates in lysosomes of afflicted individuals. The encoded protein product of *NPC1* gene is 1278 amino acids long. Within the protein are regions of homology to the transmembrane domain of the morphogen receptor PATCHED (of *Drosophila melanogaster*), the putative sterol-sensing regions of SREBP (sterol regulatory element binding protein) cleavage-activating protein (SCAP) and *HMG-CoA reductase*.

DISORDERS OF LIPID METABOLISM

Fats (lipids) are an important source of energy for the body. The body's store of fat is constantly broken down and reassembled to balance the body's energy needs with the food available. Groups of specific enzymes help the body break down and process fats. Certain abnormalities in these enzymes can lead to the buildup of specific fatty substances that normally would have been broken down by the enzymes. Over time, accumulations of these substances can be harmful to many organs of the body. Disorders caused by the accumulation of lipids are called lipidoses. Other enzyme abnormalities prevent the body from converting fats into energy normally. These abnormalities are called fatty acid oxidation disorders.

GAUCHER'S DISEASE

Gaucher's disease is caused by a buildup of glucocerebrosides in tissues. Children who have the infantile form usually die within a year, but children and adults who develop the disease later in life may survive for many years.

In Gaucher's disease, glucocerebrosides, which are a product of fat metabolism, accumulate in tissues. Gaucher's disease is the most common lipidosis. The disease is most common among Ashkenazi (Eastern European) Jews. Gaucher's disease leads to an enlarged liver and spleen and a brownish pigmentation of the skin. Accumulations of glucocerebrosides in the eyes cause yellow spots called pingueculae to appear. Accumulations in the bone marrow can cause pain and destroy bone.

RARE HEREDITARY DISORDERS OF LIPID METABOLISM

Wolman's disease results when specific types of cholesterol and glycerides accumulate in tissues. This disease causes enlargement of the spleen and liver.

Calcium deposits in the adrenal glands cause them to harden, and fatty diarrhea (steatorrhea) also occurs. Infants with Wolman's disease usually die by 6 months of age. Cerebrotendinous xanthomatosis occurs when cholestanol, a product of cholesterol metabolism, accumulates in tissues. This disease eventually leads to uncoordinated movements, dementia, cataracts, and fatty growths (xanthomas) on tendons. The disabling symptoms often appear after age 30. If started early, the drug chenodiol helps prevent progression of the disease, but it cannot undo any damage already done.

In sitosterolemia, fats from fruits and vegetables accumulate in blood and tissues. The buildup of fats leads to atherosclerosis, abnormal red blood cells, and xanthomas on tendons. Treatment consists of reducing the intake of foods that are rich in plant fats, such as vegetable oils, and takingcholestyramine resin. In Refsum's disease, phytanic acid, which is a product of fat metabolism, accumulates in tissues. A buildup of phytanic acid leads to nerve and retinal damage, spastic movements, and changes in the bone and skin. Treatment involves avoiding eating green fruits and vegetables that contain chlorophyll. Plasmapheresis, in which phytanic acid is removed from the blood, may be helpful.

- Type 1, the chronic form of Gaucher's disease, is the most common. It results in an enlarged liver and spleen and bone abnormalities. Most commonly diagnosed during adulthood, type 1 Gaucher's disease may lead to severe liver disease, including increased risk of bleeding from the stomach and esophagus and liver cancer. Nervous system problems can also occur.
- Type 2, the infantile form, usually causes death in the first year of life. Affected infants have an enlarged spleen and severe nervous system problems.
- Type 3, the juvenile form, can begin at any time during childhood. Children with type 3 disease have an enlarged liver and spleen, bone abnormalities, and slowly progressive nervous system problems. Children who survive to adolescence may live for many years.

Many people with Gaucher's disease can be treated with enzyme replacement therapy, in which enzymes are given by vein, usually every 2 weeks. Enzyme replacement therapy is most effective for people who do not have nervous system complications.

TAY-SACHS DISEASE

Tay-Sachs disease is caused by a buildup of gangliosides in the tissues. This disease results in early death. In Tay-Sachs disease, gangliosides, which are products of fat metabolism, accumulate in tissues. The disease is most common among families of Eastern European Jewish origin. At a very early age, children with this disease become progressively intellectually disabled and appear to have floppy muscle tone. Spasticity develops and is followed by paralysis, dementia, and blindness. These children usually die by age 3 or 4. The disease cannot be treated or cured. Before conception, parents can find out whether they carry the gene that causes the disease. During pregnancy, Tay-Sachs disease can be identified in the fetus by chorionic villus sampling or amniocentesis.

NIEMANN-PICK DISEASE

Niemann-Pick disease is caused by a buildup of sphingomyelin or cholesterol in the tissues. This disease causes many neurologic problems. In Niemann-Pick disease, the deficiency of a specific enzyme results in the accumulation of sphingomyelin (a product of fat metabolism) or cholesterol. Niemann-Pick disease has several forms, depending on the severity of the enzyme deficiency, which determines how much sphingomyelin or cholesterol accumulates. The most severe forms tend to occur in Jewish people. The milder forms occur in all ethnic groups.

In the most severe form (type A), children fail to grow normally and have several neurologic problems. These children usually die by age 3. Children with type B disease develop fatty growths in the skin, areas of dark pigmentation, and an enlarged liver, spleen, and lymph nodes. They may be intellectually disabled. Children with type C disease develop symptoms during childhood, with seizures and neurologic deterioration. Some forms of Niemann-Pick disease can be diagnosed in the fetus by chorionic villus sampling or amniocentesis. After birth, the diagnosis can be made by a liver biopsy (removal of a tissue specimen for examination under a microscope). None of the types of Niemann-Pick disease can be cured, and children tend to die of infection or progressive dysfunction of the central nervous system. Currently, some therapies that may slow or halt the progression of symptoms in types B and C are being studied.

FABRY'S DISEASE

Fabry's disease is caused by a buildup of glycolipid in tissues. This disease causes skin growths, pain in the extremities, poor vision, recurrent episodes of fever, and kidney or heart failure. In Fabry's disease, glycolipid, which is a product of fat metabolism, accumulates in tissues. Because the defective gene for this rare disorder is carried on the X chromosome, the full-blown disease occurs only in males (Genetics: X-Linked Inheritance). The accumulation of glycolipid causes noncancerous (benign) skin growths (angiokeratomas) to form on the lower part of the trunk. The corneas become cloudy, resulting in poor vision. A burning pain may develop in the arms and legs, and children may have episodes of fever. Children with Fabry's disease eventually develop kidney failure and heart disease, although most often, they live into adulthood.

Kidney failure may lead to high blood pressure, which may result in stroke. Fabry's disease can be diagnosed in the fetus by chorionic villus sampling or amniocentesis. The disease cannot be cured or even treated directly, but researchers are investigating a treatment in which the deficient enzyme is replaced by transfusion. Treatment consists of taking analgesics to help relieve pain and fever or anticonvulsants. People with kidney failure may need a kidney transplant.

FATTY ACID OXIDATION DISORDERS

Fatty acid oxidation disorders are caused by a lack or deficiency of the enzymes needed to break down fats, resulting in delayed mental and physical development. Several enzymes help break down fats so that they may be turned into energy. An inherited defect or deficiency of one of these enzymes leaves the body short of energy and allows breakdown products, such as acyl-CoA, to accumulate. The enzyme most commonly deficient is medium chain acyl-CoA dehydrogenase (MCAD). Other enzyme deficiencies include short chain acyl-CoA-dehydrogenase deficiency (SCAD), long chain-3-hydroxyacyl-CoA-deficiency (LCHAD), and trifunctional protein deficiency (TFP). MCAD Deficiency: This disorder is one of the most common inherited disorders of metabolism, particularly among people of Northern European descent.

Symptoms usually develop between birth and age 3. Children are most likely to develop symptoms if they go without food for a period of time (which depletes other sources of energy) or have an increased need for calories because of exercise or illness. The level of sugar in the blood drops significantly, causing confusion or coma. Children become weak and may have vomiting or seizures. Over the long term, children have delayed mental and physical development, an enlarged liver, heart muscle weakness, and an irregular heartbeat. Sudden death may occur.

METABOLISM OF THE EICOSANOIDS

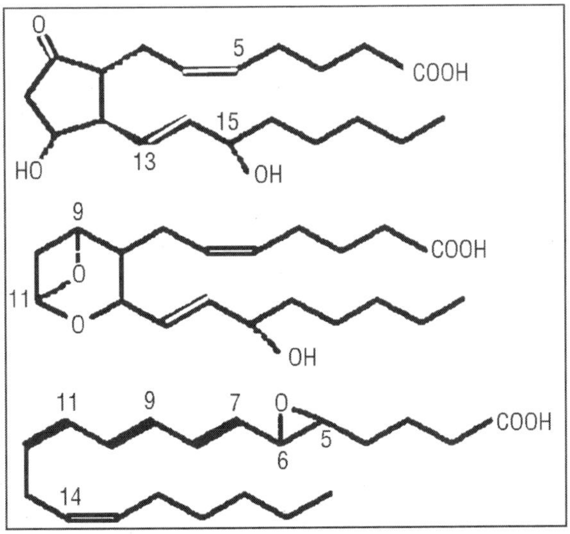

The eicosanoids consist of the prostaglandins (PGs), thromboxanes (TXs) and leukotrienes (LTs). The PGs and TXs are collectively identified as prostanoids. Prostaglandins were originally shown to be synthesized in the prostate gland, thromboxanes from platelets (thrombocytes) and leukotrienes from leukocytes, hence the derivation of their names.

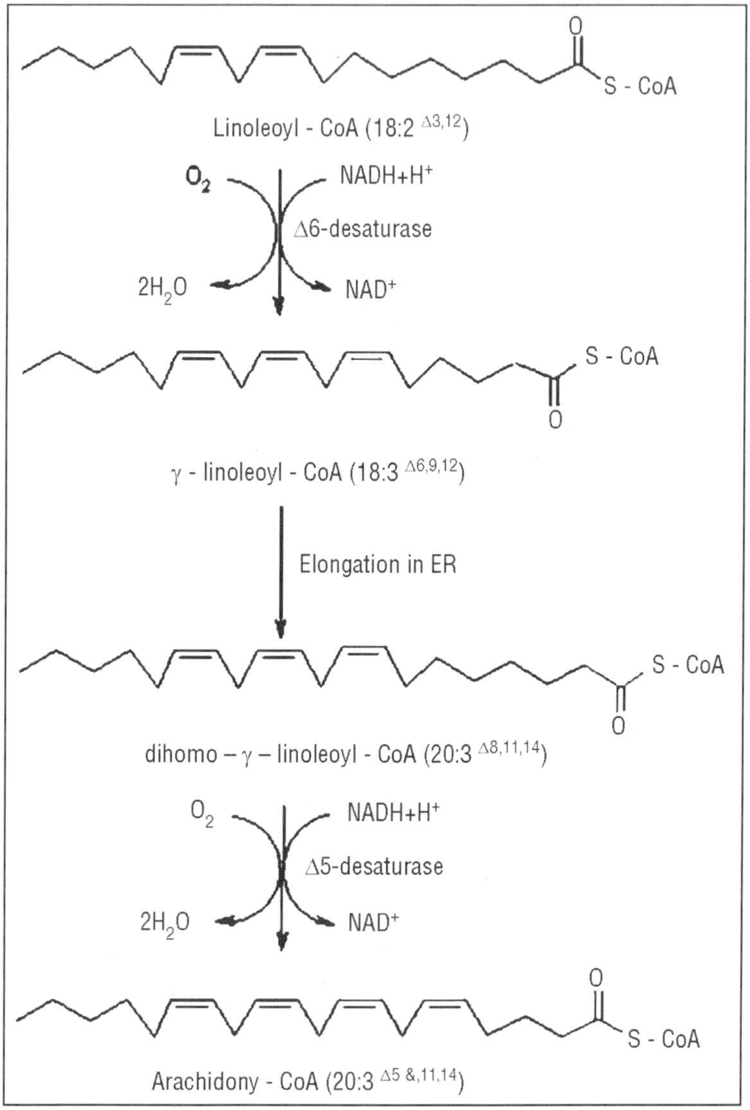

The eicosanoids produce a wide range of biological effects on inflammatory responses (predominantly those of the joints, skin and eyes), on the intensity and duration of pain and fever, and on reproductive function (including the induction of labour). They also play important roles in inhibiting gastric acid secretion, regulating blood pressure through vasodilation or constriction, and inhibiting or activating platelet aggregation and thrombosis. The principal eicosanoids of biological significance to humans are a group of molecules derived from the C_{20} fatty acid, arachidonic acid. Minor eicosanoids are derived from dihomo-g-linoleic acid and eicosopentaenoic acid. The major source of arachidonic acid is through its release from cellular stores. Within the cell, it resides predominantly at the C-2 position of membrane phospholipids and is released from there upon the activation of *phospholipase* A_2.

The immediate dietary precursor of arachidonate is linoleate. Linoleate is also the precursor for dihomo-g-linoleic acid and eicosopentaenoic acid synthesis. Therefore, the absence of linoleic acid from the diet would seriously threaten the body's ability to synthesize eicosanoids. Pathway from linoleic acid to arachidonic acid. Numbers in parentheses refer to the fatty acid length and the number and positions of unsaturations All mammalian cells except erythrocytes synthesize eicosanoids. These molecules are extremely potent, able to cause profound physiological effects at very dilute concentrations.

All eicosanoids function locally at the site of synthesis, through receptor-mediated G-protein linked signaling pathways leading to an increase in cAMP levels. Two main pathways are involved in the biosynthesis of eicosanoids. The prostaglandins and thromboxanes are synthesized by the cyclic pathway, the leukotrienes by the linear pathway. Synthesis of the clinically relevant prostaglandins and thromboxanes from arachidonic acid. Numerous stimuli (e.g. epinephrine, thrombin and bradykinin) activate *phospholipase* A_2 which hydrolyzes arachidonic acid from membrane phospholipids. The prostaglandins are identified as PG and the thromboxanes as TX. Prostaglandin PGI_2 is also known as prostacyclin. The subscript 2 in each molecule refers to the number of –C=C- present.

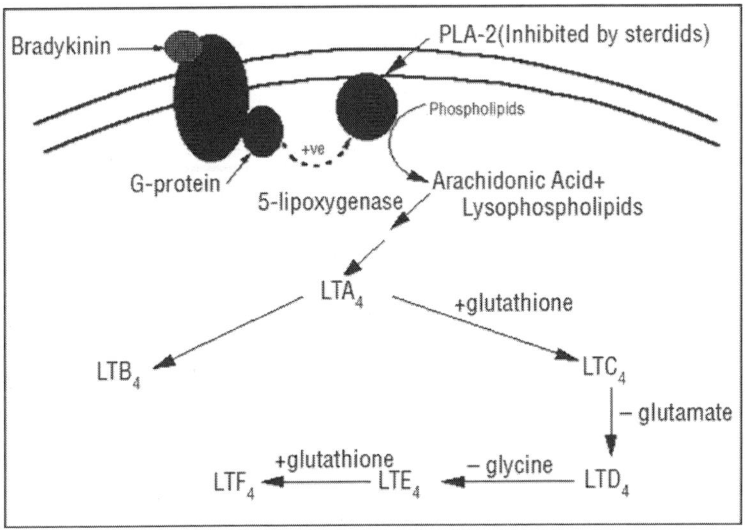

Synthesis of the clinically relevant leukotrienes from arachidonic acid. Numerous stimuli (e.g. epinephrine, thrombin and bradykinin) activate *phospholipase* A_2 which hydrolyzes arachidonic acid from membrane phospholipids. The leukotrienes are identified as LT. The leukotrienes, LTC_4, LTD_4, LTE_4 and LTF $_4$ are known as the peptidoleukotrienes because of the presence of amino acids. The peptidol-eukotrienes, LTC $_4$, LTD_4 and LTE_4 are components of slow-reacting substance of anaphylaxis The subscript 4 in each molecule refers to the number of –C=C- present.

The cyclic pathway is initiated through the action of *prostaglandin endoperoxide synthetase*. This enzyme possesses two activities, *cyclooxygenase* and *peroxidase*. The linear pathway is initiated through the action of *lipoxygenases*. It is the enzyme, *5-lipoxygenase* that gives rise to the leukotrienes. A widely used class of drugs, the non-steroidal anti-inflammatory drugs (NSAIDs such as ibuprofen, indome-thacin, naproxen, phenylbutazone and aspirin, all act upon the cyclooxygenase activity. Another class, the corticosteroidal drugs, act to inhibit *phospholipase* A_2, thereby inhibiting the release of arachidonate from membrane phospholipids and the subsequent synthesis of eicosinoids.

METABOLISM OF THE SPHINGOLIPIDS AND PHOSPHOLIPIDS

The sphingolipids, like the phospholipids, are composed of a polar head group and two nonpolar tails. The core of sphingolipids is the long-chain amino alcohol, sphingosine. The sphingolipids include sphingomyelins and glycos-phingolipids (cerebrosides, sulfatides, globosides and gangliosides). Sphingomyelins are the only sphingolipid that are also phospholipids. Sphingolipids are a component of all membranes but are particularly abundant in the myelin sheath. Ceramide,the molecule present both in sphingomyelins and glycosphingolipids derives from the condensation of serine and palmitoyl-CoA the following the pathway :

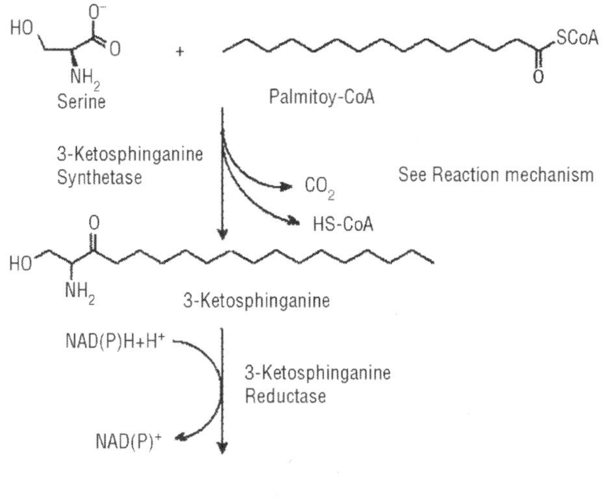

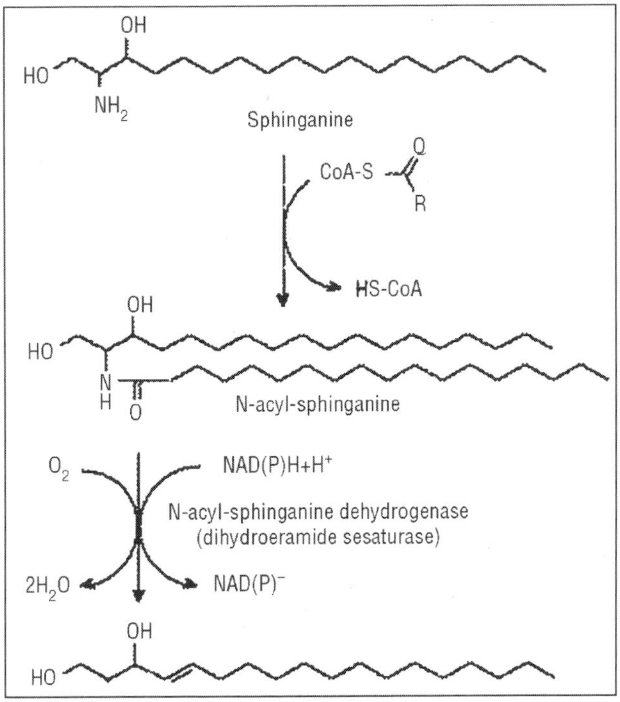

Sphingomyelins are sphingolipids that are also phospholipids. Sphingomyelins are important structural lipid components of nerve cell membranes. The predominant sphingomyelins contain palmitic or stearic acid N-acylated at carbon 2 of sphingosine. The sphingomyelins are synthesized by the transfer of phosphorylcholine from phosphatidylcholine to a ceramide in a reaction catalyzed by *sphingomyelin synthase.*

Defects in the enzyme *acid sphingomyelinase* result in the lysosomal storage disease known as Niemann-Pick disease. There are at least 4 related disorders identified as Niemann-Pick disease Type A and B (both of which result from defects in *acid sphingomyelinase*), Type C1 and a related C2 and Type D. Types C1, C2 and D do not result from defects in *acid sphingomyelinase.* Glycosphingolipids, or glycolipids, are composed of a ceramide backbone with a wide variety of carbohydrate groups (mono- or oligosaccharides) attached to carbon 1 of sphingosine. The four principal classes of glycosphingolipids are the cerebrosides, sulfatides, globosides and gangliosides.

Cerebrosides have a single sugar group linked to ceramide. The most common of these is galactose (galactocerebrosides), with a minor level of glucose (glucocerebrosides). Galactocerebrosides are found predominantly in neuronal cell membranes. By contrast glucocerebrosides are not normally found in membranes, especially neuronal membranes; instead, they represent intermediates in the synthesis or degradation of more complex glycosphingolipids.

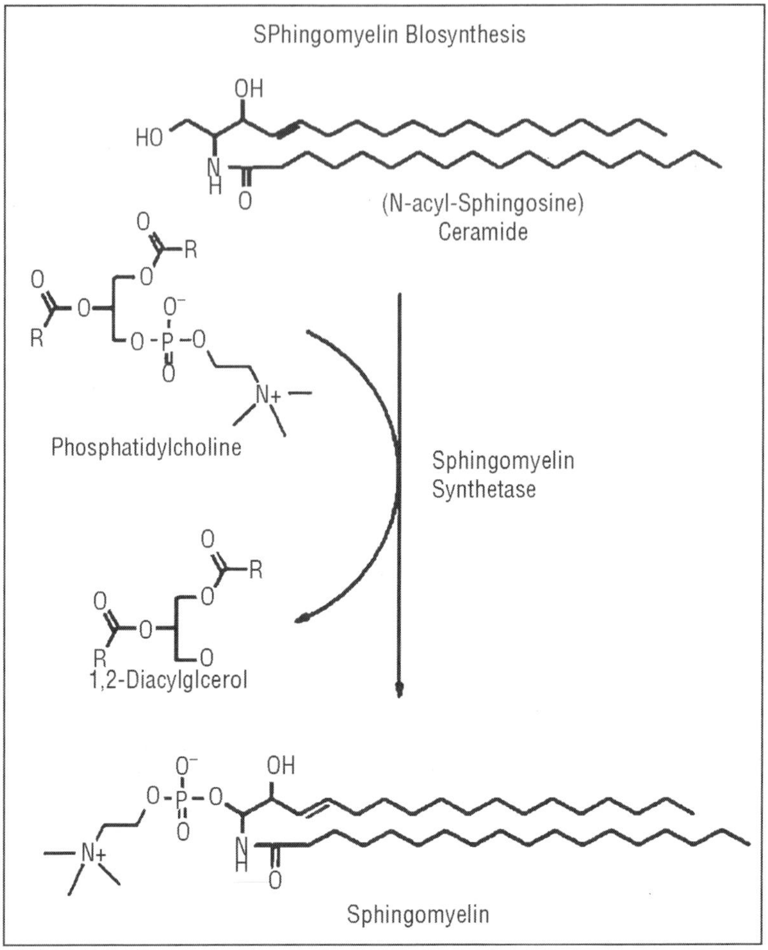

Galactocerebrosides are synthesized from ceramide and UDP-galactose. Excess accumulation of glucocerebrosides is observed in Gaucher's disease.

Sulfatides: The sulfuric acid esters of galactocerebrosides are the sulfatides. Sulfatides are synthesized from galactocerebrosides and activated sulfate, 3'-phosphoadenosine 5'-phosphosulfate (PAPS). Excess accumulation of sulfatides is observed in sulfatide lipidosis (metachromatic leukodystrophy).

Globosides: Globosides represent cerebrosides that contain additional carbohydrates, predominantly galactose, glucose or GalNAc. Lactosyl ceramide is a globoside found in erythrocyte plasma membranes. Globotriaosylceramide (also called ceramide trihexoside) contains glucose and two moles of galactose and accumulates, primarily in the kidneys, of patients suffering from Fabry's disease. *Gangliosides*: Gangliosides are very similar to globosides except that they also contain NANA in varying amounts. The specific names for gangliosides are a key to their structure. The letter G refers to "ganglioside", and the subscripts M, D, T and Q indicate that the molecule contains mono-, di-, tri and quatra(tetra)-sialic acid. The numerical subscripts 1, 2 and 3 refer to the carbohydrate sequence that is attached to ceramide;1 stands for Gal-GalNAc-Gal-Glc-ceramide, 2 for GalNAc-Gal-Glc-ceramide and 3 for Gal-Glc-ceramide.

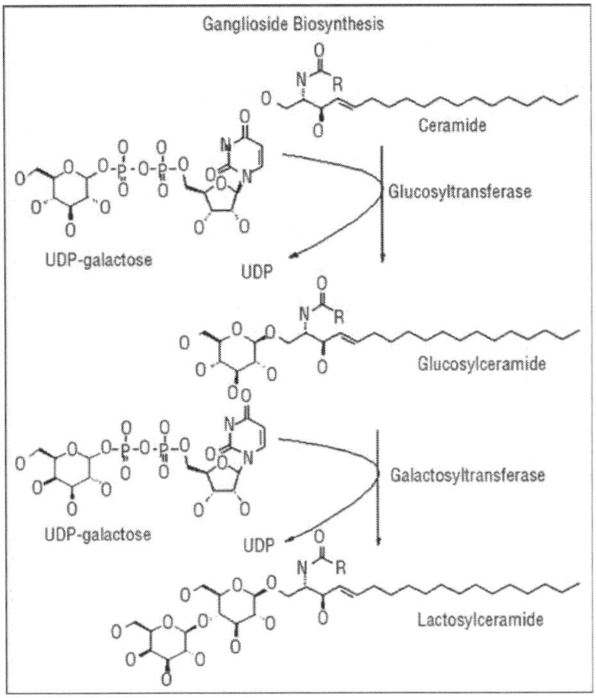

Deficiencies in lysosomal enzymes, which normally are responsible for the degradation of the carbohydrate portions of various gangliosides, underlie the symptoms observed in rare autosomally inherited diseases termed lipid storage diseases.

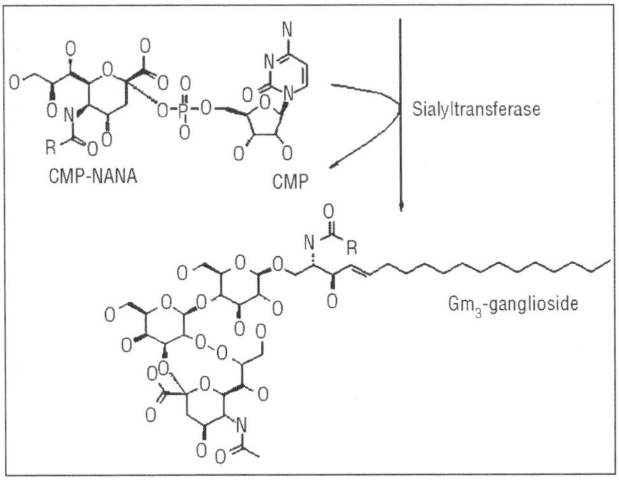

PLASMALOGENS

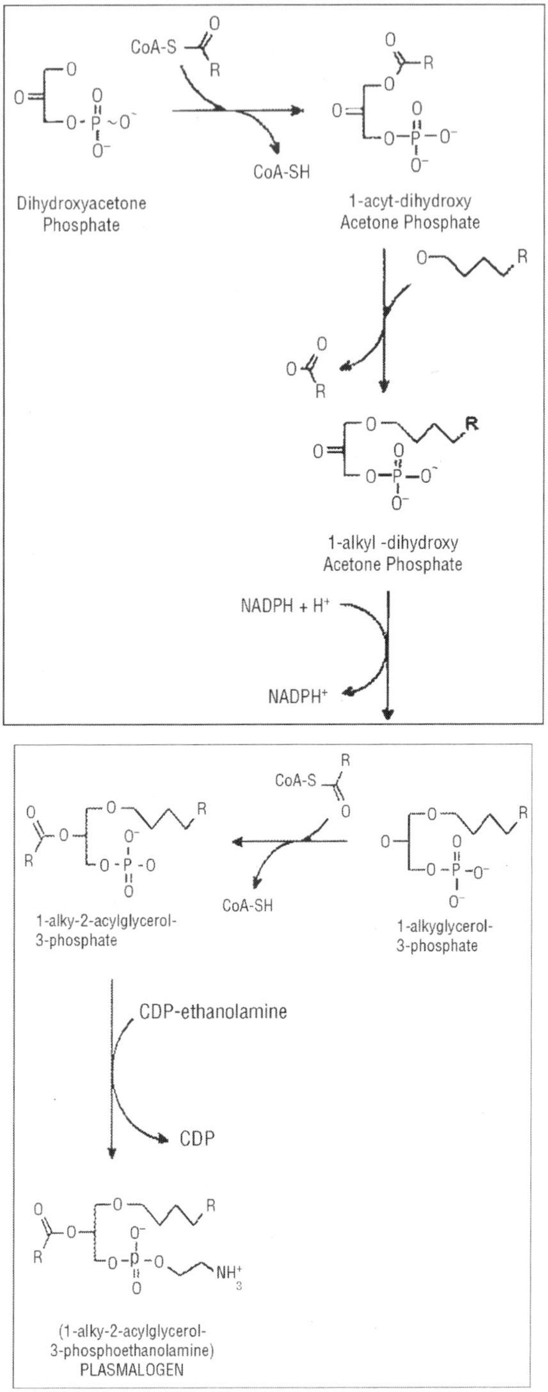

Plasmalogens are glycerol ether phospholipids. They are of two types, alkyl ether and alkenyl ether. Dihydroxyacetone phosphate serves as the glycerol precursor for the synthesis of glycerol ether phospholipids. Three major classes of plasmalogens have been identified: choline, ethanolamine and serine plasmalogens. Ethanolamine plasmalogen is prevalent in myelin. Choline plasmalogen is abundant in cardiac tissue.

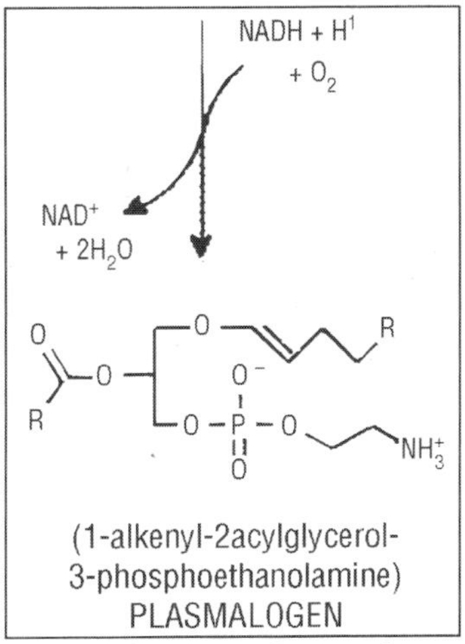

One particular choline plasmalogen (1-alkyl, 2-acetyl phosphatidylcholine) has been identified as an extremely powerful biological mediator, capable of inducing cellular responses at concentrations as low as 10^{-11} M. This molecule is called platelet activating factor, PAF.

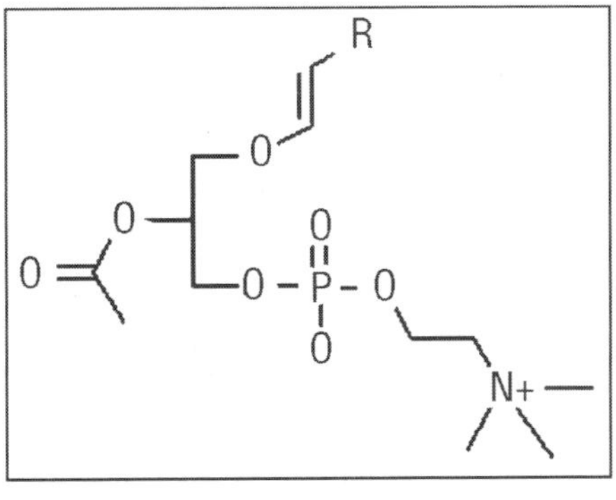

Fig. Platelet Activating Factor

PAF functions as a mediator of hypersensitivity, acute inflammatory reactions and anaphylactic shock. PAF is synthesized in response to the formation of antigen-IgE complexes on the surfaces of basophils, neutrophils, eosinophils, macrophages and monocytes. The synthesis and release of PAF from cells leads to platelet aggregation and the release of serotonin from platelets. PAF also produces responses in liver, heart, smooth muscle, and uterine and lung tissues.

EXTRACTION OF LIPID

The synthesis of malonyl-CoA is the first committed step of fatty acid synthesis and the enzyme that catalyzes this reaction, *acetyl-CoA carboxylase* (ACC), is the major site of regulation of fatty acid synthesis. Like other enzymes that transfer CO_2 to substrates, ACC requires a biotin co-factor.

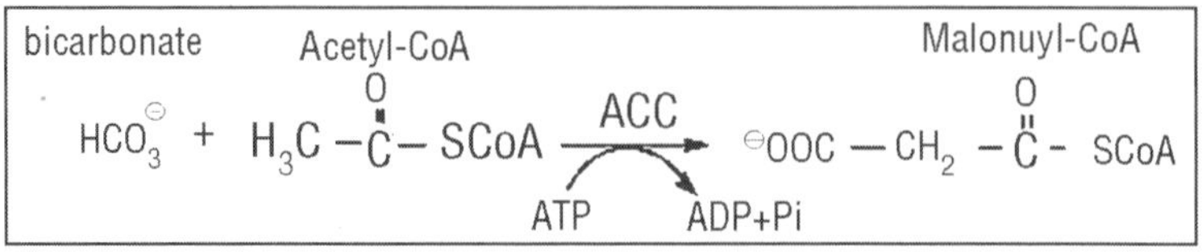

The rate of fatty acid synthesis is controlled by the equilibrium between monomeric ACC and polymeric ACC. The activity of ACC requires polymerization. This conformational change is enhanced by citrate and inhibited by long-chain fatty acids. ACC is also controlled through hormone mediated phosphorylation. The acetyl groups that are the products of fatty acid oxidation are linked to CoASH. As you should recall, CoA contains a phosphopantetheine group coupled to AMP. The carrier of acetyl groups (and elongating acyl groups) during fatty acid synthesis is also a phosphopantetheine prosthetic group, however, it is attached a serine hydroxyl in the synthetic enzyme complex. The carrier portion of the synthetic complex is called acyl carrier protein, ACP. This is somewhat of a misnomer in eukaryotic fatty acid synthesis since the ACP portion of the synthetic complex is simply one of many domains of a single polypeptide. The acetyl-CoA and malonyl-CoA are transferred to ACP by the action of *acetyl-CoA transacylase* and *malonyl-CoA transacylase,* respectively.

The synthesis of fatty acids from acetyl-CoA and malonyl-CoA is carried out by *fatty acid synthase, FAS.*

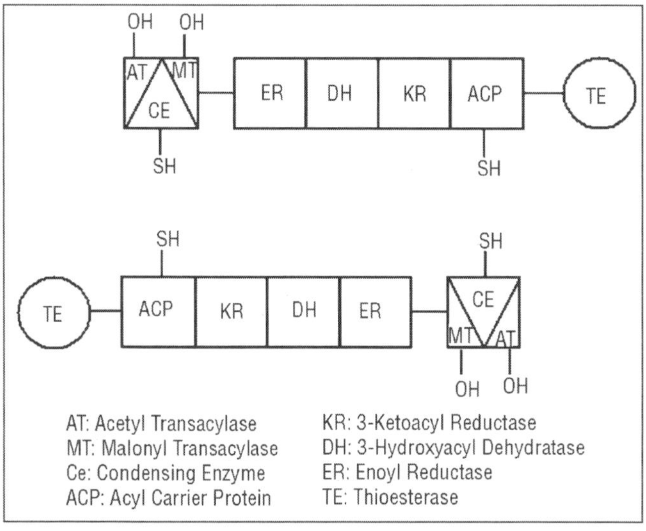

Fig. The Active Enzyme is a Dimer of Identical Subunits.

All of the reactions of fatty acid synthesis are carried out by the multiple enzymatic activities of FAS. Like fat oxidation, fat synthesis involves 4 enzymatic activities. These are, *b-keto-ACP synthase (condensing enzyme), β-keto-ACP reductase, 3-OH acyl-ACP dehydratase* and *enoyl-ACP reductase*(give a look to the FAS reaction mechanism). The two reduction reactions require NADPH oxidation to $NADP^+$. The primary fatty acid synthesized by FAS is palmitate. Palmitate is then released from the enzyme by *thioesterase* reaction and can then undergo separate elongation and/or unsaturation to yield other fatty acid molecules.

ORIGIN OF CYTOPLASMIC ACETYL

Acetyl-CoA is generated in the mitochondria primarily from two sources, the pyruvate dehydrogenase (PDH) reaction and fatty acid oxidation. In order for these acetyl units to be utilized for fatty acid synthesis they must be present in the cytoplasm. The shift from fatty acid oxidation and glycolytic oxidation occurs when the need for energy diminishes. This results in reduced oxidation of acetyl-CoA in the TCA cycle and the oxidative phosphorylation pathway. Under these conditions the mitochondrial acetyl units can be stored as fat for future energy demands. Acetyl-CoA enters the cytoplasm in the form of citrate via the tricarboxylate transport system as diagrammed. In the cytoplasm, citrate is converted to oxaloacetate and acetyl-CoA by the ATP driven *ATP-citrate lyase* reaction. This reaction is essentially the reverse of that catalyzed by the TCA enzyme *citrate synthase* except it requires the energy of ATP hydrolysis to drive it forward. The resultant oxaloacetate is converted to malate by *malate dehydrogenase* (MDH).

The malate produced by this pathway can undergo oxidative decarboxylation by *malic enzyme*. The co-enzyme for this reaction is $NADP^+$ generating NADPH. The advantage of this series of reactions for converting mitochondrial acetyl-CoA into cytoplasmic acetyl-CoA is that the NADPH produced by the *malic enzyme* reaction can be a major source of reducing co-factor for the *fatty acid synthase* activities.

REGULATION OF FATTY ACID METABOLISM

One must consider the global organismal energy requirements in order to effectively understand how the synthesis and degradation of fats (and also carbohydrates) needs to be exquisitely regulated. The blood is the carrier of triacylglycerols in the form of VLDLs and chylomicrons, fatty acids bound to albumin, amino acids, lactate, ketone bodies and glucose. The pancreas is the primary organ involved in sensing the organisms dietary and energetic states via glucose concentrations in the blood. In response to low blood glucose, glucagon is secreted, whereas, in response to elevated blood glucose insulin is secreted. The regulation of fat metabolism occurs via two distinct mechanisms. One is short term regulation which is regulation effected by events such as substrate availability, allosteric effectors and/or enzyme modification. ACC is the rate limiting (committed) step in fatty acid synthesis. This enzyme is activated by citrate and inhibited by palmitoyl-CoA and other long chain fatty acyl-CoAs. ACC activity also is affected by phosphorylation. Glucagon stimulated increases in PKA activity result in phosphorylation of certain serine residues in ACC leading to decreased activity of the enzyme. On the other hand insulin leads to phosphorylation of ACC at sites distinct from glucagon that result in increased ACC activity. These forms of regulation are all defined as short term regulation.

Control of a given pathways' regulatory enzymes can also occur by alteration of enzyme synthesis and turn-over rates. These changes are long term regulatory effects. Insulin stimulates ACC and FAS synthesis, whereas, starvation leads to decreased synthesis of these enzymes. Adipose tissue lipoprotein lipase levels also are increased by insulin and decreased by starvation. However, in contrast to the effects of insulin and starvation on adipose tissue, their effects on heart lipoprotein lipase are just the inverse. This allows the heart to absorb any available fatty acids in the blood in order to oxidize them for energy production. Starvation also leads to increases in the levels of fatty acid oxidation enzymes in the heart as well as a decrease in FAS and related enzymes of synthesis.

Adipose tissue contains *hormone-sensitive lipase,* that is activated by PKA-dependent phosphorylation leading to increased fatty acid release to the blood. This leads to increased fatty acid oxidation in other tissues such as muscle and liver. In the liver the net result (due to increased acetyl-CoA levels) is the production of ketone bodies. This would occur under conditions where insufficient carbohydrate stores and gluconeogenic precursors were available in liver for increased glucose production. The increased fatty acid availability in response to glucagon or epinephrine is assured of being completely oxidized since PKA also phosphorylates (and as a result inhibits) ACC, thus inhibiting fatty acid synthesis. Insulin, on the other hand, has the opposite effect to glucagon and epinephrine leading to increased glycogen and triacylglyceride synthesis. One of the many effects of insulin is to lower cAMP levels which leads to increased dephosphorylation through the enhanced activity of protein phosphatases such as PP-1. With respect to fatty acid metabolism this yields dephosphorylated and inactive hormone sensitive lipase. Insulin also stimulates certain phosphorylation events. This occurs through activation of several cAMP-independent kinases. Insulin stimulated phosphorylation of ACC activates this enzyme. Regulation of fat metabolism also occurs through malonyl-CoA induced inhibition of *carnitine acyltransferase I.* This functions to prevent the newly synthesized fatty acids from entering the mitochondria and being oxidized.

ELONGATION AND DESATURATION

The fatty acid product released from FAS is palmitate (via the action of *palmitoyl thioesterase*) which is a 16:0 fatty acid, i.e. 16 carbons and no sites of unsaturation. Elongation and unsaturation of fatty acids occurs in both the mitochondria and endoplasmic reticulum (microsomal membranes). The predominant site of these processes is in the ER membranes. Elongation involves condensation of acyl-CoA groups with malonyl-CoA. The resultant product is two carbons longer (CO_2 is released from malonyl-CoA as in the FAS reaction) which undergoes reduction, dehydration and reduction yielding a saturated fatty acid. The reduction reactions of elongation require NADPH as co-factor just as for the similar reactions catalyzed by FAS. Mitochondrial elongation involves acetyl-CoA units and is a reversal of oxidation except that the final reduction utilizes NADPH instead of $FADH_2$ as co-factor.

Desaturation occurs in the ER membranes as well and in mammalian cells involves 4 broad specificity *fatty acyl-CoA desaturases* (non-heme iron containing enzymes). These enzymes introduce unsaturation at C4, C5, C6 or C9.

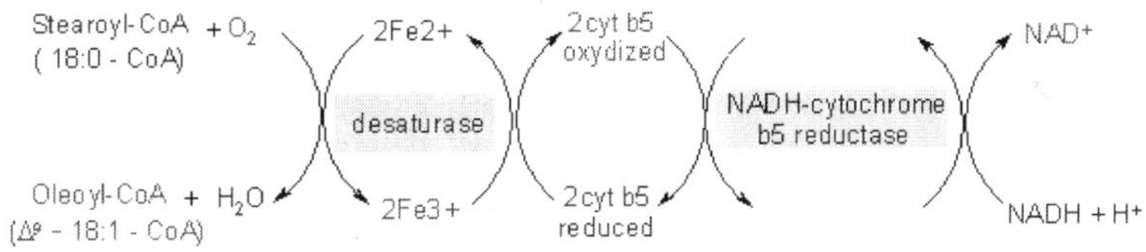

The electrons from the reduced fatty acid (saturated) and from the desaturase are transferred to molecular oxygen to form water and oxydized fatty acid (unsaturated). NADH is the donor of two electrons which keep the desaturase in the reduced form through the action of the enzyme *NADH-cytochrome b5 reductase.* These electrons are un-coupled from mitochondrial oxidative-phosphorylation and, therefore, do not yield ATP. Since these enzymes cannot introduce sites of unsaturation beyond C9 they cannot synthesize either linoleate ($18{:}2^{\Delta 9,\ 12}$) or linolenate ($18{:}3^{\Delta 9,\ 12,\ 15}$). These fatty acids must be acquired from the diet and are, therefore, referred to as essential fatty acids. Linoleic is especially important in that it required for the synthesis of arachidonic acid.

As we shall encounter later, arachindonate is a precursor for the eicosanoids (the prostaglandins and thromboxanes). It is this role of fatty acids in eicosanoid synthesis that leads to poor growth, wound healing and dermatitis in persons on fat free diets. Also, linoleic acid is a constituent of epidermal cell sphingolipids that function as the skins water permeability barrier.

SYNTHESIS OF TRIGLYCERIDES

Fatty acids are stored for future use as triacylglycerols in all cells, but primarily in adipocytes of adipose tissue. Triacylglycerols constitute molecules of glycerol to which three fatty acids have been esterified. The fatty acids present in triacylglycerols are predominantly saturated.

The major building block for the synthesis of triacylglycerols, in tissues other than adipose tissue,is glycerol-3-phosphate. Adipocytes lack *glycerol kinase*, therefore, dihydroxyacetone phosphate (DHAP), produced during glycolysis, is the precursor for triacylglycerol synthesis in adipose tissue. This means that adipoctes must have glucose to oxidize in order to store fatty acids in the form of triacylglycerols. DHAP can also serve as a backbone precursor for triacylglycerol synthesis in tissues other than adipose, but does so to a much lesser extent than glycerol.

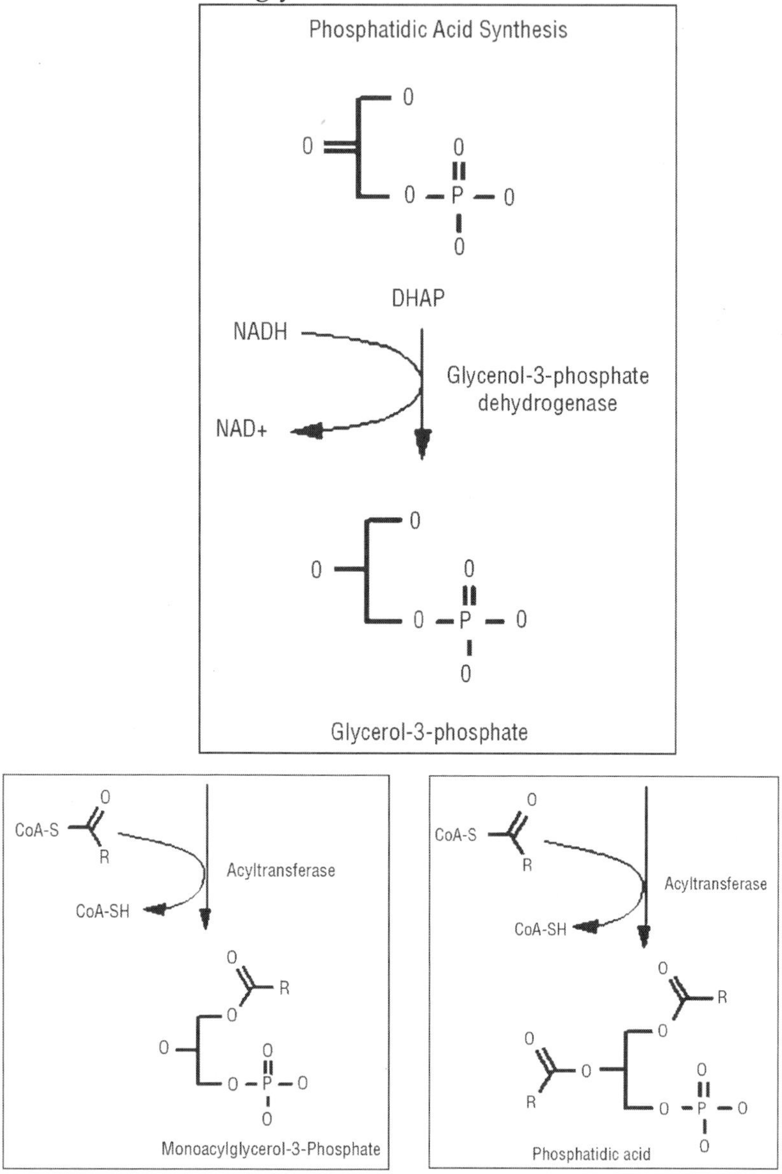

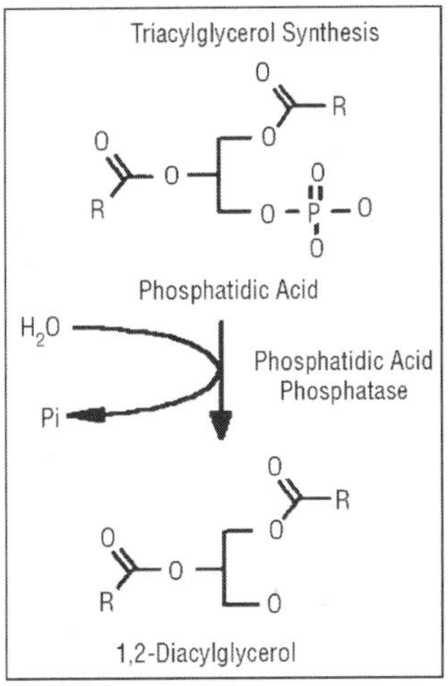

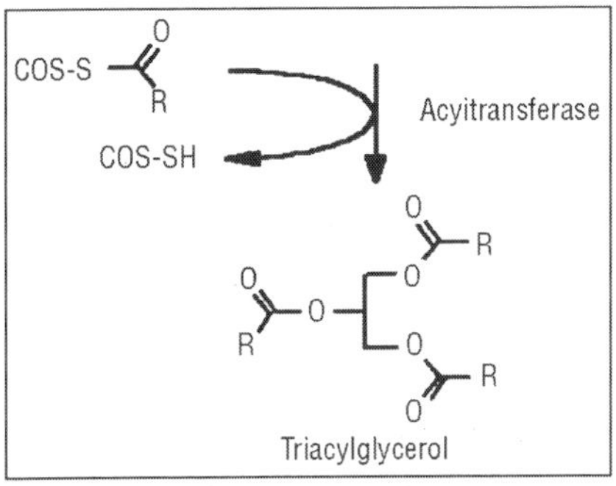

The glycerol backbone of triacylglycerols is activated by phosphorylation at the C-3 position by *glycerol kinase*. The utilization of DHAP for the backbone is carried out through the action of *glycerol-3-phosphate dehydrogenase,* a reaction that requires NADH (the same reaction as that used in the glycerol-phosphate shuttle).

The fatty acids incorporated into triacylglycerols are activated to acyl-CoAs through the action of *acyl-CoA synthetases*. Two molecules of acyl-CoA are esterified to glycerol-3-phosphate to yield 1,2-diacylglycerol phosphate (commonly identified as phosphatidic acid). The phosphate is then removed, by *phosphatidic acid phosphatase,* to yield 1,2-diacylglycerol, the substrate for addition of the third fatty acid. Intestinal monoacylglycerols, derived from the hydrolysis of dietary fats, can also serve as substrates for the synthesis of 1,2-diacylglycerols.

PHOSPHOLIPID STRUCTURES

Phospholipids are synthesized by the addition of a basic group (predominantly a nitrogenous base) to phosphatidic acid or 1,2-diacylglycerol. Most phospholipids have a saturated fatty acid on C-1 and an unsaturated fatty acid on C-2 of the glycerol backbone.

The major classifications of phospholipids are:

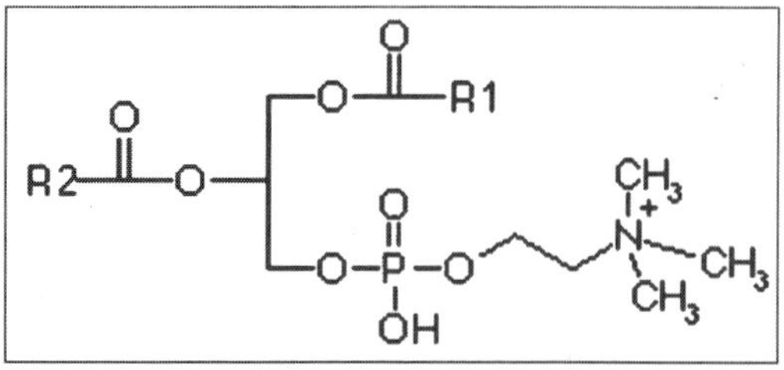

Phosphatidylcholine (PC)

(1-stearoyl,2-oleoyl)

Top of Form
Bottom of Form

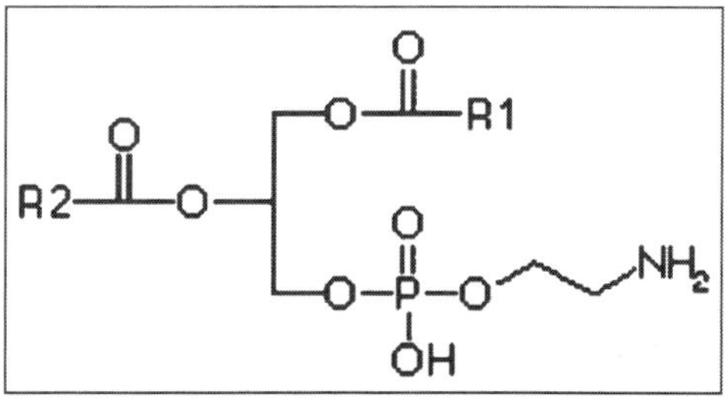

Phosphatidylethanolamine (PE)

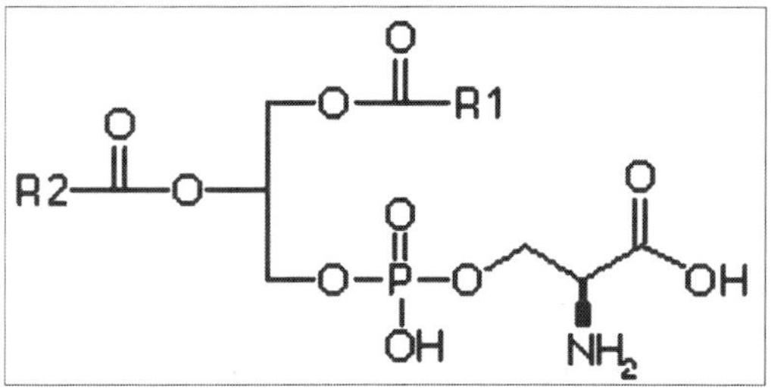

Phosphatidylserine (PS)

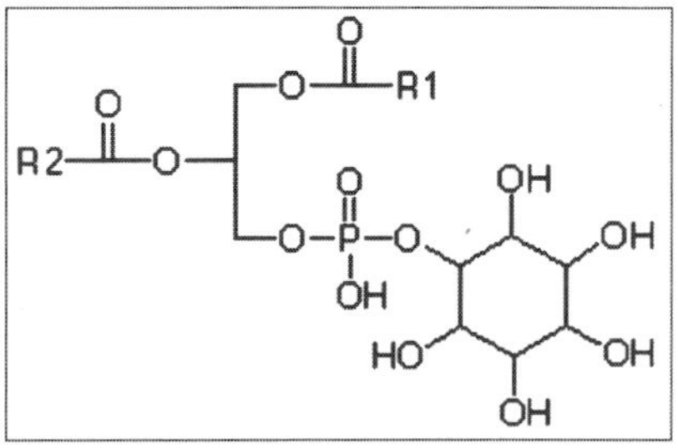

Phosphatidylinositol (PI)

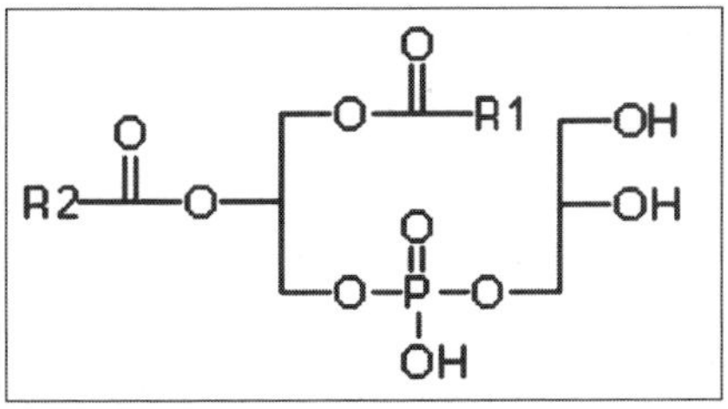

Phosphatidylglycerol (PG)

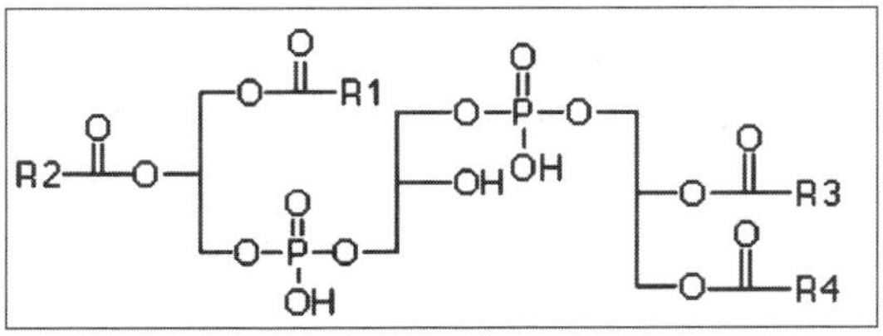

Diphosphatidylglycerol (DPG)

PHOSPHOLIPID SYNTHESIS

Phospholipids can be synthesized by two mechanisms. One utilizes a CDP-activated polar head group for attachment to the phosphate of phosphatidic acid. The other utilizes CDP-activated 1,2-diacylglycerol and an inactivated polar head group.

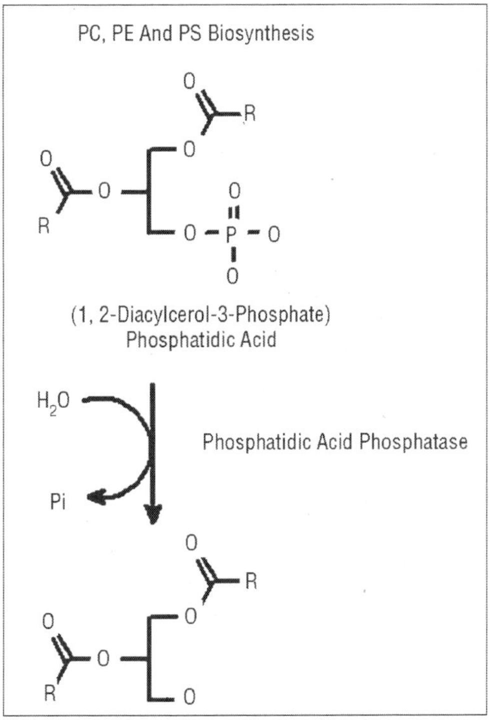

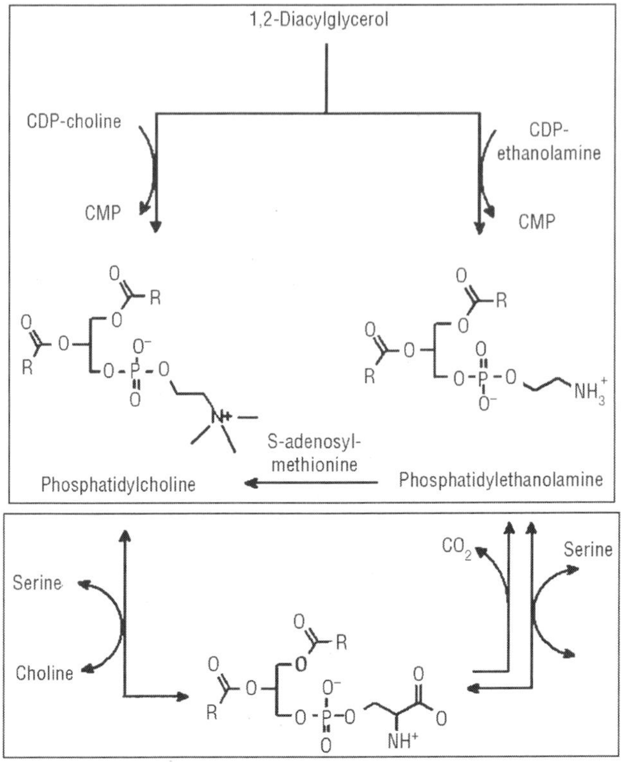

PC:This class of phospholipids is also called the lecithins. At physiological pH, phosphatidylcholines are neutral zwitterions. They contain primarily palmitic or stearic acid at carbon 1 and primarily oleic, linoleic or linolenic acid at carbon 2. The lecithin dipalmitoyllecithin is a component of lung or pulmonary surfactant. It contains palmitate at both carbon 1 and 2 of glycerol and is the major (80%) phospholipid found in the extracellular lipid layer lining the pulmonary alveoli. Choline is activated first by phosphorylation and then by coupling to CDP prior to attachment to phosphatidic acid. An other pathway to PC synthesis, involves the conversion of either PS or PE to PC. The conversion of PS to PC first requires decarboxylation of PS to yield PE; this then undergoes a series of three methylation reactions utilizing S-adenosylmethionine (SAM) as methyl group donor.

PE:These molecules are neutral zwitterions at physiological pH. They contain primarily palmitic or stearic acid on carbon 1 and a long chain unsaturated fatty acid (e.g. 18:2, 20:4 and 22:6) on carbon 2. Synthesis of PE can occur by two pathways. The first requires that ethanolamine be activated by phosphorylation and then by coupling to CDP. The ethanolamine is then transferred from CDP-ethanolamine to phosphatidic acid to yield PE. The second involves the decarboxylation of PS. PS:Phosphatidylserines will carry a net charge of -1 at physiological pH and are composed of fatty acids similar to the phosphatidylethanolamines.The pathway for PS synthesis involves an exchange reaction of serine for ethanolamine in PE and choline in PC. These exchange occur when PE and PC are in the lipid bilayer of the a membrane. As indicated above, PS can serve as a source of PE through a decarboxylation reaction.

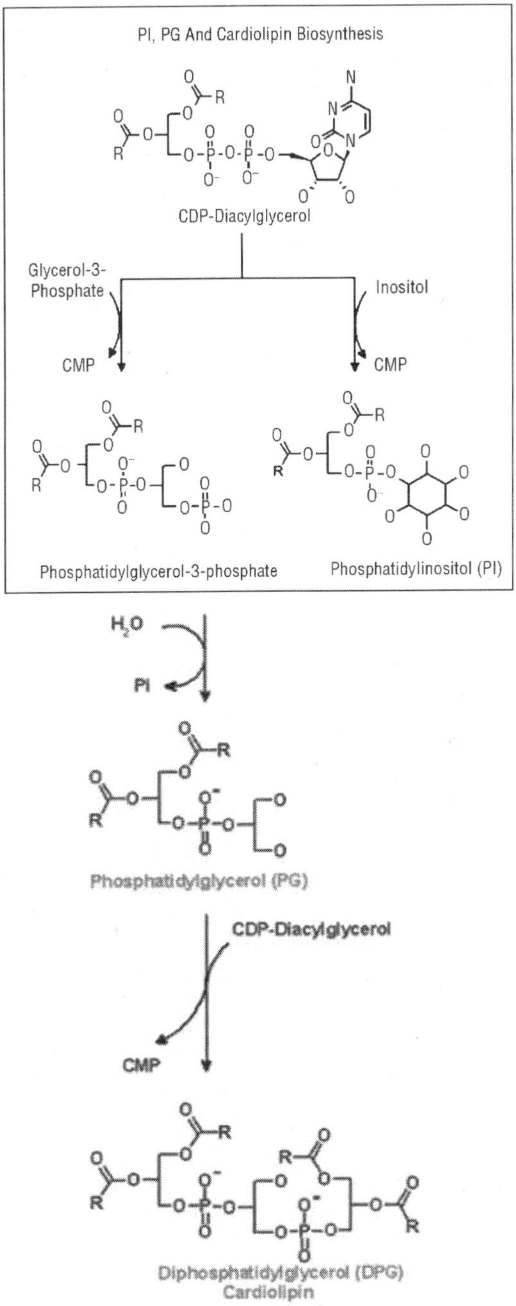

PI:These molecules contain almost exclusively stearic acid at carbon 1 and arachidonic acid at carbon 2. Phosphatidy-linositols composed exclusively of non-phosphorylated inositol exhibit a net charge of -1 at physiological pH. These molecules exist in membranes with various levels of phosphate esterified to the hydroxyls of the inositol. Molecules with phosphorylated inositol are termed polyphosphoinositides. The polyphosphoinositides are important intracellular transducers of signals emanating from the plasma membrane.

The synthesis of PI involves CDP-activated 1,2-diacylglycerol condensation with *myo*-inositol. PI subsequently undergoes a series of phosphorylations of the hydroxyls of inositol leading to the production of polyphosphoinositides. One polyphosphoinositide (phosphatidylinositol 4,5-bisphosphate, PIP_2) is a critically important membrane phospholipid involved in the transmission of signals for cell growth and differentiation from outside the cell to inside. PG:Phosphatidylglycerols exhibit a net charge of -1 at physiological pH. These molecules are found in high concentration in mitochondrial membranes and as components of pulmonary surfactant. PG is synthesized from CDP-diacylglycerol and glycerol-3-phosphate. The vital role of PG is to serve as the precursor for the synthesis of diphosphatidylglycerols (DPGs).

DPG:These molecules are very acidic, exhibiting a net charge of -2 at physiological pH. They are found primarily in the inner mitochondrial membrane and also as components of pulmonary surfactant.DPG,also known as cardiolipins are synthesized by the condensation of CDP-diacylglycerol with PG. The fatty acid distribution at the C-1 and C-2 positions of glycerol within phospholipids is continually in flux, owing to phospholipid degradation and the continuous phospholipid remodeling that occurs while these molecules are in membranes. Phospholipid degradation results from the action of *phospholipases*.

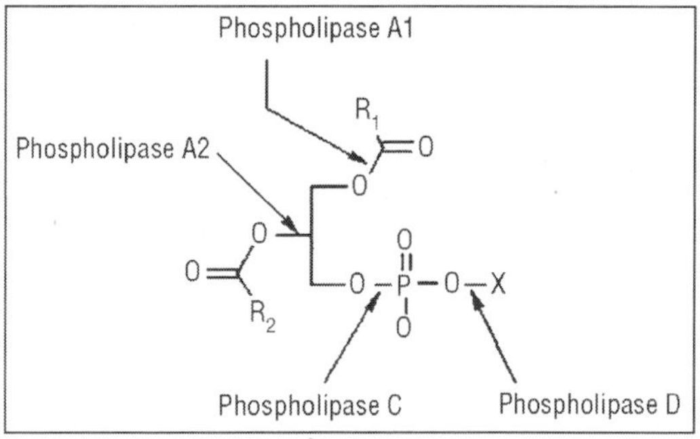

Fig. Sites of Action of the Phospholipases A_1, A_2, C and D.

There are various phospholipases that exhibit substrate specificities for different positions in phospholipids. In many cases the acyl group which was initially transferred to glycerol, by the action of the acyl transferases, is not the same acyl group present in the phospholipid when it resides within a membrane. The remodeling of acyl groups in phospholipids is the result of the action of *phospholipase* A_1 and *phospholipase* A_2.

The products of these phospholipases are called lysophospholipids and can be substrates for acyl transferases utilizing different acyl-CoA groups. Lysophospholipids can also accept acyl groups from other phospholipids in an exchange reaction catalyzed by lysolecithin:lecithin acyltransferase (LLAT).

Phospholipase A_2 is also an important enzyme, whose activity is responsible for the release of arachidonic acid from the C-2 position of membrane phospholipids. The released arachidonate is then a substrate for the synthesis of the prostaglandins and leukotrienes.

3

Clinical Significance of Fatty Acids

The majority of clinical problems related to fatty acid metabolism are associated with processes of oxidation.

These disorders fall into four main groups:

- Deficiencies in Carnitine: Deficiencies in carnitine lead to an inability to transport fatty acids into the mitochondria for oxidation. This can occur in newborns and particularly in pre-term infants. Carnitine deficiencies also are found in patients undergoing hemodialysis or exhibiting organic aciduria. Carnitine deficiencies may manifest systemic symptomology or may be limited to only muscles. Symptoms can range from mild occasional muscle cramping to severe weakness or even death. Treatment is by oral carnitine administration.
- Carnitine Palmitoyltransferase I (CPT I) Deficiency: Deficiencies in this enzyme affect primarily the liver and lead to reduced fatty acid oxidation and ketogenesis. Carnitine Palmitoylransferase II (CPT II) deficiency results in recurrent muscle pain and fatigue and myoglobinuria following strenuous exercise. Carnitine acyltransferases may also be inhibited by sulfonylurea drugs such as tolbutamide and glyburide.
- Deficiencies in Acyl-CoA Dehydrogenases: A group of inherited diseases that impair b-oxidation result from deficiencies in acyl-CoA dehydrogenases. The enzymes affected may belong to one of four categories:
  - Very long-chain acyl-CoA dehydrogenase (VLCAD)
  - Long-chain acyl-CoA dehydrogenase (LCAD)
  - Medium-chain acyl-CoA dehydrogenase (MCAD)
  - Short-chain acyl-CoA dehydrogenase (SCAD)

MCAD deficiency is the most common form of this disease. In the first years of life this deficiency will become apparent following a prolonged fasting period. Symptoms include vomiting, lethargy and frequently coma. Excessive urinary excretion of medium-chain dicarboxylic acids as well as their glycine and carnitine esters is diagnostic of this condition. In the case of this enzyme deficiency. taking care to avoid prolonged fasting is sufficient to prevent clinical problems.

- Refsum's Disease: Refsum's disease is a rare inherited disorder in which patients lack the mitochondrial *a*-oxidizing enzyme. As a consequence, they accumulate large quantities of phytanic acid in their tissues and serum. This leads to severe symptoms, including cerebellar ataxia, retinitis pigmentosa, nerve deafness and peripheral neuropathy. As expected, the restriction of dairy products and ruminant meat from the diet can ameliorate the symptoms of this disease.

KETOGENESIS

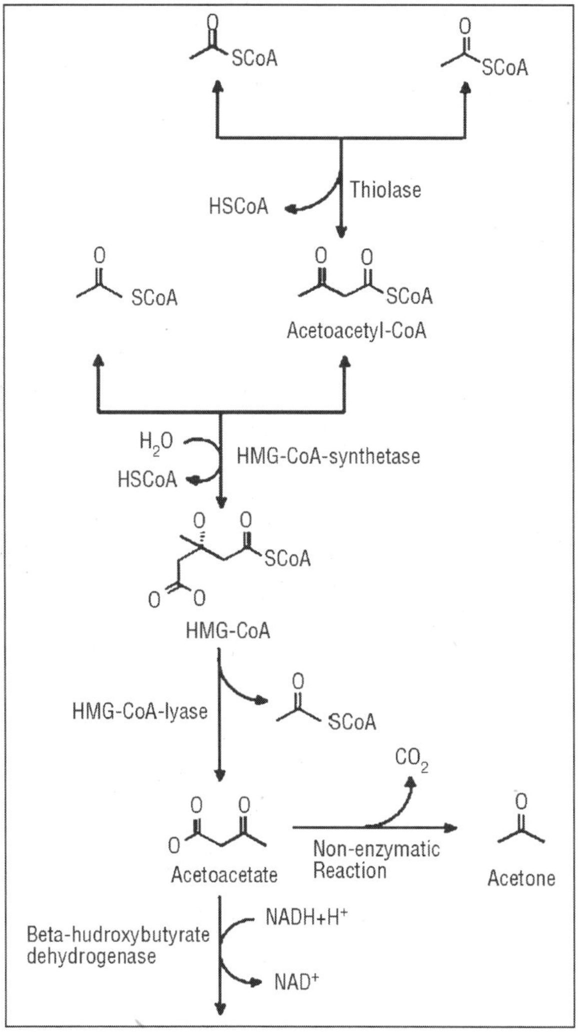

During high rates of fatty acid oxidation, primarily in the liver, large amounts of acetyl-CoA are generated. These exceed the capacity of the TCA cycle, and one result is the synthesis of ketone bodies, or ketogenesis. The ketone bodies are acetoacetate, β-hydroxybutyrate, and acetone.

The formation of acetoacetyl-CoA occurs by condensation of two moles of acetyl-CoA through a reversal of the *thiolase* catalyzed reaction of fat oxidation. Acetoacetyl-CoA and an additional acetyl-CoA are converted to β-hydroxy-β-methylglutaryl-CoA (HMG-CoA) by *HMG-CoA synthase,* an enzyme found in large amounts only in the liver. Some of the HMG-CoA leaves the mitochondria, where it is converted to mevalonate (the precursor for cholesterol synthesis) by *HMG-CoA reductase.* HMG-CoA in the mitochondria is converted to acetoacetate by the action of *HMG-CoA lyase.* Acetoacetate can undergo spontaneous decarboxylation to acetone, or be enzymatically converted to β-hydroxybutyrate through the action of *β-hydroxybutyrate dehydrogenase.* When the level of glycogen in the liver is high the production of β-hydroxybutyrate increases.

When carbohydrate utilization is low or deficient, the level of oxaloacetate will also be low, resulting in a reduced flux through the TCA cycle. This in turn leads to increased release of ketone bodies from the liver for use as fuel by other tissues. In early stages of starvation, when the last remnants of fat are oxidized, heart and skeletal muscle will consume primarily ketone bodies to preserve glucose for use by the brain. Acetoacetate and β-hydroxybutyrate, in particular, also serve as major substrates for the biosynthesis of neonatal cerebral lipids. Ketone bodies are utilized by extrahepatic tissues through the conversion of β-hydroxybutyrate to acetoacetate and of acetoacetate to acetoacetyl-CoA. The first step involves the reversal of the β-hydroxybutyrate dehydrogenase reaction, and the second involves the action of *acetoacetate:succinyl-CoA transferase,* also called *ketoacyl-CoA-transferase.*

$$\text{Acetoacetate + Succinyl–CoA} \longleftrightarrow \text{Acetoacetyl–CoA + succinate}$$

The latter enzyme is present in all tissues except the liver. Importantly, its absence allows the liver to produce ketone bodies but not to utilize them. This ensures that extrahepatic tissues have access to ketone bodies as a fuel source during prolonged starvation. Acetoacetyl-CoA is converted into two molecules of acetyl-CoA by a *thiolase*:

$$\text{Acetoacetyl-CoA + HS-CoA} \longleftrightarrow \text{2 Acetyl-CoA}$$

REGULATION OF KETOGENESIS

The fate of the products of fatty acid metabolism is determined by an individual's physiological status. Ketogenesis takes place primarily in the liver and may by affected by several factors:

- Control in the release of free fatty acids from adipose tissue directly affects the level of ketogenesis in the liver. This is, of course, substrate-level regulation.

- Once fats enter the liver, they have two distinct fates. They may be activated to acyl-CoAs and oxidized, or esterified to glycerol in the production of triacylglycerols. If the liver has sufficient supplies of glycerol-3-phosphate, most of the fats will be turned to the production of triacylglycerols.
- The generation of acetyl-CoA by oxidation of fats can be completely oxidized in the TCA cycle. Therefore, if the demand for ATP is high the fate of acetyl-CoA is likely to be further oxidation to CO_2.
- The level of fat oxidation is regulated hormonally through phosphorylation of ACC, which may activate it (in response to glucagon) or inhibit it (in the case of insulin).

STRUCTURE OF FATTY ACIDS

The"tail" of a fatty acid is a long hydrocarbon chain, making it hydrophobic. The"head" of the molecule is a carboxyl group which is hydrophilic. Fatty acids are the main component of soap, where their tails are soluble in oily dirt and their heads are soluble in water to emulsify and wash away the oily dirt. However, when the head end is attached to glycerol to form a fat, that whole molecule is hydrophobic.

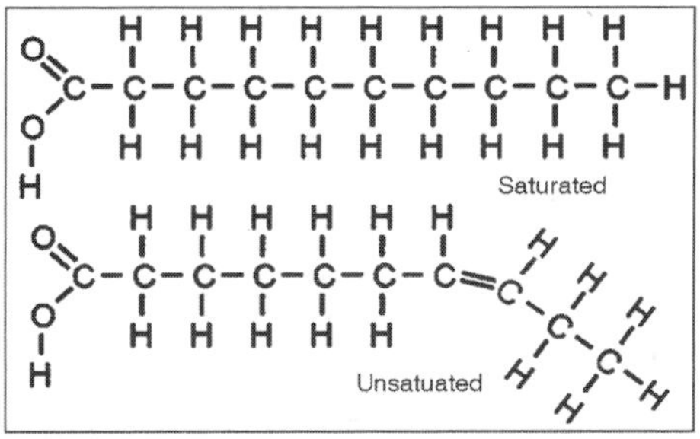

The terms saturated, mono-unsaturated, and poly-unsaturated refer to the number of hydrogens attached to the hydrocarbon tails of the fatty acids as compared to the number of double bonds between carbon atoms in the tail. Fats, which are mostly from animal sources, have all single bonds between the carbons in their fatty acid tails, thus all the carbons are also bonded to the maximum number of hydrogens possible. Since the fatty acids in these triglycerides contain the maximum possible amouunt of hydrogens, these would be calledsaturated fats.

The hydrocarbon chains in these fatty acids are, thus, fairly straight and can pack closely together, making these fats solid at room temperature. Oils, mostly from plant sources, have some double bonds between some of the carbons in the hydrocarbon tail, causing bends or"kinks" in the shape of the molecules. Because some of the carbons share double bonds, they're not bonded to as many hydrogens as they could if they weren't double bonded to each other. Therefore these oils are called unsaturated fats.

Because of the kinks in the hydrocarbon tails, unsaturated fats can't pack as closely together, making them liquid at room temperature. Many people have heard that the unsaturated fats are"healthier" than the saturated ones. Hydrogenated vegetable oil (as in shortening and commercial peanut butters where a solid consistency is sought) started out as"good" unsaturated oil. However, this commercial product has had all the double bonds artificially broken and hydrogens artificially added (in a chemistry lab-type setting) to turn it into saturated fat that bears no resemblance to the original oil from which it came (so it will be solid at room temperature).

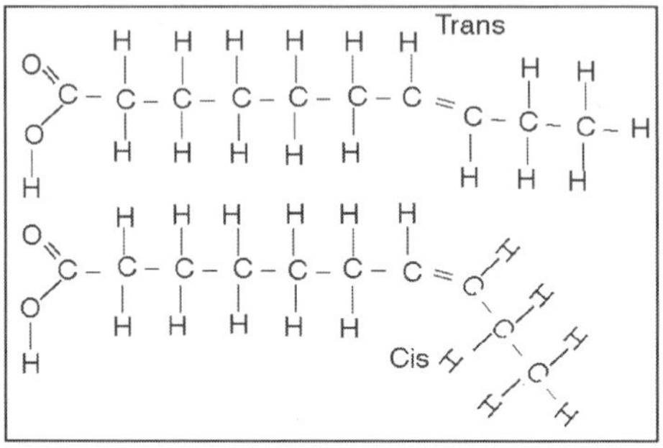

In unsaturated fatty acids, there are two ways the pieces of the hydrocarbon tail can be arranged around a C=C double bond. In cis bonds, the two pieces of the carbon chain on either side of the double bond are either both"up" or both"down," such that both are on the same side of the molecule. In trans bonds, the two pieces of the molecule are on opposite sides of the double bond, that is, one"up" and one"down" across from each other. Naturally-occurring unsaturated vegetable oils have almost all cis bonds, but using oil for frying causes some of the cis bonds to convert to trans bonds. If oil is used only once like when you fry an egg, only a few of the bonds do this so it's not too bad. However, if oil is constantly reused, like in fast food French fry machines, more and more of the cis bonds are changed to trans until significant numbers of fatty acids with trans bonds build up. The reason this is of concern is that fatty acids with trans bonds are carcinogenic, or cancer-causing.

The levels of trans fatty acids in highly-processed, lipid-containing products such as margarine are quite high, and I have heard that the government is considering requiring that the amounts of trans fatty acids in such products be listed on the labels. We need fats in our bodies and in our diet. Animals in general use fat for energy storage because fat stores 9 KCal/g of energy. Plants, which don't move around, can afford to store food for energy in a less compact but more easily accessible form, so they use starch (a carbohydrate, NOT A LIPID) for energy storage. Carbohydrates and proteins store only 4 KCal/g of energy, so fat stores over twice as much energy/gram as fat. By the way, this is also related to the idea behind some of the high-carbohydrate weight loss diets. The human body burns carbohydrates and fats for fuel in a given proportion to each other. The theory behind these diets is that if they supply carbohydrates but not fats, then it is hoped that the fat needed to balance with the sugar will be taken from the dieter's body stores. Fat is also is used in our bodies to a) cushion vital organs like the kidneys and b) serve as insulation, especially just beneath the skin.

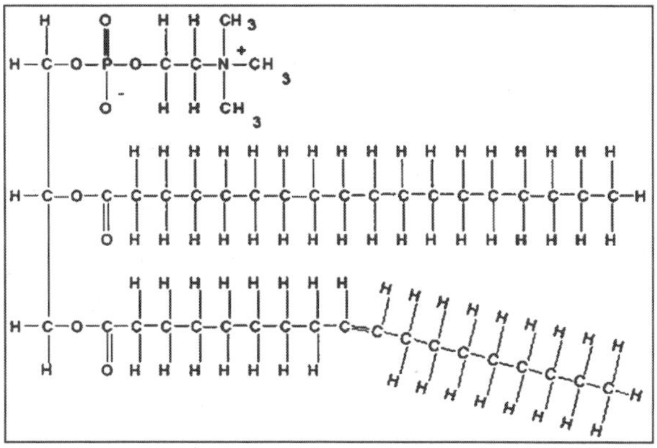

PHOSPHOLIPIDS

Phospholipids are made from glycerol, two fatty acids, and (in place of the third fatty acid) a phosphate group with some other molecule attached to its other end. The hydrocarbon tails of the fatty acids are still hydrophobic, but the phosphate group end of the molecule is hydrophilic because of the oxygens with all of their pairs of unshared electrons. This means that phospholipids are soluble in both water and oil. An emulsifying agent is a substance which is soluble in both oil and water, thus enabling the two to mix. A"famous" phospholipid is lecithin which is found in egg yolk and soybeans. Egg yolk is mostly water but has a lot of lipids, especially cholesterol, which are needed by the developing chick. Lecithin is used to emulsify the lipids and hold them in the water as an emulsion. Lecithin is the basis of the classic emulsion known as mayonnaise.

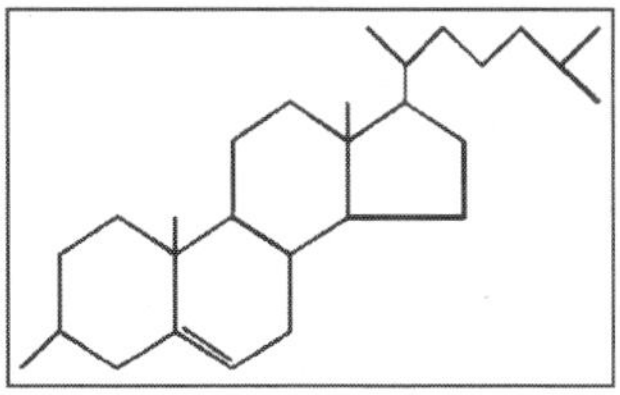

Our cell membranes are made mostly of phospholipids arranged in a double layer with the tails from both layers"inside" (facing towards each other) and the heads facing"out" (towards the watery environment) on both surfaces.

STEROIDS

The general structure of cholesterol consists of two six-membered rings side-by-side and sharing one side in common, a third six-membered ring off the top corner of the right ring, and a five-membered ring attached to the right side of that. The central core of this molecule, consisting of four fused rings, is shared by all steroids, including estrogen (estradiol), progesterone, corticosteroids such as cortisol (cortisone), aldosterone, testosterone, and Vitamin D. In the various types of steroids, various other groups/molecules are attached around the edges. Know how to draw the four rings that make up the central structure.

Cholesterol is not a"bad guy!" Our bodies make about 2 g of cholesterol per day, and that makes up about 85% of blood cholesterol, while only about 15% comes from dietary sources. Cholesterol is the precursor to our sex hormones and Vitamin D. Vitamin D is formed by the action of UV light in sunlight on cholesterol molecules that have"risen" to near the surface of the skin. At least one source I read suggested that people not shower immediately after being in the sun, but wait at least ½ hour for the new Vitamin D to be absorbed deeper into the skin. Our cell membranes contain a lot of cholesterol (in between the phospholipids) to help keep them"fluid" even when our cells are exposed to cooler temperatures.

Many people have hear the claims that egg yolk contains too much cholesterol, thus should not be eaten. An interesting study was done at Purdue University a number of years ago to test this. Men in one group each ate an egg a day, while men in another group were not allowed to eat eggs. Each of these groups was further subdivided such that half the men got"lots" of exercise while the other half were"couch potatoes." The results of this experiment showed no significant difference in blood cholesterol levels between egg-eaters and non-egg-eaters while there was a very significant difference between the men who got exercise and those who didn't.

Lipoproteins are clusters of proteins and lipids all tangled up together. These act as a means of carrying lipids, including cholesterol, around in our blood. There are two main categories of lipoproteins distinguished by how compact/dense they are. LDL or low density lipoprotein is the"bad guy," being associated with deposition of"cholesterol" on the walls of someone's arteries. HDL or high density lipoprotein is the"good guy," being associated with carrying"cholesterol" out of the blood system, and is more dense/more compact than LDL.

Triglycerides

The majority of consumed lipids are triglycerides, which contain three fatty acids and a glycerol molecule. Fat is stored in the form of triglycerides, which we manufacture when excess energy is consumed. We store triglycerides in adipose tissues (groups of fat cells) and inside muscle cells (intramuscular tri glyceride), both of which are available as an energy source when needed. When fat is burned as a source of energy, the stored triglycerides are taken out of storage, and each molecule is cleaved into its component fatty acids and glycerol molecule. Each fatty acid can then be broken apart (two carbon units at a time) and thrown into the cellular furnaces for the creation of ATP to form heat and provide the energy for muscular work. This process is referred to as the beta-oxidative metabolic pathway because burning fat, besides requiring some carbohydrate for its complete oxidation, also requires oxygen.

Glycerol is a unique lipid that is burned like a carbohydrate rather than a fat and is also an effective humectant (it holds water). Some long-endurance athletes find that adding glycerol to water helps them retain more water (i.e., to superhydrate) than if they consumed water alone. In an extremely hot and humid environment, water loss is likely to be higher than the athlete's fluid-replacement capacity, so beginning a competition in a superhydrated state may provide some advantages.

In studies of tennis and Olympic distance triathlon, athletes consuming glycerol in water prior to competition experienced some protective hyperhydration benefits when exercising in high heat.17,18 However, being superhydrated is likely to impart a degree of discomfort that requires some adaptation. Athletes who consume fluids containing glycerol often describe the feeling of holding extra body water as making them feel"like a water bag," or"heavy," or"stiff." Still, they maintain that how they feel at the end of a race is more important than how they feel at the beginning, so adding glycerol to precompetition fluids has become a standard protocol for some athletes.

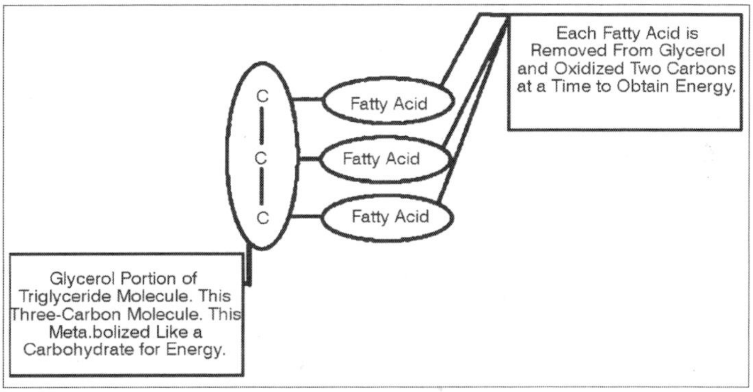

FATTY ACIDS

Linoleic (omega-6) and linolenic (omega-3) fatty acids are the essential fatty acids; although they are needed for metabolic processes, we are incapable of synthesizing them. The"omega-6" classification means that these fatty acids, which are polyunsaturated, have the last double bond six carbons from the end of the carbon chain. The"omega-3" classification means that these fatty acids have the last double bond three carbons from the end of the carbon chain. Linoleic acid is an essential part of lipid membranes and is required for normal skin health. Linolenic acid is necessary for neural function and growth. The AMDR for omega-6 andomega-3 fatty acids is 5 to 10 and.6 to 1.2 grams per day, respectively.11Both fatty acids are easily obtained from vegetable oils (corn, safflower, canola, and so on) and the oils of cold-water fish. Fish liver oils that are high in omega-3 acids have recently received much attention. These oils have been shown to reduce the ability of red blood cells to congregate, thereby decreasing the chance for an unwanted blood clot to form. This reduces the risk of a heart attack, which is most commonly caused by a clot formation in one of the major heart arteries. The oils from cold-water fish are the main source of the omega-3 fatty acids eicosapentanoic acid (EPA) and docosohexanoic acid (DHA).

Even a once-weekly consumption of cold-water fish (salmon, albacore tuna, Atlantic herring) is sufficient to significantly reduce the risk of heart attack and stroke.11Despite these findings, excessive intake of these fish oils may cause problems, including an increase in cellular oxidative damage. The best rule of thumb is to make fish consumption a regular part of the weekly diet so that supplemental intake of omega-3 fatty acids is unnecessary.

FAT REQUIREMENTS

From an exercise standpoint, there is little reason to believe that increasing fat consumption results in improving athletic performance, unless the increase in fat intake is the only reasonable means for the athlete to obtain sufficient energy. For the athlete who needs more than 4,000 calories each day to meet the combined demands of growth, exercise, and maintenance, moderate increases in dietary fat (preferably from plant and fish sources) may be needed. Since fat is a more concentrated form of energy than either carbohydrate or protein, more energy can be consumed in a smaller food package if the foods contain more fat.

LIPIDS AND PHYSICAL ACTIVITY

Even the leanest, healthiest athletes have a substantial energy pool of stored lipids. The average storage in adipose tissue ranges between 50,000 and 100,000 calories, or enough energy to walk or run 500 to 1,000 miles19 (800 to 1,600 kilometres) without a refueling stop. In addition, athletes store approximately 2,000 to 3,000 calories of lipids inside the muscle tissue.20 These lipids, which are stored in the form of triglycerides, are available as a fuel under the proper conditions of oxygen availability. Maximal fat oxidation occurs at 60 to 65 Per cent VO_2 max, but at higher levels of VO2max there is insufficient oxygen to derive the majority of energy used from fat catabolism.

Triglycerides stored in adipose tissue are cleaved into their component molecules of glycerol and fatty acids and transported to the blood plasma. The glycerol is available to all tissues for energy metabolism, and the free fatty acids are transported to working muscles where they are oxidized for energy. The triglycerides stored in working muscles are cleaved into glycerol and fatty acids, and the fatty acids can be oxidized for energy where they are resident. The glycerol can also be burned for energy in the working muscle or can be transported to the blood plasma as a source of energy for other tissues. The lower the exercise intensity, the greater the proportion of fat burned to satisfy energy needs. As exercise intensity increases, the proportion of fat burned decreases and the proportion of carbohydrate burned increases. It is this basic reality that is behind why so many people do low-intensity activity to burn fat and lower body fat levels. However, the proportion of fat burned should not be confused by the total amount of fat burned at different intensities of physical activity.

As exercise intensity increases, the total number of calories burned per unit of time also increases. Although there may be a decrease in the proportion of fat burned to satisfy total energy needs in higher-intensity activity, the total volume of fat burned is greater because the total energy requirement is higher. The take-away message from this metabolic reality is that athletes interested in lowering body fat should exercise at least as high as 65 Per cent of VO2 max for the duration of their workouts to optimize the total mass of fat that is burned. Exercising at lower intensities burns a greater proportion of fat but less total fat than exercising at higher intensities.

BIOSYNTHESIS OF FATTY ACID

Identification of Genes in Biosynthesis of Fatty Acids and Cholesterol Pathways Using Microarrays Sterol regulatory binding proteins (SREBP) are a family of basic helix-loop-helix leucine zipper transcription factors that regulate biosynthesis of fatty acids and cholesterol. To identify candidate genes in the fatty acid synthesis pathway, transgenic mice over expressing SREBP-1a and SREBP-2 isofroms in liver were used. Labeled cRNA was prepared from liver and hybridized to mouse expression microarrays interrogating ~30,000 genes/ESTs. Some of the known genes, ACC, FAS, and ATP citrate lyase regulated by over expression of SREBP were identified by microarray analysis.

A novel gene, long chain fatty acyl elongase (LCE) regulated by the SREBP1a and 2 was also identified. LCE was over expressed with a fold change of 25 and 9 in SREBP-1a and SREBP-2 mice respectively. The confirmation of LCE was done by northern blot analysis, sequence homology searches and biochemical tests. This enzyme specifically catalyzes the elongation of 12, 14 and 16 carbons but does not elongate the C-18 fatty acids or very long chain fatty acids.

DEGRADATION OF FATTY ACIDS

Multi-million pounds of viscera are generated from the catfish filleting process. Our previous study showed that catfish viscera contained 30-35%

crude fat, and catfish visceral oil contained health-promoting fatty acids. Catfish visceral oils show potential for use as nutrient supplement and as cooking oil. Information on thermal stability and degradation of catfish visceral oils is not available. The objective of this study was to determine weight loss (%) of catfish visceral oils and fatty acids after having been subjected to heat treatment. Oil samples included crude, degummed, neutralized, bleached, and deodorized oils.

The oils and individual fatty acids, FAs (C14:0, C16:0, C18:0, C20:0, C16:1, C18:1, C18:2, C18:3, C20:2, C20:4, C22:6) were subjected to the Thermogravimetric Analyser. Samples (0.5-1 mg) was placed in the furnace and heated up to 600°C at the rate of 5°C/min. Weight loss (%) of oils and fatty acids were calculated based on the original weight of non-heated samples. Duplicate experiments were conducted.

Over 90% weight loss of all FAs (except C18:1 and C20:2) was observed at 250°C. Less than 2% remaining weight of FAs (except C20:2) was observed at 350°C. Saturated FAs completely decomposed above 400°C. Comparing C18:1 with C18:2 and C18:3, weight loss increased with increased double bonds. Between 200-550°C, weight loss of oils slightly increased with increased temperature, regardless of the purification process. There was, however, no distinct weight loss of all oils below 300°C. Oils with less impurity had greater weight loss especially at 450-500°C. At 550°C, all oil samples completely decomposed. Knowledge of thermal stability or degradation (weight loss through eVapouration or decomposition) of catfish oils during purification or cooking is essential for the design of processing units, analysis of production cost, and cooking quality if to be used as cooking or seasoning oil.

MOBILIZATION OF FAT STORES

The primary sources of fatty acids for oxidation are dietary and mobilization from cellular stores. Fatty acids from the diet can are delivered from the gut to cells via transport in the blood. Fatty acids are stored in the form of triacylglycerols primarily within adipocytes of adipose tissue. In response to energy demands, the fatty acids of stored triacylglycerols can be mobilized for use by peripheral tissues. The release of metabolic energy, in the form of fatty acids, is controlled by a complex series of interrelated cascades that result in the activation of *hormone-sensitive lipase.*

The stimulus to activate this cascade, in adipocytes, can be glucagon, epinephrine or b-corticotropin. These hormones bind cell-surface receptors that are coupled to the activation of *adenylate cyclase* upon ligand binding. The resultant increase in cAMP leads to activation of PKA, which in turn phosphorylates and activates hormone-sensitive lipase. This enzyme hydrolyzes fatty acids from carbon atoms 1 or 3 of triacylglycerols. The resulting diacylglycerols are substrates for either *hormone-sensitive lipase* or for the non-inducible enzyme *diacylglycerol lipase*. Finally the monoacylglycerols are substrates for *monoacylglycerol lipase*. The net result of the action of these enzymes is three moles of free fatty acid and one mole of glycerol. The free fatty acids diffuse from adipose cells, combine with albumin in the blood, and are thereby transported to other tissues, where they passively diffuse into cells.

Hormone Induced Fatty Acid Mobilization in Adipocytes

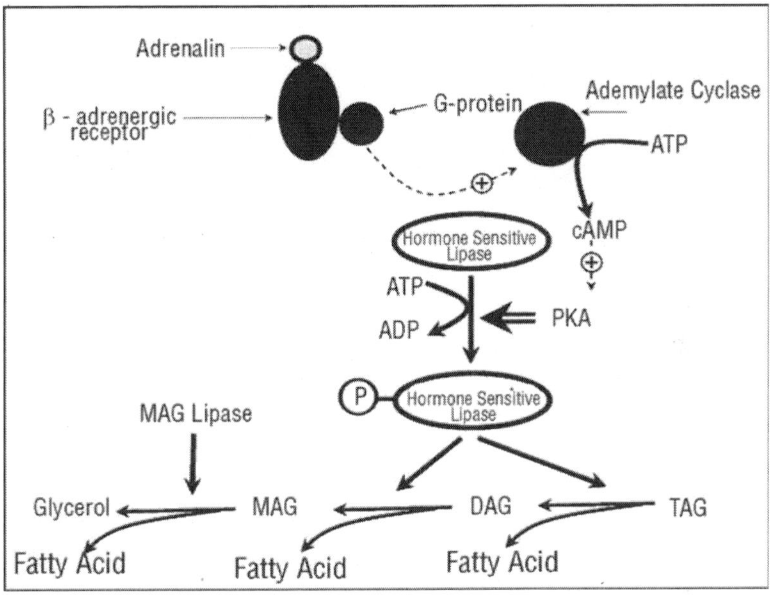

Model for the activation of *hormone-sensitive lipase* by epinephrine. Epinephrine binds its' receptor and leads to the activation of *adenylate cyclase*. The resultant increase in cAMP activates PKA which then phosphorylates and activates *hormone-sensitive lipase*. *Hormone-sensitive lipase* hydrolyzes fatty acids from triacylglycerols and diacylglycerols. The final fatty acid is released from monoacylglycerols through the action of *monoacylglycerol lipase*, an enzyme active in the absence of hormonal stimulation. In contrast to the hormonal activation of *adenylate cyclase* and (subsequently) *hormone-sensitive lipase* in adipocytes, the mobilization of fat from adipose tissue is inhibited by numerous stimuli. The most significant inhibition is that exerted upon *adenylate cyclase* by insulin. When an individual is well fed state, insulin released from the pancreas prevents the inappropriate mobilization of stored fat. Instead, any excess fat and carbohydrate are incorporated into the triacylglycerol pool within adipose tissue.

Activation and Transfer of Fatty Acid

Fatty acids must be activated in the cytoplasm before being oxidized in the mitochondria. Activation is catalyzed by *fatty acyl-CoA ligase* (also called *acyl-CoA synthetase* or *thiokinase*). The net result of this activation process is the consumption of 2 molar equivalents of ATP.

Fatty acid + ATP + CoA $\longrightarrow$ Acyl-CoA + PP_i + AMP

Oxidation of fatty acids occurs in the mitochondria. The transport of fatty acyl-CoA into the mitochondria is accomplished via an acyl-carnitine intermediate, which itself is generated by the action of *carnitine acyltransferase I*, an enzyme that resides in the outer mitochondrial membrane, from acyl-CoA and carnitine.

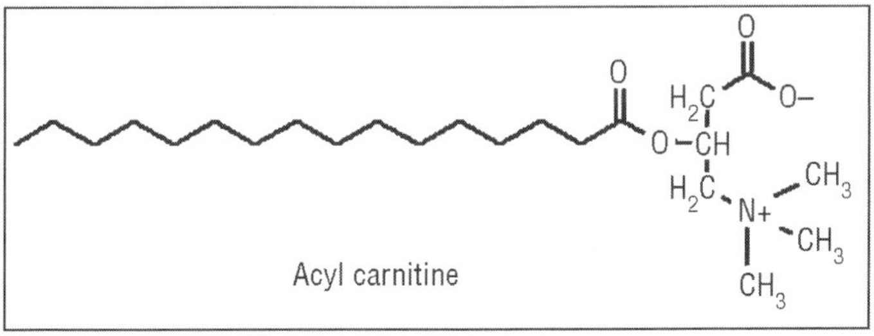

The acyl-carnitine molecule then is transported into the mitochondria where *carnitine acyltransferase II* catalyzes the regeneration of the fatty acyl-CoA molecule.

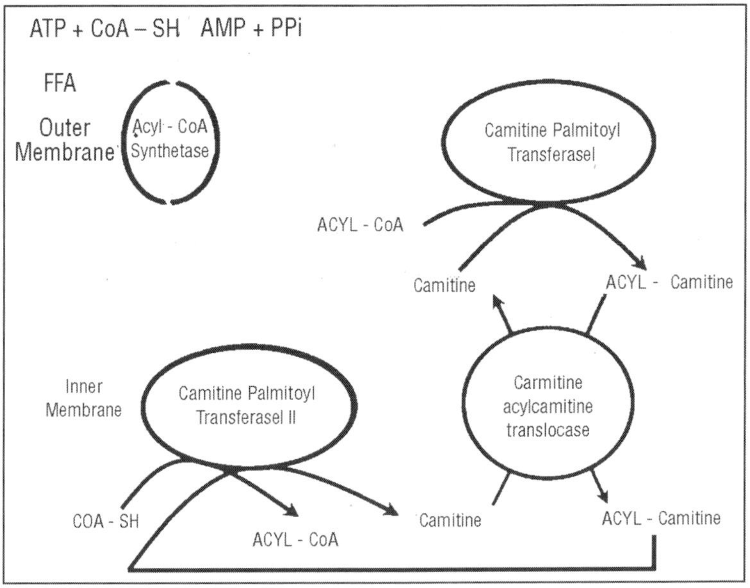

Transport of fatty acids from the cytoplasm to the inner mitochondrial space for oxidation. Following activation to a fatty-CoA, the CoA is exchanged for carnitine by *carnitine-palmitoyltransferase I*. The fatty-carnitine is then transported to the inside of the mitochondrion where a reversal exchange takes place through the action of *carnitine-palmitoyltransferase II*. Once inside the mitochondrion the fatty-CoA is a substrate for the b-oxidation machinery. The human body produces carnitine through an enzymatically catalyzed biosynthetic pathway out of lysine and methionine.

REACTION SEQUENCE IN THE β-OXIDATION

The process of fatty acid oxidation is termed β-oxidation since it occurs through the sequential removal of 2-carbon units by oxidation at the β-carbon position of the fatty acyl-CoA molecule.

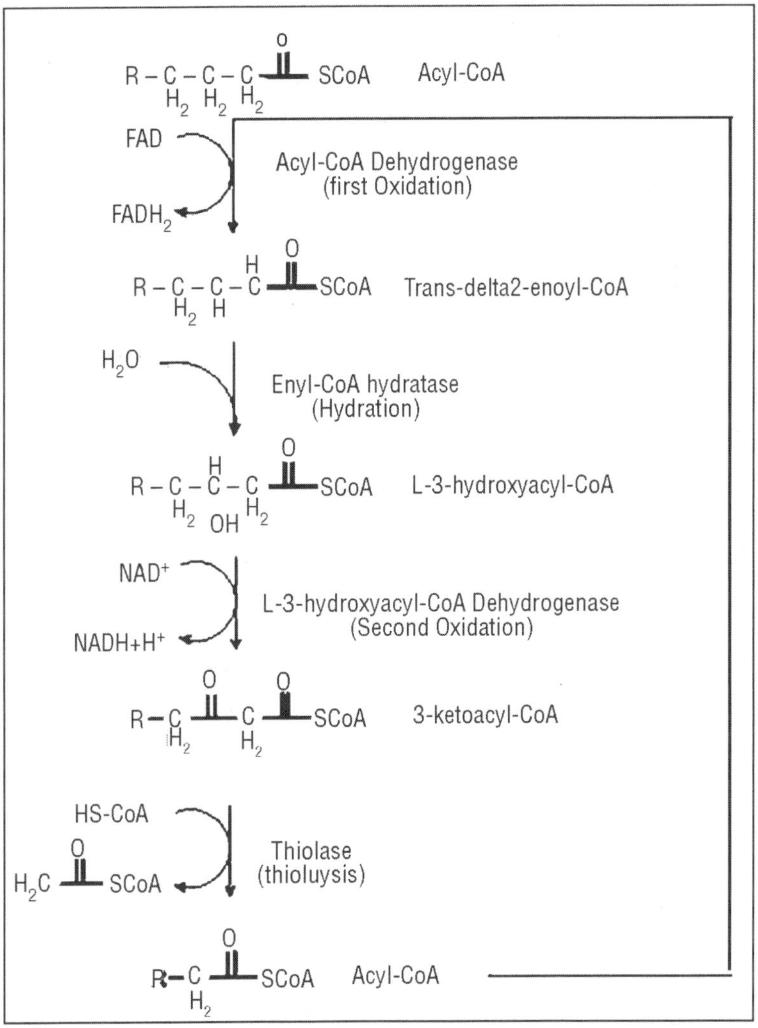

Each round of β-oxidation produces one mole of NADH, one mole of $FADH_2$ and one mole of acetyl-CoA. The acetyl-CoA— the end product of each round of β-oxidation— then enters the TCA cycle, where it is further oxidized to CO_2 with the concomitant generation of three moles of NADH, one mole of $FADH_2$ and one mole of ATP.

The NADH and $FADH_2$ generated during the fat oxidation and acetyl-CoA oxidation in the TCA cycle then can enter the respiratory pathway for the production of ATP. The oxidation of fatty acids yields significantly more energy per carbon atom than does the oxidation of carbohydrates. The net result of the oxidation of one mole of oleic acid (an 18-carbon fatty acid) will be 146 moles of ATP (2 mole equivalents are used during the activation of the fatty acid), as compared with 114 moles from an equivalent number of glucose carbon atoms.

ALTERNATIVE OXIDATION PATHWAYS

The majority of natural lipids contain an even number of carbon atoms. A small proportion that contain odd numbers; upon complete β-oxidation, these yield acetyl-CoA units plus a single mole of propionyl-CoA.

The propionyl-CoA is converted, in an ATP-dependent pathway, to succinyl-CoA. The succinyl-CoA can then enter the TCA cycle for further oxidation. The oxidation of unsaturated fatty acids is essentially the same process as for saturated fats, except when a double bond is encountered.

In such a case, the bond is isomerized by a specific *enoyl-CoA isomerase* and oxidation continues. In the case of linoleate, the presence of the D-12 unsaturation results in the formation of a dienoyl-CoA during oxidation. This molecule is the substrate for an additional oxidizing enzyme, the NADPH requiring *2,4-dienoyl-CoA reductase*. Phytanic acid is a fatty acid present in the tissues of ruminants and in dairy products and is, therefore, an important dietary component of fatty acid intake.

Because phytanic acid is methylated, it cannot act as a substrate for the first enzyme of the β-oxidation pathway (acyl-CoA dehydrogenase). An additional mitochondrial enzyme, *a-hydroxylase*, adds a hydroxyl group to the α-carbon of phytanic acid, which then serves as a substrate for the remainder of the normal oxidative enzymes. This process is termed α-oxidation.

REGULATION OF FATTY ACID METABOLISM

In order to understand how the synthesis and degradation of fats needs to be exquisitely regulated, one must consider the energy requirements of the organism as a whole. The blood is the carrier of triacylglycerols in the form of VLDLs and chylomicrons, fatty acids bound to albumin, amino acids, lactate, ketone bodies and glucose. The pancreas is the primary organ involved in sensing the organism's dietary and energetic states by monitoring glucose concentrations in the blood. Low blood glucose stimulates the secretion of glucagon, whereas, elevated blood glucose calls for the secretion of insulin.

The metabolism of fat is regulated by two distinct mechanisms. One is short-term regulation, which can come about through events such as substrate availability, allosteric effectors and/or enzyme modification. The other mechanism, long-term regulation, is achieved by alteration of the rate of enzyme synthesis and turn-over.

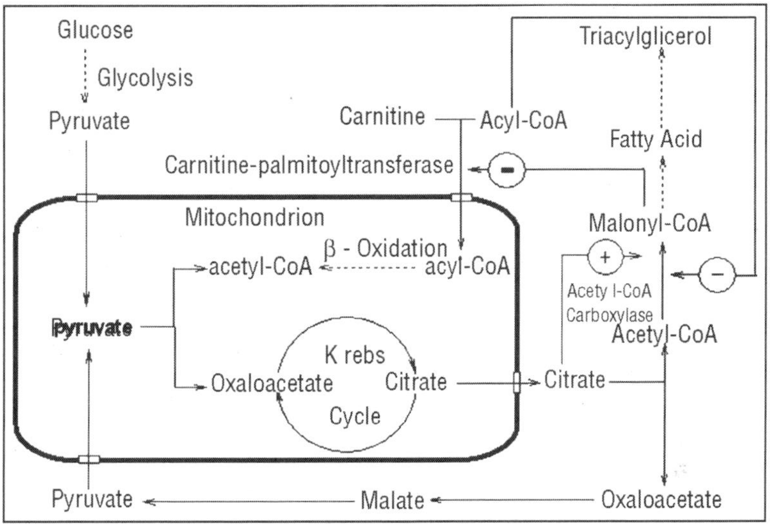

ACC (acetyl-CoA carboxylase) is the rate-limiting (committed) step in fatty acid synthesis. This enzyme is activated by citrate and inhibited by palmitoyl-CoA and other long-chain fatty acyl-CoAs. ACC activity can also be affected by phosphorylation.

For instance, glucagon-stimulated increases in PKA activity result in the phosphorylation of certain serine residues in ACC leading to decreased activity of the enzyme. By contrast, insulin leads to PKA-independent phosphorylation of ACC at sites distinct from glucagon, which bring about increased ACC activity. Both of these reaction chains are examples of short-term regulation.

Insulin, a product of the well-fed state, stimulates ACC and FAS synthesis, whereas starvation leads to a decrease in the synthesis of these enzymes. Adipose tissue levels of lipoprotein lipase also are increased by insulin and decreased by starvation. However, the effects of insulin and starvation on lipoprotein lipase in the heart are just the inverse of those in adipose tissue.

This sensitivity allows the heart to absorb any available fatty acids in the blood in order to oxidize them for energy production. Starvation also leads to increases in the levels of cardiac enzymes of fatty acid oxidation, and to decreases in FAS and related enzymes of synthesis.

Adipose tissue contains *hormone-sensitive lipase,* which is activated by PKA-dependent phosphorylation; this activation increases the release of fatty acids into the blood. This in turn leads to the increased oxidation of fatty acids in other tissues such as muscle and liver. In the liver, the net result (due to increased acetyl-CoA levels) is the production of ketone bodies. This would occur under conditions in which the carbohydrate stores and gluconeogenic precursors available in the liver are not sufficient to allow increased glucose production. The increased levels of fatty acid that become available in response to glucagon or epinephrine are assured of being completely oxidized, because PKA also phosphorylates ACC; the synthesis of fatty acid is thereby inhibited. Insulin has the opposite effect to glucagon and epinephrine: it increases the synthesis of triacylglycerols (and glycogen).

One of the many effects of insulin is to lower cAMP levels, which leads to increased dephosphorylation through the enhanced activity of protein phosphatases such as PP-1. With respect to fatty acid metabolism, this yields dephosphorylated and inactive hormone-sensitive lipase. Insulin also stimulates certain phosphorylation events. This occurs through activation of several cAMP-independent kinases, one of which phosphorylates and thereby stimulates the activity of ACC. Fat metabolism can also be regulated by malonyl-CoA-mediated inhibition of carnitine acyltransferase I. Such regulation serves to prevent *de novo* synthesized fatty acids from entering the mitochondria and being oxidized.

TRANS FATTY ACIDS

A trans fatty acid (commonly shortened to trans fat) is an unsaturated fatty acid molecule that contains a trans double bond between carbon atoms, which makes the molecule less'kinked' in comparison to fatty acids with cis double bonds. These bonds are characteristically produced during industrial hydrogenation of plant oils. Since they are also produced in bacterial metabolism, ruminant fats (e.g. in milk) also contain about 4% trans fatty acids. Research suggests that amounts of trans fats correlate with circulatory diseases such as atherosclerosis and coronary heart disease more than the same amount of cis fats, for reasons that are not fully understood. It is known, however, that trans fats, just like saturated fats, raise the LDL ("bad") cholesterol and lowers the HDL ("good") cholesterol. They have also been shown to have other harmful effects such as increasing triglycerides and Lp(a) lipoproteins. They are also thought to cause more inflammation, which is thought to occur through damage to the cells lining of blood vessels.

Long and Short

In addition to saturation, fatty acids are short, medium, or long.

- Short-chain fatty acids (SCFA) are fatty acids with aliphatic tails of fewer than six carbons.
- Medium-chain fatty acids (MCFA) are fatty acids with aliphatic tails of 6-12 carbons, which can form medium-chain triglycerides.
- Long-chain fatty acids (LCFA) are fatty acids with aliphatic tails longer than 12 carbons.
- Very-Long-chain fatty acids (VLCFA) are fatty acids with aliphatic tails longer than 22 carbons

When discussing essential fatty acids (EFA), a slightly different terminology applies. Short-chain EFA are 18 carbons long; long-chain EFA have 20 or more carbons.

Free Fatty Acids

Fatty acids can be bound or attached to other molecules, such as in triglycerides or phospholipids. When they are not attached to other molecules, they are known as"free" fatty acids. The uncombined fatty acids or free fatty acids may come from the breakdown of a triglyceride into its components (fatty acids and glycerol). However as fats are insoluble in water they must be bound to appropriate regions in the plasma protein albumin for transport around the body. The levels of"free fatty acid" in the blood are limited by the number of albumin binding sites available.

Free fatty acids are an important source of fuel for many tissues since they can yield relatively large quantities of ATP. Many cell types can use either glucose or fatty acids for this purpose. In particular, heart and skeletal muscle prefer fatty acids. The brain cannot use fatty acids as a source of fuel; it relies on glucose, or on ketone bodies. Ketone bodies are produced in the liver by fatty acid metabolism during starvation, or during periods of low carbohydrate intake.

Acidity

Short-chain carboxylic acids such as formic acid and acetic acid are miscible with water and dissociate to form reasonably strong acids (pKa 3.77 and 4.76, respectively). Longer-chain fatty acids do not show a great change in pKa. Nonanoic acid, for example, has a pKa of 4.96. However, as the chain length increases the solubility of the fatty acids in water decreases very rapidly, so that the longer-chain fatty acids have very little effect on the pH of a solution. The significance of their pKa values therefore has relevance only to the types of reactions in which they can take part.

Even those fatty acids that are insoluble in water will dissolve in warm ethanol, and can be titrated with sodium hydroxide solution using phenolphthalein as an indicator to a pale-pink endpoint. This analysis is used to determine the free fatty acid content of fats; i.e., the proportion of the triglycerides that have been hydrolyzed.

Reaction of Fatty Acids

Fatty acids react just like any other carboxylic acid, which means they can undergo esterification and acid-base reactions. Reduction of fatty acids yields fatty alcohols. Unsaturated fatty acids can also undergo addition reactions, most commonly hydrogenation, which is used to convert vegetable oils into margarine. With partial hydrogenation, unsaturated fatty acids can be isomerized from cis to trans configuration. In the Varrentrapp reaction certain unsaturated fatty acids are cleaved in molten alkali, a reaction at one time of relevance to structure elucidation.

Auto-oxidation and Rancidity

Fatty acids at room temperature undergo a chemical change known as auto-oxidation. The fatty acid breaks down into hydrocarbons, ketones, aldehydes, and smaller amounts of epoxides and alcohols. Heavy metals present at low levels in fats and oils promote auto-oxidation. Fats and oils often are treated with chelating agents such as citric acid.

CIRCULATION

Digestion and Intake

Short- and medium-chain fatty acids are absorbed directly into the blood via intestine capillaries and travel through the portal vein just as other absorbed nutrients do. However, long-chain fatty acids are too large to be directly released into the tiny intestine capillaries. Instead they are absorbed into the fatty walls of the intestine villi and reassembled again into triglycerides. The triglycerides are coated with cholesterol and protein (protein coat) into a compound called a chylomicron.

Within the villi, the chylomicron enters a lymphatic capillary called a lacteal, which merges into larger lymphatic vessels. It is transported via the lymphatic system and the thoracic duct up to a location near the heart (where the arteries and veins are larger). The thoracic duct empties the chylomicrons into the bloodstream via the left subclavian vein. At this point the chylomicrons can transport the triglycerides to where they are needed.

Utilization of dietary lipids requires that they first be absorbed through the intestine. As these molecules are oils they would be essentially insoluble in the aqueous intestinal environment. Solubilization (emulsification) of dietary lipid is accomplished via bile salts that are synthesized in the liver and secreted from the gallbladder.

The emulsified fats can then be degraded by pancreatic lipases (*lipase* and *phospholipase* A_2). These enzymes, secreted into the intestine from the pancreas, generate free fatty acids and a mixtures of mono- and diacylglycerols from dietary triacylglycerols. Pancreatic lipase degrades triacylglycerols at the 1 and 3 positions sequentially to generate 1,2-diacylglycerols and 2-acylglycerols. Phospholipids are degraded at the 2 position by pancreatic *phospholipase* A_2 releasing a free fatty acid and the lysophospholipid.

Following absorption of the products of pancreatic lipase by the intestinal mucosal cells, the resynthesis of triacylglycerols occurs. The triacylglycerols are then solubilized in lipoprotein complexes (complexes of lipid and protein) called chylomicrons. A chylomicron contains lipid droplets surrounded by the more polar lipids and finally a layer of proteins. Triacylglycerols synthesized in the liver are packaged into VLDLs and released into the blood directly. Chylomicrons from the intestine are then released into the blood via the lymph system for delivery to the various tissues for storage or production of energy through oxidation.

The triacylglycerol components of VLDLs and chylomicrons are hydrolyzed to free fatty acids and glycerol in the capillaries of adipose tissue and skeletal muscle by the action of *lipoprotein lipase*. The free fatty acids are then absorbed by the cells and the glycerol is returned via the blood to the liver (and kidneys). The glycerol is then converted to the glycolytic intermediate DHAP. The classification of blood lipids is distinguished based upon the density of the different lipoproteins. As lipid is less dense than protein, the lower the density of lipoprotein the less protein there is.

CLINICAL SIGNIFICANCE OF KETOGENESIS

The production of ketone bodies occurs at a relatively low rate during normal feeding and under conditions of normal physiological status. Normal physiological responses to carbohydrate shortages cause the liver to increase the production of ketone bodies from the acetyl-CoA generated from fatty acid oxidation. This allows the heart and skeletal muscles primarily to use ketone bodies for energy, thereby preserving the limited glucose for use by the brain.

The most significant disruption in the level of ketosis, leading to profound clinical manifestations, occurs in untreated insulin-dependent diabetes mellitus. This physiological state, diabetic ketoacidosis (DKA), results from a reduced supply of glucose (due to a significant decline in circulating insulin) and a concomitant increase in fatty acid oxidation (due to a concomitant increase in circulating glucagon). The increased production of acetyl-CoA leads to ketone body production that exceeds the ability of peripheral tissues to oxidize them. Ketone bodies are relatively strong acids (pK_a around 3.5), and their increase lowers the pH of the blood. This acidification of the blood is dangerous chiefly because it impairs the ability of hemoglobin to bind oxygen.

CHOLESTEROL METABOLISM

Cholesterol is an extremely important biological molecule that has roles in membrane structure as well as being a precursor for the synthesis of the steroid hormones and bile acids. Both dietary cholesterol and that synthesized *de novo* are transported through the circulation in lipoprotein particles. The same is true of cholesteryl esters, the form in which cholesterol is stored in cells. The synthesis and utilization of cholesterol must be tightly regulated in order to prevent over-accumulation and abnormal deposition within the body. Of particular importance clinically is the abnormal deposition of cholesterol and cholesterol-rich lipoproteins in the coronary arteries. Such deposition, eventually leading to atherosclerosis, is the leading contributory factor in diseases of the coronary arteries.

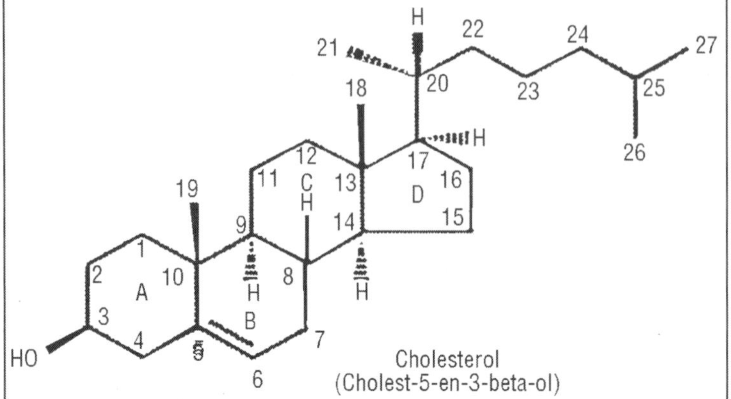

BIOSYNTHESIS OF CHOLESTEROL

Slightly less than half of the cholesterol in the body derives from biosynthesis *de novo*. Biosynthesis in the liver accounts for approximately 10%, and in the intestines approximately 15%, of the amount produced each day.

Cholesterol synthesis occurs in the cytoplasm and microsomes from the two-carbon acetate group of acetyl-CoA.

The process has nine major steps:

- Acetyl-CoAs are converted to 3-hydroxy- 3-methyl glutaryl-CoA (HMG-CoA).Unlike the HMG-CoA formed during ketone body synthesis in the mitochondria, this form is synthesized in the cytoplasm. However, the pathway and the necessary enzymes are the same as those in the mitochondria. Two moles of acetyl-CoA are condensed in a reversal of the *thiolase* reaction, forming acetoacetyl-CoA. Acetoacetyl-CoA and a third mole of acetyl-CoA are converted to HMG-CoA by the action of *HMG-CoA synthase*.
- HMG-CoA is converted to mevalonate by *HMG-CoA reductase* (this enzyme is bound to the endoplasmic reticulum). *HMG-CoA reductase* absolutely requires NADPH as a cofactor and two moles of NADPH are consumed during the conversion of HMG-CoA to mevalonate.

PATHWAY OF CHOLESTEROL BIOSYNTHESIS.

Synthesis begins with the transport of acetyl-CoA from the mitochondrion to the cytosol. The rate limiting step occurs at the *3-hydroxy-3-methylglutaryl-CoA (HMG-CoA) reducatse* catalyzed step. The phosphorylation reactions are required to solubilize the isoprenoid intermediates in the pathway. Intermediates in the pathway are used for the synthesis of prenylated proteins, dolichol, coenzyme Q and the side chain of heme *a*.

The acetyl-CoA utilized for cholesterol biosynthesis is derived from an oxidation reaction (eg, fatty acids or pyruvate) in the mitochondria and is transported to the cytoplasm by the same process as that described for fatty acid synthesis. Acetyl-CoA can also be derived from cytoplasmic oxidation of ethanol by *acetyl-CoA synthetase*. All the reduction reactions of cholesterol biosynthesis use NADPH as a cofactor.

The isoprenoid intermediates of cholesterol biosynthesis can be diverted to other synthesis reactions, such as those for dolichol (used in the synthesis of N-linked glycoproteins, coenzyme Q (of the oxidative phosphorylation) pathway or the side chain of heme *a*. Additionally, these intermediates are used in the lipid modification of some proteins.

PROTEOLYTIC REGULATION OF HMG-COA REDUCTASE

The stability of HMGR is regulated as the rate of flux through the mevalonate synthesis pathway changes. When the flux is high the rate of HMGR degradation is also high. When the flux is low, degradation of HMGR decreases. This phenomenon can easily be observed in the presence of the statin drugs. HMGR is localized to the ER and like SREBP contains a sterol-sensing domain,SSD. When sterol levels increase in cells there is an concommitant increase in the rate of HMGR degradation. The degradation of HMGR occurs within the proteosome, a multiprotein complex dedicated to protein degradation.

The primary signal directing proteins to the proteosome is ubiquitination. Ubiquitin is a 7.6kDa protein that is covalently attached to proteins targeted for degradation by ubiqitin ligases. These enzymes attach multiple copies of ubiquitin allowing for recognition by the proteosome. HMGR has been shown to be ubiquitinated prior to its degradation. The primary sterol regulating HMGR degradation is cholesterol itself. As the levels of free cholesterol increase in cells, the rate of HMGR degradation increases.

The Utilization of Cholesterol

Cholesterol is transported in the plasma predominantly as cholesteryl esters associated with lipoproteins. Dietary cholesterol is transported from the small intestine to the liver within chylomicrons. Cholesterol synthesized by the liver, as well as any dietary cholesterol in the liver that exceeds hepatic needs, is transported in the serum within LDLs. The liver synthesizes VLDLs and these are converted to LDLs through the action of endothelial cell-associated lipoprotein lipase. Cholesterol found in plasma membranes can be extracted by HDLs and esterified by the HDL-associated enzyme LCAT. The cholesterol acquired from peripheral tissues by HDLs can then be transferred to VLDLs and LDLs via the action of cholesteryl ester transfer protein (apo-D) which is associated with HDLs. Reverse cholesterol transport allows peripheral cholesterol to be returned to the liver in LDLs. Ultimately, cholesterol is excreted in the bile as free cholesterol or as bile salts following conversion to bile acids in the liver.

Regulation of Cellular Sterol Content

The continual alteration of the intracellular sterol content occurs through the regulation of key sterol synthetic enzymes as well as by altering the levels of cell-surface LDL receptors. As cells need more sterol they will induce their synthesis and uptake, conversely when the need declines synthesis and uptake are decreased.

Regulation of these events is brought about primarily by sterol-regulated transcription of key rate limiting enzymes and by the regulated degradation of HMGR. Activation of transcriptional control occurs through the regulated cleavage of the membrane-bound transcription factor sterol regulated element binding protein, SREBP. As discussed above, degradation of HMGR is controlled by the ubiquitin-mediated pathway for proteolysis.

Sterol control of transcription affects more than 30 genes involved in the biosynthesis of cholesterol, triacylglycerols, phospholipids and fatty acids. Transcriptional control requires the presence of an octamer sequence in the gene termed the sterol regulatory element, SRE-1. It has been shown that SREBP is the transcription factor that binds to SRE-1 elements. It turns out that there are 2 distinct SREBP gene, SREBP-1 and SREBP-2. In addition, the SREBP-1 gene encodes 2 proteins, SREBP-1a and SREBP-1c as a consequence of alternative exon usage.

All 3 proteins are proteolytically regulated by sterols. Full-length SREBPs have several domains and are embedded in the membrane of the endoplasmic reticulum (ER). The N-terminal domain contains a transcription factor motif of the basic helix-loop-helix (bHLH) type that is exposed to the cytoplasmic side of the ER. There are 2 transmembrane spanning domains followed by a large C-terminal domain also exposed to the cytosolic side. When sterols are scarce, cleavage of the full-length SREBP takes place with the result being that the N-terminla bHLH motif is released into the cytosol. The bHLH domian then migrates to the nucleus to direct transcription.

Conversely, when sterols are abundant, cleavage of SREBP is inhibited. To control the level of SREBP-mediated transcription, the soluble bHLH domain is itself subject to rapid proteolysis. The cleavage of SREBP is carried out by 2 distinct enzymes, one of which is regulated by sterols.

The regulated cleavage occurs in the lumenal loop between the 2 transmembrane domains. This cleavage is catalyzed by site-1 protease, S1P. High sterol content blocks the activity of S1P. The second cleavage, catalyzed by site-2 protease, S2P, occurs in the first transmembrane span, leading to release of active SREBP. In order for S2P to act on SREBP site-1 must already have been cleaved. Additional studies on sterol-regulated gene expression demonstrated that cleavage of SREBP by S1P is controlled by the level and action of an additional protein termed, SREBP cleavage-activating protein, SCAP. SCAP is a large protein also found in the ER membrane and contains at least 8 transmembrane spans. The C-terminal portion, which extends into the cytosol, has been shown to interact with the C-terminal domain of SREBP.

This C-terminal region of SCAP contains 4 motifs called WD40 repeats. The WD40 repeats are required for interaction of SCAP with SREBP. Interestingly, the N-terminus of SCAP, including membrane spans 2-6, resembles HMGR which itself is subject to sterol-stimulated degradation. This shared motif is called the sterol sensing domain, SSD. Several proteins whose functions involve streols also contain the SSD.

These include patched, an important development regulating receptor whose ligand, hedgehog, is modified by attachment of cholesterol and the Niemann Pick C1 (NPC1) protein which is involved in cholesterol transport in the secretory pathway. The function of SCAP is to positively stimulate S1P-mediated cleavage of SREBP. The function of sterols is to inhibit this positive action of SCAP. The activity of SCAP involves movement from the ER to the Golgi and back. Because the C-terminus of SCAP interacts with SREBP, movement of SCAP takes SREBP along for the ride.

When sterols are low, SCAP and SREBP move to the Golgi. This transit is required for SREBP cleavage as S1P is Golgi-localized. When sterols are high, movement of SCAP is halted. Thus, the overall effect of sterols is to regulate the ability of SCAP to present SREBP to S1P.

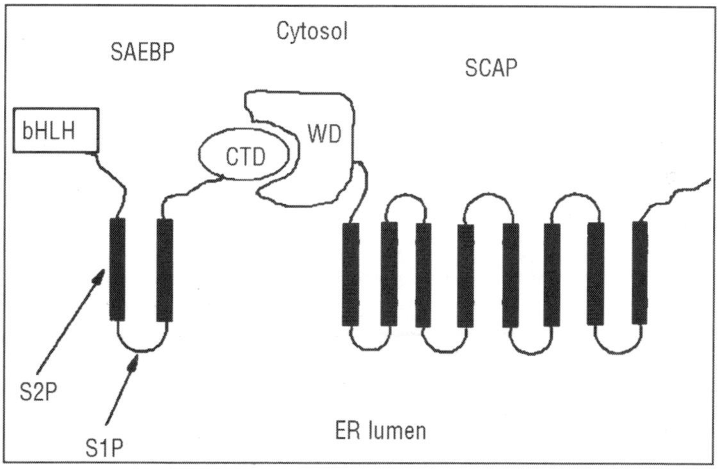

Diagramatic representation of the interactions between SREBP and SCAP in the membrane of the ER.

CTD = C-terminal Domain.

BILE ACIDS SYNTHESIS AND UTILIZATION

The end products of cholesterol utilization are the bile acids, synthesized in the liver. Synthesis of bile acids is the predominant mechanisms for the excretion of excess cholesterol. However, the excretion of cholesterol in the form of bile acids is insufficient to compensate for an excess dietary intake of cholesterol.

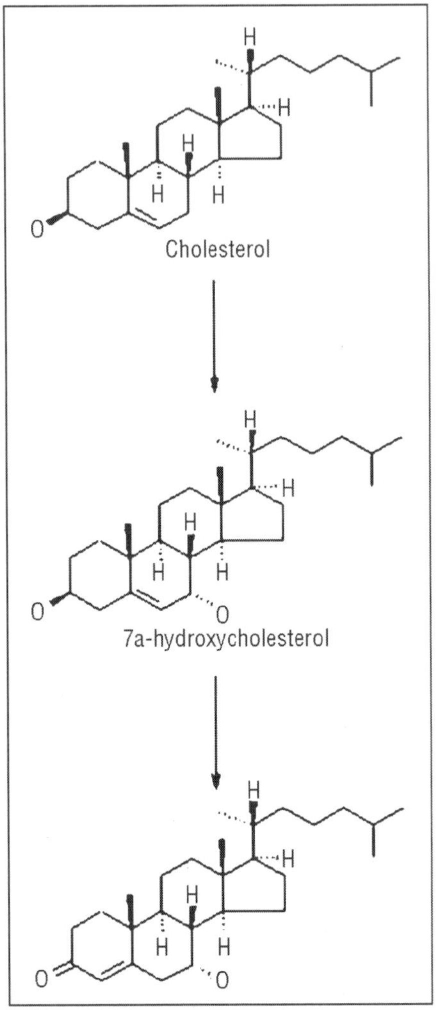

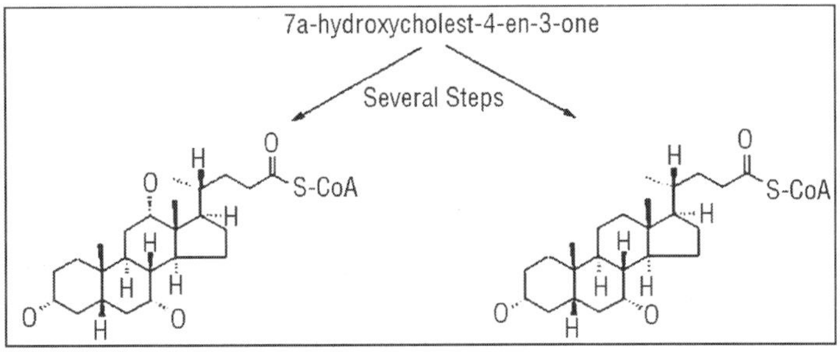

Synthesis of the 2 primary bile acids, cholic acid and chenodeoxycholic acid. The reaction catalyzed by the 7a-hydroxylase is the rate limiting step in bile acid synthesis. Conversion of 7a-hydroxycholesterol to the bile acids requires several steps not shown in detail in this image.

The most abundant bile acids in human bile are chenodeoxycholic acid (45%) and cholic acid (31%). These are referred to as the primary bile acids. Within the intestines the primary bile acids are acted upon by bacteria and converted to the secondary bile acids, identified as deoxycholate (from cholate) and lithocholate (from chenodeoxycholate). Both primary and secondary bile acids are reabsorbed by the intestines and delivered back to the liver via the portal circulation.

Within the liver the carboxyl group of primary and secondary bile acids is conjugated via an amide bond to either glycine or taurine before their being re-secreted into the bile canaliculi. These conjugation reactions yield glycoconjugates and tauroconjugates, respectively. The bile canaliculi join with the bile ductules, which then form the bile ducts. Bile acids are carried from the liver through these ducts to the gallbladder, where they are stored for future use. The ultimate fate of bile acids is secretion into the intestine, where they aid in the emulsification of dietary lipids. In the gut the glycine and taurine residues are removed and the bile acids are either excreted (only a small percentage) or reabsorbed by the gut and returned to the liver. This process of secretion from the liver to the gallbladder, to the intestines and finally re-absorbtion is termed the enterohepatic circulation.

CLINICAL SIGNIFICANCE OF BILE ACID SYNTHESIS

Bile acids perform four physiologically significant functions:

- Their synthesis and subsequent excretion in the feces represent the only significant mechanism for the elimination of excess cholesterol.
- Bile acids and phospholipids solubilize cholesterol in the bile, thereby preventing the precipitation of cholesterol in the gallbladder.
- They facilitate the digestion of dietary triacylglycerols by acting as emulsifying agents that render fats accessible to pancreatic lipases.
- They facilitate the intestinal absorption of fat-soluble vitamins.

MASTER METABOLIC REGULATOR

AMP-activated protein kinase (AMPK) was first discovered as an activity that inhibited preparations of acetyl-CoA carboxylase (ACC) and 3-hydroxy-3-methylglutaryl-CoA reductase (HMG-CoA reductase, HMGR) and was induced by AMP. AMPK induces a cascade of events within cells in response to the ever changing energy charge of the cell. The role of AMPK in regulating cellular energy charge places this enzyme at a central control point in maintaining energy homeostasis. More recent evidence has shown that AMPK activity can also be regulated by physiological stimuli, independent of the energy charge of the cell, including hormones and nutrients. Once activated, AMPK-mediated phosphorylation events switch cells from active ATP consumption (e.g. fatty acid and cholesterol biosynthesis) to active ATP production (e.g. fatty acid and glucose oxidation). These events are rapidly initiated and are referred to as short-term regulatory processes. The activation of AMPK also exerts long-term effects at the level of both gene expression and protein synthesis. Other important activities attributable to AMPK are regulation of insulin synthesis and secretion in pancreatic islet b-cells and modulation of hypothalamic functions involved in the regulation of satiety.

STRUCTURE OF AMPK

The mammalian AMPK is a trimeric enzyme composed of a catalytic α subunit and the non-catalytic β and γ subunits. There are two genes encoding isoforms of both the α and β subunits (α1, α2, β1 and β2) and three genes encoding isoforms of the g subunit (γ1-γ3). The α2 isoform is the subunit of AMPK found predominantly within skeletal and cardiac muscle, whereas, approximately equal distribution of both the α1 and α2 isoforms are present in hepatic AMPK. Within pancreatic islet β-cells the α1 isoform predominates.

The N-terminal half of the a subunits contains a typical serine/threonine kinase catalytic domain. Interaction with the β and γ subunits occurs via the C-terminal half of the α subunits. The yeast AMPK β subunits are lipid modified with myristic acid. Myristoylation may account for the membrane association mammalian AMPK. The core of the β subunits have α glycogen-binding domain (GBD). This domain is closely related to the isoamylase N domain subfamily and weakly related to domains in the glycogen-targeting phosphatase subunits and several starch-binding proteins. The close proximity of AMPK to cellular glycogen stores allows it to rapidly affect changes in glycogen metabolism in response to changes in metabolic demands. The γ subunits of AMPK have been shown to contain nucleotide binding sites with similarity to cystathionine β-synthase (CBS) domains. Indeed, direct AMP-binding studies have shown that AMP is bound to the γ subunits by a pair of CBS domains. Of clinical significance is the observation that mutations in the CBS domains of the γ2 subunit are associated with Wolff-Parkinson-White syndrome and familial hypertrophic cardiomyopathy.

REGULATION OF AMPK

AMPK is activated by phosphorylation by one or more upstream AMPK kinases (AMPKKs). In the absence of phosphorylation there is no detectable activity of AMPK towards any substrates. Phosphorylation of AMPK occurs in the a subunit at threonine 172 (Thr-172) which lies in the activation loop. One kinase activator of AMPK is Ca^{2+}-calmodulin-dependent kinase kinase b (CAMKKb) which phosphorylates and activates AMPK in response to increased calcium. Recent evidence has demonstrated that the threonine kinase, LKB1, encoded by the Peutz-Jegher syndrome tumor suppressor gene, is required for activation of AMPK in response to stress. Loss of LKB1 activity in adult mouse liver leads to near complete loss of AMPK activity and is associated with hyperglycemia. The hyperglycemia is, in part, due to an increase in the transcription of gluconeogenic genes. Of particular significance is the increased expression of the peroxisome proliferator-activated receptor-γ (PPAR-γ) coactivator 1α (PGC-1α) which drives gluconeogenesis. Reduction in PGC-1α activity results in normalized blood glucose levels in LKB1-deficient mice. As the name implies, AMPK is also regulated by AMP. The effects of AMP are two-fold: a direct allosteric activation and making AMPK a poorer substrate for dephosphorylation. Because AMP affects both the rate of AMPK phoshorylation in the positive direction and dephosphorylation in the negative direction, the cascade is ultrasensitive. This means that a very small rise in AMP levels can induce a dramatic increase in the activity of AMPK. The activity of adenylate kinase, catalyzing the reaction shown below, ensures that AMPK is highly sensitive to small changes in the intracellular [ATP]/[ADP] ratio.

$$2\ ADP \longrightarrow ATP + AMP$$

Negative allosteric regulation of AMPK also occurs and this effect is exerted by phosphocreatine. As indicated above, the β subunits of AMPK have a glycogen-binding domain, GBD. In muscle, a high glycogen content represses AMPK activity and this is likely the result of interaction between the GBD and glycogen, although this has not been shown directly. As suggested above, the GBD of AMPK allows association of the enzyme with the regulation of glycogen metabolism by placing AMPK in close proximity to one of its substrates glycogen synthase. AMPK has also been shown to be activated by receptors that are coupled to phospholipase C-γ (PLC-γ) and by cytokines secreted by adipose tissue (adipocytokines) such as leptin and adiponectin.

TARGETS OF AMPK

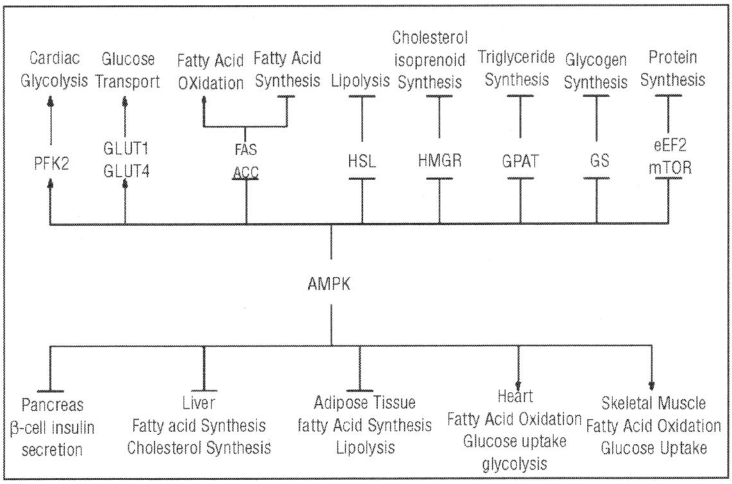

The signaling cascades initiated by the activation of AMPK exert effects on glucose and lipid metabolism, gene expression and protein synthesis. These effects are most important for regulating metabolic events in the liver, skeletal muscle, heart, adipose tissue, and pancreas.

Demonstration of the central role of AMPK in the regulation of metabolism in response to events such as nutrient- or exercise-induced stress. Several of the known physiologic targets for AMPK are included as well as several pathways whose flux is affected by AMPK activation. Arrows indicate positive effects of AMPK, whereas, T-lines indicate inhibitory effects.

The uptake, by skeletal muscle, accounts for >70% of the glucose removal from the serum in humans. Therefore, it should be obvious that this event is extremely important for overall glucose homeostasis, keeping in mind, of course, that glucose uptake by cardiac muscle and adipocytes cannot be excluded from consideration. An important fact related to skeletal muscle glucose uptake is that this process is markedly impaired in individuals with type 2 diabetes. The uptake of glucose increases dramatically in response to stress (such as ischemia) and exercise and is stimulated by insulin-induced recruitment of glucose transporters to the plasma membrane, primarily GLUT4. Insulin-independent recruitment of glucose transporters also occurs in skeletal muscle in response to contraction (exercise).

The activation of AMPK plays an important, albeit not an exclusive, role in the induction of GLUT4 recruitment to the plasma membrane. In addition, there is some demonstration that AMPK may regulate glucose transport through GLUT1. Increased glucose uptake will result in an increase in glycolysis and ATP production. Under ischemic conditions in the heart the activation of AMPK leads to the phosphorylation and activation of phosphofructo-kinase-2, PFK-2 (6-phosphofructo-2-kinase). PFK-2 is one of the most potent regulators of the rate of flux through glycolysis and gluconeogenesis. It is important to note that like many enzymes, there are multiple isoforms of PFK-2 and neither the liver nor the skeletal muscle isoforms contain the AMPK phosphorylation sites of the inducible form of PFK-2. Of particular significance is the fact that the inducible form of PFK-2 is commonly found in tumor cells and this may allow AMPK to play an important role in protecting tumor cells from hypoxic stress.

Indeed, techniques for depleting AMPK in tumor cells have shown that these cells become sensitized to nutritional stress upon loss of AMPK activity. Whereas, stress and exercise are powerful inducers of AMPK activity in skeletal muscle, additional regulators of its activity have been identified. Insulin-sensitizing drugs of the thiazolidinedione family (activators of PPAR-γ) as well as the hypoglycemia drug metformin exert a portion of their effects through regulation of the activity of AMPK.

As indicated above, the activity of the AMPK activating kinase LKB1, is critical for regulation of gluconeogenic flux and consequent glucose homeostasis. The action of metformin in reducing blood glucose levels requires the activity of LKB1 in the liver for this function. Also, several adipocytokines either stimulate or inhibit AMPK activation: leptin and adiponectin have been shown to stimulate AMPK activation, whereas, resistin inhibits AMPK activation.

Within skeletal muscle and heart activation of AMPK leads to the phosphorylation and inhibition of acetyl-CoA carboxylase (ACC). This inhibition results in a drop in the level of malonyl-CoA which itself is an inhibitor of carnitine palmitoyltransferase I (CPT I). With a drop in the inhibition of CPT I a concomitant increase in β-oxidation of fatty acid will occur within the mitochondria.

An increase in fatty acid oxidation, like increases in glycolysis, will lead to increases in ATP production. In addiiton to ACC, AMPK has been shown to phosphorylate and thus regulate the activities of HMG-CoA reductase, (HMGR); hormone-sensitive lipase, (HSL); glycerophosphate acyltransferase, (GPAT); malonyl-CoA decarboxylase, (MCD); glycogen synthase, (GS) and creatine kinase, (CK).

Therefore, the effects of AMPK activation are exerted on not only glucose homeostasis and fatty acid metabolism but overall energy homeostasis including glycogen metabolism, cholesterol metabolism and phosphocreatine metabolism. Not only does activation of AMPK exerts effects on enzyme activity through phosphorylation, there are marked effects on the expression of a number of glycolytic and lipogenic enzymes in the liver and adipose tissue. Included in this list are the genes for the liver isoform of pyruvate kinase (L-PK), fatty acid synthase (FAS), and ACC. Activation of AMPK leads to a reduction in the level of SREBP a transcription factor that is a key regulator of the expression of numerous lipogenic enzymes. Another transcription factor reduced in response to AMPK activation is hepatocyte nuclear factor 4α, HNF4α which is a member of the steroid/thyroid hormone superfamily. HNF4α in known to regulate the expression of several liver and pancreatic β-cell genes such as GLUT2, L-PK and preproinsulin. Of clinical significance is that mutations in HNF4a are responsible for maturity-onset diabetes of the young, MODY-1.

Recent evidence indicates that the newly discovered carbohydrate-response-element-binding protein (ChREBP) is a target for AMPK in the liver. ChREBP is implicated in the transcriptional regulation of L-PK. The target of the thiazolidinedione class of drugs used to treat type 2 diabetes is the peroxisome proliferator-activated receptor γ, PPARγ which itself may be a target for the action of AMPK. The transcription co-activator, p300, is phosphorylated by AMPK which inhibits interaction of p300 with not only PPARγ but also the retinoic acid receptor, retinoid X receptor, and thyroid receptor. Another transcription factor target of AMPK is the forkhead protein, FKHR (now refered to as FoxO1α). FoxO1α is involved in the activation of glucose-6-phosphatase expression and, therefore, loss of FoxO1α activity in response to AMPK activation will lead to reduced hepatic output of glucose. AMPK activation in response to hypoxia exerts effects on rates of protein synthesis. Hepatic translation elongation factor, eEF2 is a target for phosphorylation in repsonse to AMPK activation. AMPK phosphorylates and activates the kinase that phosphorylates eEF2 (eEF2K) leading to inhibition of protein synthesis. Another indirect substrate for AMPK that plays a role in protein synthesis is the mammalian target of rapamycin, mTOR. A detailed description of the role of mTOR in regulating protein synthesis

RELEVANCE OF AMPK TO TYPE 2 DIABETES

An impairment in fuel metabolism occurs in obesity and this impairment is a leading pathogenic factor in type 2 diabetes. The insulin resistance associated with type 2 diabetes is most profound at the level of skeletal muslce as this is the primary site of glucose and fatty acid utilization. Therefore, an understanding of how to activate AMPK in skeletal muscle would offer significant pharmacologic benefits in the treatment of type 2 diabetes. As indicated above, it has already been shown that metformin and the thiazolidinedione drugs exert some of their effects via activation of AMPK. In the non-pharmacologic context, activation of AMPK occurs in response to exercise, an activity known to have significant benefit for type 2 diabetics. Future targets for type 2 diabetes treatments will likely be those that can effect beneficial changes in the activity of AMPK.

ESPECIALLY BIOCHEMISTRY IN FATTY ACID

In chemistry, especially biochemistry, a fatty acid is a carboxylic acid with a long unbranched aliphatic tail (chain), which is either saturated or unsaturated. Carboxylic acids as short as butyric acid (4 carbon atoms) are considered to be fatty acids, whereas fatty acids derived from natural fats and oils may be assumed to have at least eight carbon atoms, caprylic acid (octanoic acid), for example. The most abundant natural fatty acids have an even number of carbon atoms because their biosynthesis involves acetyl-CoA, a coenzyme carrying a two-carbon-atom group

Fatty acids are produced by the hydrolysis of the ester linkages in a fat or biological oil (both of which are triglycerides), with the removal of glycerol. See oleochemicals. Fatty acids are aliphatic monocarboxylic acids derived from, or contained in esterified form in, an animal or vegetable fat, oil, or wax. Natural fatty acids commonly have a chain of four to 28 carbons (usually unbranched and even numbered), which may be saturated or unsaturated. By extension, the term is sometimes used to embrace all acyclic aliphatic carboxylic acids.

This would include acetic acid, which is not usually considered a fatty acid because it is so short that the triglyceride triacetin made from it is substantially miscible with water and is thus not a lipid. It is proposed that the blend of fatty acids exuded by mammalian skin, together with lactic acid and pyruvic acid, are distinctive and enable animals with a keen sense of smell to differentiate individuals.

TYPES

Fatty acids can be saturated and unsaturated, depending on double bonds. They differ in length as well.

Unsaturated Fatty Acids

Unsaturated fatty acids are of similar form, except that one or more alkenyl functional groups exist along the chain, with each alkene substituting a single-bonded" -CH_2-CH_2-" part of the chain with a double-bonded"-CH=CH-" portion (that is, a carbon double-bonded to another carbon).

The two next carbon atoms in the chain that are bound to either side of the double bond can occur in a cis or trans configuration.

- *Cis*: A cis configuration means that adjacent hydrogen atoms are on the same side of the double bond. The rigidity of the double bond freezes its conformation and, in the case of the cis isomer, causes the chain to bend and restricts the conformational freedom of the fatty acid. The more double bonds the chain has in the cis configuration, the less flexibility it has. When a chain has many cis bonds, it becomes quite curved in its most accessible conformations. For example, oleic acid, with one double bond, has a"kink" in it, whereas linoleic acid, with two double bonds, has a more pronounced bend. Alpha-linolenic acid, with three double bonds, favors a hooked shape. The effect of this is that, in restricted environments, such as when fatty acids are part of a phospholipid in a lipid bilayer, or triglycerides in lipid droplets, cis bonds limit the ability of fatty acids to be closely packed, and therefore could affect the melting temperature of the membrane or of the fat.
- Trans: A trans configuration, by contrast, means that the next two hydrogen atoms are bound to opposite sides of the double bond. As a result, they do not cause the chain to bend much, and their shape is similar to straight saturated fatty acids.

In most naturally-occurring unsaturated fatty acids, each double bond has three n carbon atoms after it, for some n, and all are cis bonds. Most fatty acids in the trans configuration (trans fats) are not found in nature and are the result of human processing (e.g., hydrogenation). The differences in geometry between the various types of unsaturated fatty acids, as well as between saturated and unsaturated fatty acids, play an important role in biological processes, and in the construction of biological structures (such as cell membranes).

Examples of unsaturated fatty acids:

Common name	Chemical structure
Myristoleic acid	$CH_3(CH_2)_3CH=CH(CH_2)_7COOH$
Palmitoleic acid	$CH_3(CH_2)_5CH=CH(CH_2)_7COOH$
Sapienic acid	$CH_3(CH2)_8CH=CH(CH_2)_4COOH$
Oleic acid	$CH_3(CH_2)_7CH=CH(CH_2)_7COOH$
Linoleic acid	$CH_3(CH_2)_4CH=CHCH_2CH = CH(CH_2)_7COOH$
?-Linolenic acid	$CH_3CH_2CH=CHCH_2CH$ $=CHCH_2CH=CH(CH_2)_7COOH$
Arachidonic acid	$CH_3(CH_2)_4CH=CHCH_2CH=$ $CHCH_2CH=CHCH_2CH=CH(CH_2)_3COOH$
Eicosapentaenoic acid	$CH_3CH_2CH=CHCH_2CH=CHCH_2CH$ $=CHCH_2CH=CHCH_2CH =CH(CH_2)_3COOH$
Erucic acid	$CH_3(CH_2)_7CH=CH(CH_2)_{11}COOH$
Docosahexaenoic acid	$CH_3CH_2CH=CHCH_2CH=CHCH_2CH =CHCH_2CH$ $=CHCH_2CH=CHCH_2CH =CH(CH_2)_2COOH$

Saturated Fatty Acids

Saturated fatty acids are a long-chain carboxylic acid that usually has between 12 and 24 carbon atoms that has no double bonds. Thus, saturated fatty acids are saturated with hydrogen (since double bonds reduce the number of hydrogens on each carbon).

Example:

- Lauric acid (12 C)
- Myristic acid (14 C)
- Palmitic acid (16 C)
- Stearic acid (18 C)
- Arachidic acid (20 C)

Essential Fatty Acids

The human body can produce all but two of the fatty acids it needs. These two, linoleic acid (LA) and alpha-linolenic acid (ALA), are widely distributed in plant oils. In addition, fish oils contain the longer-chain omega-3 fatty acids eicosapentaenoic acid (EPA) and docosahexaenoic acid (DHA). Other marine oils, such as from seal, also contain significant amounts of docosapentaenoic acid (DPA), which is also an omega-3 fatty acid. Although the body to some extent can convert ALA into these longer-chain omega-3 fatty acids, the omega-3 fatty acids found in marine oils help fulfill the requirement of essential fatty acids (and have been shown to have wholesome properties of their own).

Since they cannot be made in the body from other substrates and must be supplied in food, they are called essential fatty acids. Mammals lack the ability to introduce double bonds in fatty acids beyond carbons 9 and 10. Hence linoleic acid and alpha-linolenic acid are essential fatty acids for humans. In the body, essential fatty acids are primarily used to produce hormone-like substances that regulate a wide range of functions, including blood pressure, blood clotting, blood lipid levels, the immune response, and the inflammation response to injury infection.

Essential fatty acids are polyunsaturated fatty acids and are the parent compounds of the omega-6 and omega-3 fatty acid series, respectively. They are essential in the human diet because there is no synthetic mechanism for them. Humans can easily make saturated fatty acids or monounsaturated fatty acids with a double bond at the omega-9 position, but do not have the enzymes necessary to introduce a double bond at the omega-3 position or omega-6 position.

The essential fatty acids are important in several human body systems, including the immune system and in blood pressure regulation, since they are used to make compounds such as prostaglandins. The brain has increased amounts of linoleic and alpha-linolenic acid derivatives. Changes in the levels and balance of these fatty acids due to a typical Western diet rich in omega-6 and poor in omega-3 fatty acids is alleged to be associated with depression and behavioral change, including violence. The actual connection, if any, is still under investigation.

Further, changing to a diet richer in omega-3 fatty acids, or consumption of supplements to compensate for a dietary imbalance, has been associated with reduced violent behaviour and increased attention span, but the mechanisms for the effect are still unclear. So far, at least three human studies have shown results that support this: two school studies as well as a double blind study in a prison.

Fatty acids play an important role in the life and death of cardiac cells because they are essential fuels for mechanical and electrical activities of the heart.

FATTY ACID COMPOSITION

ENVIRONMENTAL INFLUENCES

Salinity

The difference between fatty acid compositions of marine and freshwater fish has been noted by several authors. Some examples of fatty acid patterns. Although these fish lipids are higher in ω 3 fatty acids, it is clear that freshwater fish have higher levels of ω 6 fatty acids than marine species. The average ω 6/ω 3 ratios are 0.37 and 0.16 for freshwater and marine fish, respectively. Fish in general contain more ω 3 than ω 6 polyunsaturated fatty acids (PUFA) and should have a higher dietary requirement for ω 3 PUFA; thus the dietary EFA requirement of marine fish for ω 3 PUFA may be higher than that of freshwater fish.

The same type of difference in the ω 6/ω 3 ratio between freshwater and seawater is seen when some species of fish migrate from oceans to streams or vice versa. The PUFA ratio of sweet smelt (Plecoglosus altivelis) changes drastically in only one month as they migrate from the sea to a freshwater river. A similar but reverse change occurs in the masu salmon (Oncorhynchus masu) as they migrate from freshwater to seawater. Even within the same species of fish, the salinity of the water seems to cause a dramatic change in the fatty acid pattern. The difference between marine and freshwater fish may be due simply to differences in the fatty acid content in the diet or it may be related to a specific requirement in fish related to physiological adaptations to the environments. The phospholipids are generally considered to be structural or functional lipid, being incorporated to a large extent in the membrane structure of cell and subcellular particles. The triglycerides are more often storage lipids and reflect the fatty acid composition of the diet to a greater extent than do the phospholipids. The fatty acid compositions of the triglyceride and phospholipid fractions of fish lipids are presented. It can be seen that the effect of changing environment on the fatty acid composition of the phospholipid is as great in the case of salmon, and considerably greater in the case of the sweet smelt, than it is on the triglyceride composition. Rainbow trout on diets containing either corn oil, which is high in ω 6 but low in ω 6 PUFA, showed a higher mortality and growth reduction in seawater than in freshwater over the twelve-week feeding period.

Temperature

There are several other factors besides the salinity of the water which affect the fatty acid composition and especially the PUFA of fish. In Tables 1 and 2 it can be seen that the salmonids, even in freshwater, tend to have a higher total PUFA of the 20 and 22 carbon chain length, and a lower ω 6/ω 3 ratio than the other fish. The salmonids are mostly cold-water fish.

The fatty acids from a number of marine animals from temperate and arctic waters show some significant differences in the general pattern; unfortunately analysis included fatty acids longer than 20:1. There are a number of other experiments demonstrating the effect of environmental temperature on fatty acid composition of aquatic animals. The general trend toward higher content of long chain PUFA at lower temperatures is quite clear. The ω 6/ω 3 ratio decreases with a decrease in temperature. If the trends in fatty acid composition can be taken as clues to the EFA requirements of fish, the ω 3 requirement would be greater for fish raised at lower temperatures. Fish raised in warmer waters, such as common carp, channel catfish, and tilapia may do better with a mixture of ω 6 and ω 3 fatty acids.

Effects of Diet

The mosquito fish and guppies were fed trout pellets which had an ω 6/

ω 3 ratio of 2.75. The catfish were fed diets supplemented with either beef tallow or menhaden oil, with ω 6/ω 3 ratios of 18.13 and 0.15, respectively. These fish were able to alter the dietary ω 6/ω 3 ratio in favour of ω 3 fatty acid incorporation into the flesh lipids even at the highest temperature.

Commercially available trout pellets are often low in ω 3 PUFA and high in ω 6 fatty acids. It is important not to ignore the effect of dietary lipid composition on fatty acid composition of fish fed artificial diets. It is clear from the data in Table that the ω 6/ω 3 ratio of the fish lipids is greatly affected by the ω 6/ω 3 ratio of the dietary lipids.

When the dietary ratio is very high in ω 6 fatty acids supplied by animal lard or vegetable oils, there is a tendency for fish to alter the ratio of PUFA incorporated in favour of ω 3 fatty acids. When the dietary oil is a fish oil high in to3 fatty acids, there is little change in the ω 6/ω 3 ratio of lipids incorporated into the fish. This is further suggestive evidence of an EFA requirement of fish for ω 3 PUFA.

Seasonal Variation

Seasonal variations in the fatty acid composition of fish species have often been reported. Seasonal changes have been observed in total lipid and iodine values of herring oils. The iodine value or degree of unsaturation of the oil was minimal in April and maximal in June.

The great increase in unsaturation corresponded to the onset of feeding in spring. The absence of a gas liquid chromatograph (GLC) at the time precluded identification of changes in individual fatty acids. Flesh and viscera lipid content of the sardine Sardinops melanosticta vary from 3.9 to 10.77 per cent and from 10.9 to 38.3 per cent, respectively. The fatty acids of principal interest with respect to EFA metabolism are 20:4ω 6, 20:5ω 3, and 22:6ω 3.

There was considerable variation in all of these fatty acids in both neutral and polar lipid from both tissues. In the flesh, the 20:4 ω 6 was consistently higher in the neutral lipid than in the polar lipid. The total 20:5ω 3 plus 22:6ω 3 was consistently higher in polar lipid than in the neutral lipid. Thus, in spite of the major fluctuations in fatty acids caused by changes in diet and temperature throughout the seasons, there was a consistent preferential incorporation of PUFA of the ω 3 series into the polar or phospholipid fraction of the lipids.

One of the best clues to the EFA requirements of a species can be gained from the fatty acid composition of the lipids incorporated into the offspring or egg. The act of reproduction or spawning also has a significant effect on the seasonal fluctuation of lipids in fish. Fatty acid composition of fish egg lipids is probably distinctive for each species and contains increased levels of 16:0, 20:4 ω 6, 20:5 ω 3 and 22:6ω 3 compared to the liver lipids of the same female fish. Elevated levels of 16:0, 20:5ω 3, and 22:6 ω 3 and reduced 18:1 in the ovary occurred compared to mesenteric fat of Pacific sardine fed a natural copepod diet. The blood fatty acids of the sardine fed the natural diet were similar to those of the ovary. When the sardines were fed trout food, both the blood and mesenteric fat responded to the diet with elevated 18:2ω 6 and reduced 20:5ω 3 arid 22:6ω 3.

The effect of the diet on ovary fatty acid content was considerably less, as relatively high levels of 20:5ω 3 and 22:6ω 3 were retained. The ovary lipids of the sweet smelt show an increase in 16:0, and a reduction in the PUFA, especially in the phospholipids, compared to the lipids from the flesh of fish caught at the same time of year. The ω 6/ω 3 ratio of the ovary was lower than that of the flesh lipids, 0.21 and 0.17 for the ovary compared to 0.31 and 0.20 for the triglycerids and phospholipids of the flesh, respectively.

The hatchability of eggs from common carp fed several different formulated feeds is greatly reduced when the 22:6ω 3 of the egg lipids is less than 10 per cent. Further, the muscle, plasma, and erythrocyte fatty acid compositions are more affected by dietary lipid than those of the eggs. The EFA requirements of a number of species of fish have been investigated in nutritional studies. The fish themselves have given ample evidence for EFA preference by the types of fatty acids they incorporate into their lipids.

Fish, in general, tend to utilize ω 3 over ω 6. This is especially observed when the dietary lipids are high in ω 6, as the fish tend to alter the ω 6/ω 3 ratio toward the ω 3 fatty acids in the tissue lipids. The lipids of the egg must satisfy the EFA requirement of the embryo until it is able to feed. The fatty acid composition data suggest that the ω 3 requirement is greater in seawater than in freshwater and higher in cold water than in warm water.

BODY LIPID COMPOSITION AND DIETARY LIPID REQUIREMENTS

Detailed information is still lacking on the dietary lipid requirements of many species of fish, but there is an abundance of information on the fatty acid composition of fish oils. Information on the lipid composition of fish can be used to make some guesses about dietary lipid requirements. Linolenic acid (18:3ω 3) resulted in some sparing action and growth promotion in rats, and fatty acids of the ω 6 EFA prevented all of the EFA-deficiency symptoms.

Research with homeothermic land-dwelling animals showed that the ω 6 series of fatty acids are the "essential fatty acids", while the ω 3 series are considered to be non-essential or only have a partial sparing action on EFA-deficiency. The ω 6 series of fatty acids have been shown to be essential to enough species that it began to become accepted that these are the essential fatty acids for all animals. It was assumed by many that fish also required ω 6 fatty acids. Many researchers began by supplementing fish diets with vegetable oils, such as corn, peanut, or sunflower oil, which were rich in linoleic acid. The main sympton observed during the development of EFA deficiency in chinook salmon fed fat-free diets was a marked depigmentation that can be prevented by addition of 1 per cent trilinolein, but not by 0.1 per cent linolenic acid.

Although the ω 6 fatty acids are considered to be essential, one of the general characteristics of fish oils is the low levels of ω 6 series fatty acids and the higher levels of ω 3 type fatty acids. There is evidence that polyunsaturated fatty acids (PUPA) of the ω 3 series, which are present in relatively large concentrations in fish oil, play the role of essential fatty acid for fish.

POLYUNSATURATED FATTY ACID REQUIREMENTS OF FISH

When a test diet containing 13 per cent corn oil and 2 per cent cod-liver oil was fed to rainbow trout, subsequent deletion of the cod-liver oil from the diet depressed growth and produced some kidney degeneration which might be attributed to a lack of sufficient ω 3 PUFA present in significant quantities in cod-liver oil.

Dietary fish oil is superior to corn oil in promoting growth of rainbow trout (Salmo gairdneri) and the yellow-tail (Seriola guingueradiata). Dietary linolenic acid or ethyl linolenate (18:3 ω 3) gives a positive growth response for rainbow trout which may be attributed to a dietary requirement for ω 3 fatty acids.

ESSENTIAL FATTY ACID REQUIREMENTS OF FISH

One of the most widely accepted theories explaining the presence of such high levels of 20:5ω 3 and 22:6ω 3 fatty acids in fish oils is related to the effect of unsaturation on the melting point of a lipid. The greater degree of unsaturation of fatty acids in the fish phospholipids allows for flexibility of cell membrane at lower temperatures. The ω 3 structure allows a greater degree of unsaturation than the ω 6 or ω 9. This theory is consistent with the fact that cold water fish have a greater nutritional requirement for ω 3 fatty acids, while the EFA requirement of some warm water fish can be satisfied by a mixture of ω 6 plus ω 3.

Rainbow Trout

Rainbow trout, a cold water fish, requires ω 3 fatty acids as EFA in the diet. The EFA requirement can be met by 1 per cent 18:3ω 3 in the diet. Inclusion of 18:2ω 6 in the diet may result in some improvement in growth and feed conversion compared to EFA deficient diets; however, the ω 6 fatty acids will not prevent some EFA deficiency symptoms such as the "shock syndrome". Although it is clear that rainbow trout require ω 3 fatty acids, it remains to be shown conclusively whether some dietary level of ω 6 fatty acid is essential. In all the above studies with rainbow trout, dietary 18:2ω 6 or 18:3ω 3 were readily converted to C-20 and C-22 PUFA of the same series, and 18:3ω 3 or 22:6ω 3 had similar EFA value for rainbow trout. Either 20:5ω

3 or 22:6 ω 3 is superior to 18:3ω 3 in an EFA value for rainbow trout, and the former two fatty acids in combination are superior to either alone.

This is consistent with data for mammals, where 20:4ω 6 has higher EFA value than 18:2ω 6. The superior nutritional value of C-20 and C-22 carbon ω 3-PUFA is further supported by the excellent growth promoting effects of dietary fish oils such as pollock liver oil and salmon oil for rainbow trout.

Channel Catfish

One of the most important warm water fish in North America is the channel catfish (Ictalurus punctatus). The quantitative EFA requirement of the catfish has not yet been determined. However, the evidence is convincing that the ω 3 requirement is not as great as that of rainbow trout. Analysis of fatty acids of lipids from catfish purchased at five processing plants showed very low levels of 20:4ω 6, 20:5ω 3, and 22:6ω 3; 0.8 - 5.5, 0.2 - 1.3, and 0.6 - 6.1 per cent of the total fatty acids, respectively. It was shown that corn oil added to a semipurified casein based diet initially resulted in a positive growth response and protein sparing, but later growth inhibition was observed.

The apparent repressive effects of corn oil may be due to its 18:2 ω 6 content since 20:5 ω 3 and 22:6ω 3 present in menhaden oil had no apparent detrimental effects. The growth suppressing effects of 18:2ω 6 were also noted when 3 per cent corn oil was added to 3 per cent beef tallow and 3 per cent menhaden oil. The growth suppression caused by unsaturated fatty acids does not appear to be limited to ω 6 fatty acids. Linseed oil (high in 18:3ω 3) in the diet of catfish resulted in growth suppression similar to that caused by corn oil compared with dietary beef tallow, olive oil and menhaden oil.

The Common Carp

The picture for another warm water fish, the common carp (Cyprinus carpio) is much clearer than that for the channel catfish. This fish has an EFA requirement for both ω 3 and ω 6 fatty acids.

The best weight gains and feed conversions are obtained in fish receiving a diet containing both 1 per cent 18:2 ω 6 and 1 per cent 18:3ω 3. With the carp, 20:5ω3 and 22:6ω3 at 0.5 per cent of the diet are superior to 1 per cent of 18:3u3. Carp fed a fat-free, or EFA deficient, diet incorporated high levels of 20:3ω9 in their lipids, especially in the phospholipids.

The Eel

The eel (Anguilla japonica), another warm water fish, has a requirement for both ω 3 and ω 6 fatty acids. Corn oil (high in ω 6) and cod liver oil (high in ω 3)in a 2:1 mixture are most favourable for the growth of eels. The eel requires ω 6 and ω 3 in the same proportion as the carp, but at a lower level in the diet; namely, 0.5 per cent of each, rather than 1.0 per cent of each PUFA.

The Plaice

The plaice becomes depleted of both ω 3 and ω 6 PUFA when fed a fat-free diet. The addition of 12:0 and 14:0 to the diet result in the synthesis of saturated and monoenoic fatty acids of chain lengths up to C18; however, increased levels of 20:3w 9 noted in trout and mammals have not been reported in plaice. Plaice fed dietary 18:2w 6 and 18:3w 3 will not produce significant amounts of 20:4w 6, 20:5w 3, or 22:6w 3.

The Turbot

The growth of turbot (Scophthalmus matimus) is much better with ω 3 PUFA than with ω 6 or saturated fat (hydrogenated coconut oil) in the diet. The turbot also appears to be unable to convert dietary 18:2 ω 6 to 20:4ω6 when fed corn oil, or to convert endogenous 18:1ω9 to 20:3ω9 when fed the EFA deficient diet.

Although it appears to have an EFA requirement for ω 3 fat acids such as are present in cod liver oil, this requirement is not satisfied by 18:3ω3. The chain elongation and desaturation of 18:lω9, 18:2 ω 6, or 18:3ω3 has been found to be very limited (3-15 per cent) in turbot compared to the rainbow trout where 70 per cent of the 18:3ω3 was converted to 22:6ω3. The required level of long-chain ω3 fatty acids for turbot is at least 0.8 per cent of the diet.

The Red Sea Bream

The red sea bream (Chrysophyrys major) grows better when the dietary lipid is of marine origin (pollock residual oil) rather than a vegetable oil (such as corn oil). The EFA requirement of the red sea bream is not satisfied by either linoleic acid of corn oil or supplemented linolenate. A mixture of 20:5ω3 and 22:6ω3 supplemented to the corn oil diet has been shown to be effective in improving growth and condition of these fish. Thus, even in warm water, marine fish seem to require not just ω 3 fatty acids but 0)3 fatty acids of 20 to 22 carbon-chain length. A direct correlation between feed efficiency and the 18:1 level in the lipids of the red sea bream has been postulated.

Other Species

Among warm water marine fish, mullet and fundulus possess the ability to chain, elongate, and desaturate 18:2ω6 or 18:3ω3 PUFAs. The process is, however, inhibited in fundulus by high levels (about 5 per cent) of these PUFAs of 18:2ω6 or 18:3 ω 3 in the diet.

FATTY ACID METABOLISM IN FISH

It appears that high levels of 18-carbon ω 6 or ω 3 fatty acids inhibit the synthesis and metabolism of 18:lω9. It is interesting to note that the channel catfish, which also exhibits negative growth response to dietary 18:2ω6 or 18:3ω3, incorporates very high levels of 18:1 into its body lipids. The inclusion of either 18:2ω6 or 18:3ω3 in the diet reduces the levels of 18:1 fatty acids in body lipids. A similar reduction has also been observed in red sea bream liver phospholipid when either of the PUFAs is added to the diet.

The competitive inhibition of chain elongation and desaturation of members from one series of fatty acids for members of another series is well established, with ω 3 > ω 6 > ω 9 being the usual order of potency for inhibition.

The pathways of fatty acid metabolism have been reviewed by Mead and Kayama. Fish are able to synthesize, de novo from acetate, the even-chain, saturated fatty acids. Radio tracer studies have shown that fish can convert 16:0 to the ω 7 monoene and 18:0 to the ω 9 monoene. The ω 5, ω 11 and ω 3 monoenes are proposed based on the identification of these isomers in the monoenes of herring oil.

Fish are unable to synthesize any fatty acids of the ω 6 and u3 series unless a precursor with this ω structure is present in the diet. Fish are able to desaturate and elongate fatty acids of the ω 9, ω 6, or ω 3. There is competitive inhibition of the elongation desaturation of fatty acids of one series by members of the other series. The ω 3 fatty acids are the most potent inhibitors, the ω 9 are the least. The ability to elongate and desaturate fatty acids is not the same in all species of fish, as was noted earlier. The turbot was able to desaturate and elongate only 3-15 per cent of 18:1ω 9, 18:2ω6, or 18:3 ω 3, when given the C^{14} labelled fatty acid; in the rainbow trout, 70 per cent of the label from 18:3ω3 (C^{14}) was found in 22:6 ω 3.

The essential fatty acids are not unique in their ability to supply energy. The β -oxidation of fatty acids in fish is basically the same as in mammals. The EFA and saturated and monoenoic fatty acids are all equally utilized by fish for energy production.

Increased swelling rates of liver mitochondria occur in rainbow trout fed diets deficient in ω 3 fatty acids. It is possible that EFA plays an important role in the permeability as well as the plasticity of membranes. The role of ω 3 fatty acids in membrane permeability may be one of the factors accounting for differences in content of this family of fatty acids between freshwater and marine fish.

Fish mitochondria with high levels of the ω 3 PUFA and very low levels of ω 6 fatty acids are very similar to mammalian mitochondria with respect to cytochrome content, β -oxidation of fatty acids, operation of the tricarboxylic acid cycle, electron transport, and oxidative phosphorylation. The ω 3 PUFA may play the same role in fish that the ω 6 fatty acids play in rats. The EFA play another role in the mitochon-dria. In addition to their importance in membrane structure, the EFA are important in some enzyme systems.

Unsaturated fatty acids play an important role in the transportation of other lipids. It has been repeatedly shown that feeding PUFA will lower the cholesterol levels in animals with above-normal blood lipid and cholesterol levels. Fish oils are more effective in lowering cholesterol levels than are most dietary lipids.

The major portion of the fatty acids absorbed across the intestinal mucosa are transported as protein-lipid complexes stabilized by phospholipids. The low body temperature in fish probably results in a greater importance for unsaturation in transport of lipids than in homeothermic animals.

NEGATIVE ASPECTS OF LIPIDS IN FISH NUTRITION

The requirement by fish for PUFA of the ω 3 series creates problems with respect to feed storage. These types of fatty acids are very labile on oxidation. The products of lipid oxidation may react with other nutrients such as proteins, vitamins, etc., and reducing the available dietary levels or the oxidation products may be toxic. The effect of oxidized lipids on dietary proteins, enzymes and amino acids have been demonstrated.

The use of oxidized menhaden oil in the diets of swine and rats caused decreased appetite, reduced growth, yellowish-brown pigmentation of depot fat, and decreased haemoglobin and haematocrit levels. The negative effects of the oxidized fish oils were reversed by the addition of alpha-tocopherol acetate or ethoxyoquin to the diet.

Much of the use of vegetable oils in fish diets in the 1950s and 1960s might, in part, have been based on their greater stability in prepared diets. It has been demonstrated that rancid herring and hake meals in fish feeds caused dark colouration, anaemia, lethargy, brown-yellow pigmented liver, abnormal kidneys, and small gill clubbing in chinook salmon. The symptons can be alleviated by addition of alpha-tocopherol to the diets containing rancid fish meals. The addition of vitamin E would prevent the toxic or negative effects of adding 5 per cent highly oxidized salmon oil to the diet of rainbow trout. This same sparing effect of alpha-tocopherol can also apply to rancid carp feed.

The positive nutritional value of ω 3 fatty acids in fish lipids for fish feeds can become a negative factor if adequate care is not taken in the preparation and storage of feeds. Only fresh oils with low peroxide values should be used in feeds. Fish feed ingredients such as fish meals should be protected against oxidation. The level of vitamin E added to the diet should be increased as the PUFA level is increased. The finished feed, if possible, should be stored in air tight containers at reduced temperatures with minimum exposure to UV radiation and other factors accelerating the rate of lipid oxidation. The problems of rancidity or antioxidation of lipids in fish feeds should not be ignored.

LIPOPROTEINS NTESTINAL UPTAKE OF LIPIDS

In order for the body to make use of dietary lipids, they must first be absorbed from the small intestine. Since these molecules are oils, they are essentially insoluble in the aqueous environment of the intestine. The solubilization (or emulsification) of dietary lipids is therefore accomplished by means of bile salts, which are synthesized from cholesterol in the liver and then stored in the gallbladder; they are secreted following the ingestion of fat.

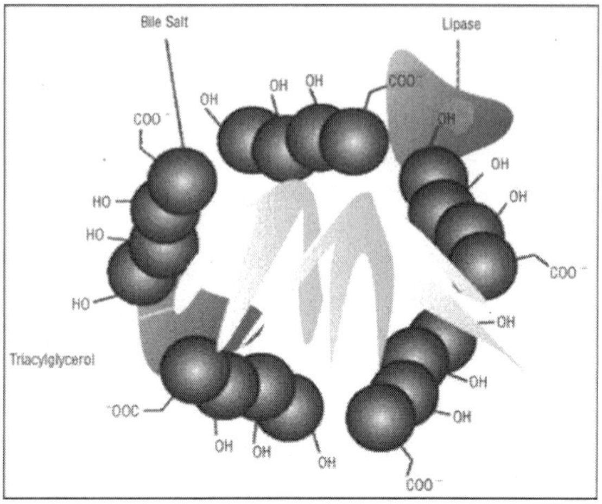

Fig. Mixed Micelle Formed by Bile Salts, Triacylglycerols and Pancreatic Lipase.

The emulsification of dietary fats renders them accessible to pancreatic lipases (primarily *lipase* and *phospholipase* A_2). These enzymes, secreted into the intestine from the pancreas, generate free fatty acids and a mixtures of mono- and diacylglycerols from dietary triacylglycerols. Pancreatic lipase degrades triacylglycerols at the 1 and 3 positions sequentially to generate 1,2-diacylglycerols and 2-acylglycerols. Phospholipids are degraded at the 2 position by pancreatic phospholipase A_2 releasing a free fatty acid and the lysophospholipid. The products of pancreatic lipases then diffuse into the intestinal epithelial cells, where the re-synthesis of triacyglycerols occurs.

Dietary triacylglycerols and cholesterol, as well as triacylglycerols and cholesterol synthesized by the liver, are solubilized in lipid-protein complexes. These complexes contain triacylglycerol lipid droplets and cholesteryl esters surrounded by the polar phospholipids and proteins identified as apolipoproteins. These lipid-protein complexes vary in their content of lipid and protein.

INTERMEDIATE DENSITY LIPOPROTEINS, IDLS

IDLs are formed as triacylglycerols are removed from VLDLs. The fate of IDLs is either conversion to LDLs or direct uptake by the liver. Conversion of IDLs to LDLs occurs as more triacylglycerols are removed. The liver takes up IDLs after they have interacted with the LDL receptor to form a complex, which is endocytosed by the cell. For LDL receptors in the liver to recognize IDLs requires the presence of both apo-B-100 and apo-E (the LDL receptor is also called the apo-B-100/apo-E receptor).

The importance of apo-E in cholesterol uptake by LDL receptors has been demonstrated in transgenic mice lacking functional apo-E genes. These mice develop severe atherosclerotic lesions at 10 weeks of age.

LOW DENSITY LIPOPROTEINS, LDLS

The cellular requirement for cholesterol as a membrane component is satisfied in one of two way: either it is synthesized de novo within the cell, or it is supplied from extra-cellular sources, namely, chylomicrons and LDLs. As indicated above, the dietary cholesterol that goes into chylomicrons is supplied to the liver by the interaction of chylomicron remnants with the remnant receptor. In addition, cholesterol synthesized by the liver can be transported to extra-hepatic tissues if packaged in VLDLs. In the circulation VLDLs are converted to LDLs through the action of lipoprotein lipase. LDLs are the primary plasma carriers of cholesterol for delivery to all tissues.

The exclusive apolipoprotein of LDLs is apo-B-100. LDLs are taken up by cells via LDL receptor-mediated endocytosis, as described above for IDL uptake. The uptake of LDLs occurs predominantly in liver (75/%), adrenals and adipose tissue. As with IDLs, the interaction of LDLs with LDL receptors requires the presence of apo-B-100.

The endocytosed membrane vesicles (endosomes) fuse with lysosomes, in which the apoproteins are degraded and the cholesterol esters are hydrolyzed to yield free cholesterol. The cholesterol is then incorporated into the plasma membranes as necessary. Excess intracellular cholesterol is re-esterified by *acyl-CoA-cholesterol acyltransferase* (ACAT), for intracellular storage. The activity of ACAT is enhanced by the presence of intracellular cholesterol. Insulin and tri-iodothyronine (T3) increase the binding of LDLs to liver cells, whereas glucocorticoids (e.g., dexamethasone) have the opposite effect. The precise mechanism for these effects is unclear but may be mediated through the regulation of apo-B degradation. The effects of insulin and T3 on hepatic LDL binding may explain the hypercholesterolemia and increased risk of athersclerosis that have been shown to be associated with uncontrolled diabetes or hypothyroidism. An abnormal form of LDL, identified as lipoprotein-X (Lp-X), predominates in the circulation of patients suffering from *lecithin-cholesterol acyl transferase* (LCAT) deficiency or cholestatic liver disease. In both cases there is an elevation in the level of circulating free cholesterol and phospholipids.

HIGH DENSITY LIPOPROTEINS, HDLS

HDLs are synthesized *de novo* in the liver and small intestine, as primarily protein-rich disc-shaped particles. These newly formed HDLs are nearly devoid of any cholesterol and cholesteryl esters. The primary apoproteins of HDLs are apo-A-I, apo-C-I, apo-C-II and apo-E. In fact, a major function of HDLs is to act as circulating stores of apo-C-I, apo-C-II and apo-E. HDLs are converted into spherical lipoprotein particles through the accumulation of cholesteryl esters. This accumulation converts nascent HDLs to HDL_2 and HDL_3. Any free cholesterol present in chylomicron remnants and VLDL remnants (IDLs) can be esterified through the action of the HDL-associated enzyme, *lecithin-cholesterol acyl transferase,* LCAT.

LCAT is synthesized in the liver and so named because it transfers a fatty acid from the C-2 position of lecithin to the C-3-OH of cholesterol, generating a cholesteryl ester and lysolecithin. The activity of LCAT requires interaction with apo-A-I, which is found on the surface of HDLs. Cholesterol-rich HDLs return to the liver, where they are endocytosed. Hepatic uptake of HDLs, or reverse cholesterol transport, may be mediated through an HDL-specific apo-A-I receptor or through lipid-lipid interactions. Macrophages also take up HDLs through apo-A-I receptor interaction. HDLs can then acquire cholesterol and apo-E from the macrophages; cholesterol-enriched HDLs are then secreted from the macrophages. The added apo-E in these HDLs leads to an increase in their uptake and catabolism by the liver.

HDLs also acquire cholesterol by extracting it from cell surface membranes. This process has the effect of lowering the level of intracellular cholesterol, since the cholesterol stored within cells as cholesteryl esters will be mobilized to replace the cholesterol removed from the plasma membrane. The cholesterol esters of HDLs can also be transferred to VLDLs and LDLs through the action of the HDL-associated enzyme, *cholesterol ester transfer protein* (CETP, also identified as apo-D). This has the added effect of allowing the excess cellular cholesterol to be returned to the liver through the LDL-receptor pathway as well as the HDL-receptor pathway.

LDL RECEPTORS

LDLs are the principal plasma carriers of cholesterol delivering cholesterol from the liver (via hepatic synthesis of VLDLs) to peripheral tissues, primarily the adrenals and adipose tissue. LDLs also return cholesterol to the liver. The cellular uptake of cholesterol from LDLs occurs following the interaction of LDLs with the LDL receptor (also called the apo-B-100/apo-E receptor). The sole apoprotein present in LDLs is apo-B-100, which is required for interaction with the LDL receptor. The LDL receptor is a polypeptide of 839 amino acids that spans the plasma membrane. An extracellular domain is responsible for apo-B-100/apo-E binding. The intracellular domain is responsible for the clustering of LDL receptors into regions of the plasma membrane termed coated pits. Once LDL binds the receptor, the complexes are rapidly internalized (endocytosed).

ATP-dependent proton pumps lower the pH in the endosomes, which results in dissociation of the LDL from the receptor. The portion of the endosomal membranes harboring the receptor are then recycled to the plasma membrane and the LDL-containing endosomes fuse with lysosomes. Acid hydrolases of the lysosomes degrade the apoproteins and release free fatty acids and cholesterol. As indicated above, the free cholesterol is either incorporated into plasma membranes or esterified (by ACAT) and stored within the cell.

The level of intracellular cholesterol is regulated through cholesterol-induced suppression of LDL receptor synthesis and cholesterol-induced inhibition of cholesterol synthesis. The increased level of intracellular cholesterol that results from LDL uptake has the additional effect of activating ACAT, thereby allowing the storage of excess cholesterol within cells.

However, the effect of cholesterol-induced suppression of LDL receptor synthesis is a decrease in the rate at which LDLs and IDLs are removed from the serum. This can lead to excess circulating levels of cholesterol and cholesteryl esters when the dietary intake of fat and cholesterol exceeds the needs of the body. The excess cholesterol tends to be deposited in the skin, tendons and (more gravely) within the arteries, leading to atherosclerosis.

CLINICAL SIGNIFICANCES OF LIPOPROTEIN METABOLISM

Fortunately, few individuals carry the inherited defects in lipoprotein metabolism that lead to hyper- or hypolipoproteinemias. Persons suffering from diabetes mellitus, hypothyroidism and kidney disease often exhibit abnormal lipoprotein metabolism as a result of secondary effects of their disorders. For example, because *lipoprotein lipase* (LPL) synthesis is regulated by insulin, LPL deficiencies leading to Type I hyperlipoproteinemia may occur as a secondary outcome of diabetes mellitus. Additionally, insulin and thyroid hormones positively affect hepatic LDL-receptor interactions; therefore, the hypercholesterolemia and increased risk of athersclerosis associated with uncontrolled diabetes or hypothyroidism is likely due to decreased hepatic LDL uptake and metabolism.

Of the many disorders of lipoprotein metabolism, familial hypercholesterolemia (FH) may be the most prevalent in the general population. Heterozygosity at the FH locus occurs in 1:500 individuals, whereas, homozygosity is observed in 1:1,000,000 individuals. FH is an inherited disorder comprising four different classes of mutation in the LDL receptor gene. The class 1 defect (the most common) results in a complete loss of receptor synthesis.

The class 2 defect results in the synthesis of a receptor protein that is not properly processed in the Golgi apparatus and therefore is not transported to the plasma membrane. The class 3 defect results in an LDL receptor that is incapable of binding LDLs. The class 4 defect results in receptors that bind LDLs but do not cluster in coated pits and are, therefore, not internalized. FH sufferers may be either heterozygous or homologous for a particular mutation in the receptor gene. Homozygotes exhibit grossly elevated serum cholesterol (primarily in LDLs). The elevated levels of LDLs result in their phagocytosis by macrophages. These lipid-laden phagocytic cells tend to deposit within the skin and tendons, leading to xanthomas. A greater complication results from cholesterol deposition within the arteries, leading to atherosclerosis, the major contributing factor of nearly all cardiovascular diseases.

PHARMACOLOGIC INTERVENTION

Drug treatment to lower plasma lipoproteins and/or cholesterol is primarily aimed at reducing the risk of athersclerosis and subsequent coronary artery disease that exists in patients with elevated circulating lipids. Drug therapy usually is considered as an option only if non-pharmacologic interventions (altered diet and exercise) have failed to lower plasma lipids.

- *Mevinolin, Mevastatin, Lovastatin*: These drugs are fungal HMG-CoA reductase inhibitors. The net result of treatment is an increased cellular uptake of LDLs, since the intracellular synthesis of cholesterol is inhibited and cells are therefore dependent on extracellular sources of cholesterol. However, since mevalonate (the product of the HMG-CoA reductase reaction) is required for the synthesis of other important isoprenoid compounds besides cholesterol, long-term treatments carry some risk of toxicity.
- *Nicotinic acid*: Nicotinic acid reduces the plasma levels of both VLDLs and LDLs by inhibiting hepatic VLDL secretion, as well as suppressing the flux of FFA release from adipose tissue by inhibiting lipolysis. Because of its ability to cause large reductions in circulating levels of cholesterol, nicotinic acid is used to treat Type II, III, IV and V hyperlipoproteinemias.
- *Clofibrate, Gemfibrozil, Fenofibrate*: These compounds are derivatives of fibric acid and promote rapid VLDL turnover by activating lipoprotein lipase. They also induce the diversion of hepatic free fatty acids from esterification reactions to those of oxidation, thereby decreasing the liver's secretion of triacylglycerol- and cholesterol-rich VLDLs.
- *Probucol*: Probucol increases the rate of LDL metabolism and may block the intestinal transport of cholesterol. The net result is a significant reduction in plasma cholesterol levels.
- *Cholestyramine or colestipol (resins)*: These compounds are nonabsorbable resins that bind bile acids which are then not reabsorbed by the liver but excreted. The drop in hepatic reabsorption of bile acids releases a feedback inhibitory mechanism that had been inhibiting bile acid synthesis. As a result, a greater amount of cholesterol is converted to bile acids to maintain a steady level in circulation. Additionally, the synthesis of LDL receptors increases to allow increased cholesterol uptake for bile acid synthesis,

and the overall effect is a reduction in plasma cholesterol. (This treatment is ineffective in homozygous FH patients, since they are completely deficient in LDL receptor

MEMBRANE LIPIDS AND BILAYER MEMBRANES

The simpler lipids are used primarily for energy storage by living organisms, but more complex lipids are used to form biological membranes. The most important and abundant class of biological membrane lipids is the class of phospholipids. Most phospholipids are derived from glycerol and so they are called phosphoglycerides. The amino alcohol sphingosine gives rise to sphingomyelin, the only significant membrane phospholipid which is not a phosphoglyceride.

In phosphoglycerides, the hydroxy groups at C-1 and C-2 of glycerol are esterified to fatty acids while the hydroxy group at C-3 of glycerol is esterified to phosphoric acid; this forms diacylglycerol-3-phosphates. The two fatty acids found in a diacylglycerol-3-phosphate molecule need not be the same. Diacylglycerol-3-phosphates are only a minor constituent of membranes. The major constituents of membranes are derivatives of diacylglycerol-3-phosphates in which the esterified phosphate also forms an ester linkage with an alcohol. The alcohols commonly found in phosphoglycerides include glycerol, ethanolamine, and choline. The glycolipids are lipids which contain sugars. Glycolipids are derived from sphingosine, as is the phospholipid sphingomyelin. Glycolipids contain no esterified phosphate.

Both phospholipids and glycolipids are amphipathic molecules, that is, molecules in which one end is hydrophobic and one end is hydrophilic. The fatty acid chains form the hydrophobic end of the molecule while the polar glycerol-phosphate-alcohol or sphingosine-sugar portion forms the hydrophilic end of the molecule.

BILAYER MEMBRANES

When amphipathic molecules are placed in aqueous solution, their hydrophobic tails attempt to orient themselves toward each other. The two ways in which they do so are to form small spherical micelles or to form a planar lipid bilayer. A micelle is a small structure, generally less than two micrometer in diameter, while a lipid bilayer typically has a thickness of about 0.5 micrometer and can have an area of several square millimeters. Sonication or other treatment of lipids can produce spherical liposomes, or lipid vesicles, which contain aqueous solution within an enclosing lipid bilayer. Liposomes are considerably larger than micelles.

Both liposomes and planar bilayer membranes can be prepared from simple solutions of phospholipids such as the diacylglycerol-3-phosphatidyl cholines. Studies of such artificially prepared membranes have given us much insight into the functions and operation of membranes in living cells. Simple bilayer membranes are highly permeable to water molecules, while ions such as sodium ion and potassium ion can traverse them only more slowly (by nine orders of magnitude!). For small molecules, the rates of permeation through simple bilayer membranes increase with their solubility in nonpolar solvents relative to their solubility in water. These observations strongly suggest that transfer of substances through a membrane requires desolvation, transfer through the anhydrous membrane interior, and then resolvation on the other side of the membrane.

The differences in transport behaviour between biological membranes and simple bilayer membranes are now believed to be due to the incorporation of protein molecules or other specifically transporting molecules directly into or onto the surface of a bilayer cell membrane. Considerable research in biology and in chemistry is being directed toward the transport properties of membranes.

LIPIDS AND FATTY ACIDS

Lipids are the generic names assigned to a group of fat soluble compounds found in the tissues of plants and animals,: and are broadly classified as:

- Fats,
- Phospholipids,
- Sphingomyelins,
- Waxes,
- Sterols.

Fats are the fatty acid esters of glycerol and are the primary energy depots of animals. These are used for long-term energy requirements during periods of extensive exercise or during periods of inadequate food and energy intake. Fish have the unique capability of metabolizing these compounds readily and, as a result, can exist for long periods of time under conditions of food deprivation. A typical example is the many weeks of migration by salmon in their return upstream to spawn; stored lipid deposits are burned for fuel to enable body processes to continue during the strenuous journey. Phospholipids are the esters of fatty acids and phosphatidic acid. These are the main constituent lipids of cellular membranes allowing the membrane surfaces to be hydrophobic or hydrophylic depending on the orientation of the lipid compounds into the intra or extracellular spaces. Sphingomyelins are the fatty acid esters of sphingosine and are present in brain and nerve tissue compounds.

Waxes are fatty acid esters of long-chain alcohols. These compounds can be metabolized for energy and to impart physical and chemical characteristics through the stored lipids of some plant and several animal compounds. Sterols are polycyclic, long-chain alcohols and function as components of several hormone systems, especially in sexual maturation and sex-related physiological functions.

Fatty acids can exist as straight chain or branch chain components; many of the fish fats contain numerous unsaturated double bonds in the fatty acid structures. A short bond designation for. fatty acids will be used throughout where the ω number identifies the position of the first double bond counting from the methyl end. Linolenic acid would be written 18:3ω 3. The first number identifies the number of carbons; the second number, the number of double bonds; and the last number, the position of the double bonds. Many reviews of fish nutrition have been published which contain information on lipid requirements.

Most work on lipid requirements of fish has been with salmonids. Rainbow trout have an essential fatty acid (EFA) requirement for the linolenic of ω 3[1] series rather than for linolenic or ω 6 as required by most mammals. The main emphasis on lipid requirements has been on EFA and on the energy value of lipids.

4

Digestion of Carbohydrates Metabolism

Carbohydrate metabolism begins with digestion in the small intestine where monosaccharides are absorbed into the blood stream. Blood sugar concentrations are controlled by three hormones: insulin, glucagon, and epinephrine. If the concentration of glucose in the blood is too high, insulin is secreted by the pancreas. Insulin stimulates the transfer of glucose into the cells, especially in the liver and muscles, although other organs are also able to metabolize glucose.

In the liver and muscles, most of the glucose is changed into glycogen by the process of glycogenesis (anabolism). Glycogen is stored in the liver and muscles until needed at some later time when glucose levels are low. If blood glucose levels are low, then eqinephrine and glucogon hormones are secreted to stimulate the conversion of glycogen to glucose. This process is called glycogeno-lysis(catabolism). If glucose is needed immediately upon entering the cells to supply energy, it begins the metabolic process calledglycoysis (catabolism). The end products of glycolysis arepyruvic acid and ATP.

Since glycolysis releases relatively little ATP, further reactions continue to convert pyruvic acid to acetyl CoA and then citric acid in the citric acid cycle. The majority of the ATP is made from oxidations in the citric acid cycle in connection with the electron transport chain. During strenuous muscular activity, pyruvic acid is converted into lactic acid rather thatn acetyl CoA. Durlng the resting period, the lactic acid is converted back to pyruvic acid. The pyruvic acid in turn is converted back to glucose by the process called gluconeogenesis(anabolism). If the glucose is not needed at that moment, it is converted into glycogen by glycogenesis. You can remember those terms if you think of"genesis" as the formation-beginning. These processes are summarized in the Metaboism Summary in the graphic on the left. Each of these processes will be developed in greater detail various pages of this module.

The activities on which the life of the body depends involve a continuous expenditure of energy and a constant exchange of material. Ultimately the body is dependent upon food for the fuel materials which supply energy and for both the substances which are transformed in, and eliminated from, the body, and those whose presence regulates and controls these transformations. The materials leaving the body are to be regarded not merely as wastes but as end products of an orderly and coordinated series of chemical reactions which occur in the body and by virtue of which its functions are performed.

Thus the chief functions of food are:

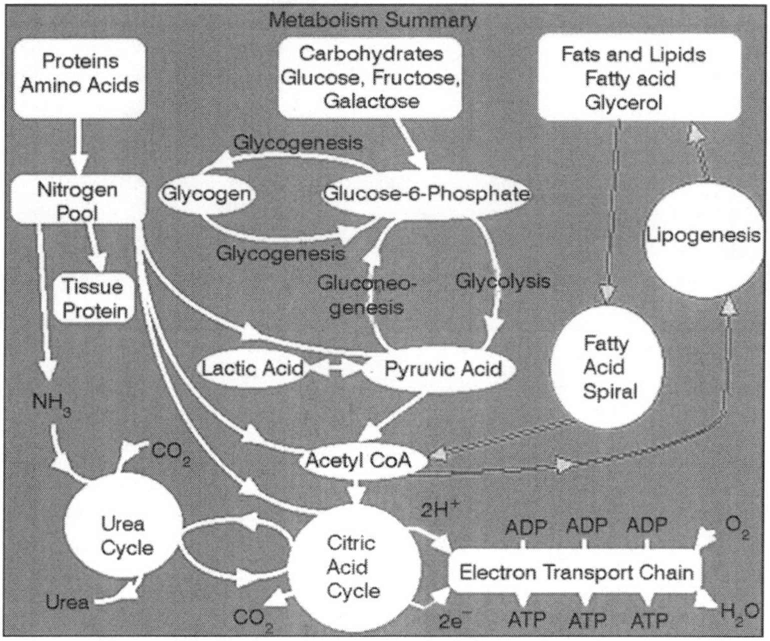

CLASSIFICATION OF CARBOHYDRATES

- To yield energy.
- To build tissue.
- To regulate body processes.

These functions involve reactions which are dependent upon the chemical composition and constitution of the food. Any food constituent which takes part in any of these functions may be regarded as having nutritive value.

Most of the nutrient material contained in food requires more or less change to bring it into the exact forms most useful in nutrition. These changes as a rule take place in the digestive tract and together constitute the process of digestion. The changes which take place in the foodstuffs, after they have been absorbed from the digestive tract, are included under the general term"metabolism." Although the chemical changes and the energy transformations are of course inseparable, it has become customary to speak of the metabolism of matter and the metabolism of energy, and to regard the extent of the metabolism of any material substance as measured by the amount of its end products eliminated, and the extent of the energy metabolism as measured by the amount of heat, or of heat and external muscular work, which the body gives off.

The metabolism of matter and the metabolism of energy are normally supported by the food; but if no food is taken, they continue at the expense of the body substance. The expenditure of energy can never cease in the living body because it includes the work involved in carrying on the internal processes which are essential to life itself; and the expenditure of matter cannot cease because the energy for this necessary work is obtained by the breaking down of the organic compounds of the food or of the body substance into simpler compounds, many of which are of no further use to the body and must be eliminated. When the food supplies sufficient energy, the body substance is protected; when the food is insufficient, body substance is burned as fuel.

In order, then, to consider intelligently the nutritive requirements of the body as regards the substances of which it is composed, it is necessary first to know whether the fuel requirements (the requirements of the energy metabolism) have been fully met. The carbohydrates, fats, and proteins of the food all serve as fuel to yield the energy required for the activities of the body, and the proteins serve also as material for the maintenance or growth of body tissue. But of the fifteen chemical elements which are essential to the structure and functions of the body, simple proteins furnish only five. The remaining ten elements are largely constituents of the ash of the food and are known as ash constituents, inorganic foodstuffs, mineral matter, or salts.

Recent investigations have developed the fact that food of sufficient energy value and containing ample amounts of each of the chemical elements known to be essential to the body is not necessarily adequate to meet all the requirements of nutrition. Thus it appears that certain substances occurring in natural foods but not yet chemically identified are also to be included among the nutritive requirements of the body and therefore among the factors which determine the nutritive values of foods. At present these unidentified substances are referred to as"vitamines" or as"fat soluble A" and"water soluble B." *The essentials of a chemically adequate food supply may therefore be summarized as follows*:

- Sufficient of the organic nutrients in digestible forms to yield the needed energy.
- Protein, sufficient in amount and appropriate in kind.
- Adequate amounts and proper proportions of the various ash constituents or inorganic foodstuffs.
- Sufficient of each of the two unidentified"vitamine" factors, the"fat soluble A" and the"water soluble B."

In attempting to give in the following pages a general view of the chemistry of food and nutrition it has seemed best to discuss first the chemical nature and nutritive functions of the substances which serve as sources of energy in nutrition, then the nutritive requirements in terms of energy, protein, the more prominent"inorganic" elements, and the"vitamines," and finally the bearing of these various factors offood value upon problems connected with the economic use of food.

Of the constituents of the ordinary mixed food of man the carbohydrates are usually the most abundant and the most economical sources of energy. They are also considered to be the first of the three great groups of foodstuffs to be formed by synthesis from simple inorganic substances in plants;"in the long run, all the energy of living matter comes from them." The synthesis of carbohydrates in nature is therefore a logical starting point for the study of the organic foodstuffs.

In the chlorophyll cells of the leaves of green plants the energy of the sun's rays brings about reaction between carbon dioxide and water which results in the liberation of oxygen and the formation of organic compounds. There is still doubt as to the exact mechanism of the process and no certainty that it is the same in all cases. It has, however, been quite generally found that the volume of oxygen liberated is equal to that of carbon dioxide consumed. The simplest possible representation of the reaction would be $CO_2 + H_20 \rightarrow CH_20 + O_2$ according to which the first product of the synthesis would be formaldehyde. There is considerable (though not conclusive) evidence that formaldehyde is thus formed and that it is rapidly built up into less reactive compounds. Whatever the steps in the process* there is normally an early production of carbohydrate. Usually the first product which can be demonstrated as accumulating in the plant as the result of the photosynthesis is a sugar (glucose or sucrose) or starch. Assuming glucose as a typical product and neglecting the intermediate stages, the photosynthesis of carbohydrate may be represented thus:

$$6\ CO_2 + 6\ H_{20} \text{ ® } C_6H_{12}O_6 + 6O_2$$

Glucose is the most familiar representative of a group of simple sugars (monosaccharides or monosaccharoses) which are in composition direct polymers of formaldehyde (CH20) and which are classified, according to the number of carbon atoms in the monosaccharide molecule, as trioses, pentoses, hexoses, etc.

CLASSIFICATION

Definitions of the term"simple sugar" vary somewhat, depending chiefly upon the views of different authors as to how simple a compound may properly be called a sugar. According to Browne, a simple sugar or monosaccharide may be defined as an aldehyde alcohol or ketone alcohol of the aliphatic series, the molecule of which contains one carbonyl and one or more alcohol groups, one of the latter being always adjacent to the carbonyl group. All simple sugars contain, therefore, as a characteristic group upon the presence of which the chief chemical properties of the sugars depend.

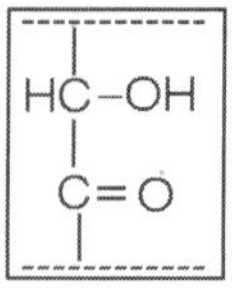

The simplest possible sugar according to this definition is glycolaldehyde, CH2OH-CHO, which (in analogy with the nomenclature of the familiar sugars) may also be called glyco-lose. The structural formulae of glucose and fructose, the most familiar representatives of the aldehyde-alcohol (aldose) and ketone-alcohol (ketose) sugars, respectively, are as follows:

Glucose

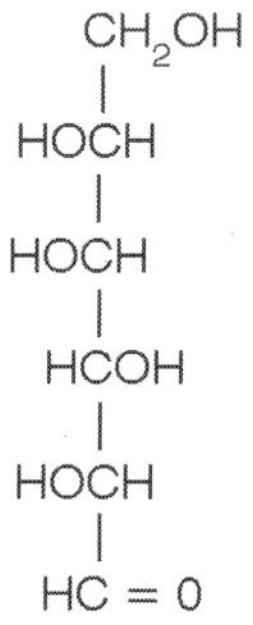

Fructose

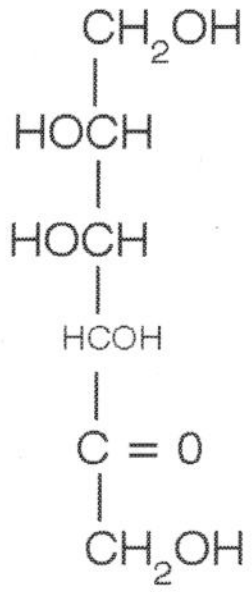

Since glucose gives aldehyde reactions but not so readily as the above structural formula would lead one to expect, it is believed that in ordinary solutions of glucose the substance exists partly in the condition indicated by the aldehyde formula and partly in a tautomeric form represented by the lactone or"oxygen bridge" formula.

Following are the aldehyde and lactone formulae written without reference to the spatial relationships of the hydrogen and hydroxyl groups:

Aldehyde form

$$CH_2OH\text{-}CHOH\text{-}CHOH\text{-}CHOH\text{-}CHOH\text{-}CHO$$

Lactone form

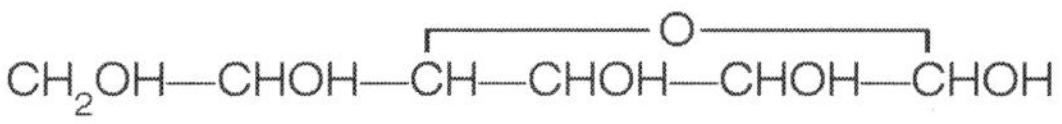

The name monosaccharide ("single sugar") implies that the monosaccharide molecule contains only one sugar radicle - that it cannot be split by hydrolysis into sugars of lower molecular weight. A substance like cane sugar which on hydrolysis splits to two molecules of simple sugar is called adisaccharide or disaccha-rose ("double sugar"). Trisaccharides and tetrasaccharides are also known.

Substances which like starch are of high molecular weight and on complete hydrolysis yield many molecules of simple sugar are called polysaccharides * or polysaccharoses. The term"carbohydrates" covers all the simple sugars and all substances which can be converted into simple sugars by hydrolysis. The term"glucosides" is applied to substances which consist of combinations of carbohydrate radicles with radicles of other kinds and which therefore yield on hydrolysis both a simple sugar and one or more products of other than carbohydrate nature.

DIAGNOSIS AND TREATMENT OF ACQUIRED DISORDERS OF CARBOHYDRATES

Diabetic ketoacidosis, hyperosmolar coma, and hypoglycemia present as obtundation or coma. The differential diagnosis is as extensive as it is for obtundation and coma in general. The first clinical task is to make sure the patient survives. Checking the airway, breathing, and circulation (ABCs) and correcting any problems with these is always the first step. The second step is to get baseline blood levels for later analysis before metabolically altering the system during treatment. The third step is to give intramuscular (IM) thiamine followed by 50 mL of 50% intravenous (IV) glucose.

Hypoglycemia can be fatal if not reversed in approximately 45 minutes. A glucose load depletes thiamine stores and precipitates Wernicke syndrome; therefore, thiamine should be given first. Then perform a rapid general examination, looking for clues of serious trouble (eg, abnormal patterns of breathing, unusual body or breath odour, jaundice, pink skin, pallor, petechiae, meningismus, abnormalities of major organs). Next, perform a neurological examination focused on variable conditions associated with coma. Fisher and Plum and Posner have written extensively about the neurological examination of comatose patients.

The neurological examination helps clarify whether the problem is focal or systemic (metabolic). If all signs point to a lesion on one side or in one location, the lesion is focal; imaging studies and IV steroids need to be considered and a neurosurgeon consulted. If the signs are diffuse or do not lateralize, the problem most likely is metabolic. Again, Fisher and Plum and Posner have published schemata that can be followed to go through the myriad possibilities in a logical manner. Subtle clues from the history or the examination can be crucial and therefore must be given serious consideration and addressed immediately.

Bhattacharya *et al*, in their pilot study compared uncooked cornstarch with uncooked physically modified cornstarch (WMHM20) in patients with glycogen storage disease types Ia, Ib, and III. The authors reported longer duration of euglycemia and better short-term metabolic control in most of the patients with WMHM20 when compared with cornstarch.4

Diabetes

The acute severe disorders in patients with diabetes are diabetic ketoacidosis (see Diabetic Ketoacidosis) and diabetic hyperosmolar coma (Hyperosmolar Coma). Hypoglycemia from lack of liver glycogen stores (as in alcoholics) or from excess insulin, whether accidental or intentional, also is an acute emergency. The chronic disorders are more common. These include mononeuritis from diabetes, multiple neuropathy (still called mononeuritis multiplex by some) in patients with diabetes, polyneuropathy in patients with diabetes, and so-called diabetic amyotrophy.

Diabetic Ketoacidosis

Even though the diagnosis may be established clinically, and even though someone experienced in the care of this problem may sometimes treat the ketoacidosis successfully without supporting Labouratory data, confirming the diagnosis by lab studies is still important. Whether the patient will respond easily or with difficulty or what hazards and problems may lie ahead cannot be known at the outset. The extent of Kussmaul breathing is a guide to the severity of acidosis; sudden relief of this breathing, as the patient is being treated, usually means the pH of the blood is back to normal. This can be checked with serial measurements of arterial blood pH (ie, blood gases) while treating the patient.

Ketones and sugars can be demonstrated in the urine and blood by dipstick, but with each repeated step of treatment, sending samples of blood for an accurate analysis by the Labouratory is advisable. Changes on the ECG may suggest hypokalemia, but, again, getting the Labouratory's ongoing help with serial assays of electrolytes, BUN, and blood pH is advisable. The workup and treatment must include any likely coincidental problem (ie, a new illness other than diabetes) that has tipped the metabolic balance.

Diabetic Hyperosmolar Coma and Hypoglycemia

These are the 2 major diagnoses to consider in any patient presenting with obtundation or coma of metabolic rather than structural etiology. Once the comatose patient's ABCs have been secured, several tubes of blood must be drawn for the host of cellular and biochemical analyses a comatose patient may need. Administer 100 mg thiamine IM and then inject an IV bolus of 50 mL of a 50% solution of glucose through a needle of largest possible diametre (the author uses a 15-gauge needle). This amount of glucose is enough to reverse hypoglycemia but represents a mere"drop in the bucket" for a patient with hyperosmolar coma or even diabetic ketoacidosis.

Rapid clinical improvement after a bolus of glucose confirms the diagnosis of hypoglycemia. Patients with hyperosmolar coma do not respond, but the initial blood glucose, as measured in the Labouratory, is very high (ie, usually well over 400 mg/dL), as is the serum osmolality. The CSF glucose corresponds to the blood glucose but with a 2-hour delay; therefore, the CSF glucose remains high even after the blood glucose and electrolytes are brought slowly back to normal. Remember the dangers of central pontine myelinolysis when correcting electrolyte imbalances due to any cause.

The chronic neurological complications of diabetes are due to damage to peripheral nerves. Mononeuritis and multiple neuropathy are both due to infarction of the tiny artery supplying a particular nerve, the vasa nervorum. Diabetes is a common cause of these problems, and it often presents with infarction of a single nerve, especially of cranial nerves III and VI and of the long mixed nerves of the limbs. When the distal motor nerves of limbs are involved, the diagnosis may be fairly clear. When the proximal motor nerves are primarily involved, the diagnosis may be difficult. The pattern can be that of weakness of the muscles of one thigh with deep pain (ie, diabetic amyotrophy).

Amyotrophy can be a presentation of diabetes. Sometimes, the opposite limb is affected much later. Amyotrophy tends to be self-limiting and improves without any treatment after a few months. The cause of the waxing and waning of symptoms of amyotrophy is unclear. In some patients, the pattern is of damage to the lumbar plexus, resulting in diabetic plexitis. Again, the reason for the differences in the extent of involvement is unknown.

The differential diagnosis of mononeuritis includes local pressure along the course of the nerve or its roots from trauma, hematoma, entrapment, abscess, or growths, and other causes of infarction of the arteria nervorum, of which a current leading cause is collagen-vascular disease, especially the peculiar form of periarteritis nodosa that is confined largely to the peripheral nervous system (ie, not clinically affecting other organ systems). These may be suggested by repeated abnormalities of the erythrocyte sedimentation rate or the C-reactive protein or by consistent elevations of antibodies specific for certain collagen-vascular diseases.

Other causes of multiple neuropathy include Lyme disease, sarcoidosis, acquired and inherited amyloidosis, and, very rarely, first attacks of tumacular neuropathy. Most cases of multiple neuropathies are treatable; therefore, this syndrome is one of the few clinical indications for a nerve biopsy should it be needed to establish the diagnosis. The sural nerve is almost always the nerve of choice. Consult an experienced surgeon to obtain the specimen. Too small a piece, or one that looked liked nerve but was collagen, does no good, and the patient usually will not consent to a second attempt at biopsy.

Many cases of plexitis, whether brachial or lumbar, unilateral or bilateral, are due to infectious or postinfectious problems. These need to be excluded carefully before diabetes is accepted as the cause of plexitis. The problem is complicated by treatment.

For example, the nondiabetic causes of plexitis often respond to high-dose steroids. Most of the author's patients have responded well to pulse-dose steroids for 5 days as currently used for other immune-mediated neurological disorders. Steroids elevate blood sugars. This means that diabetes needs to be sought and excluded before a therapeutic trial and that a patient with diabetes who is to be given even pulse-dose steroids needs careful monitoring, often with testing of blood glucose 4 times a day for monitoring excessive elevations of glucose during treatment.

In the era before CT scans of the abdomen and pelvis were available, the worsening of what appeared to be a lumbar plexitis with high-dose steroids pointed to the correct problem, pressure palsy from a pelvic abscess.

Polyneuropathy

Diffuse disorders of peripheral nerves exist, both sensory and motor, but often the sensory disorders are the most profound, clinically affecting the longest Fibres first. This is a common problem in neurology in general, and it is also a common problem in people with diabetes. The patients have numbness and tingling in the toes and feet and, later, up the legs. By the time the sensory loss reaches the knees, the fingertips begin to be involved.

Often, small-Fibre sensation (eg, superficial touch, pinprick, appreciation of hot and cold) is involved early and most predominantly. These patients are likely to scald themselves by getting into too hot a bathtub. They test the water with their feet, but their feet can no longer warn them of the high temperature. To some extent, the severity of the neuropathy correlates with the severity of the diabetes and with episodes of lack of metabolic control.

To some extent, the neuropathy may improve if control is restored. However, the correlation is not at all tight. Some patients presenting with neuropathy have mild diabetes; in others, the neuropathy occurs and does not get better readily, regardless of the degree of control of blood glucose levels. Some patients present with a marked dysesthetic pain, others do not. In many patients, later rather than sooner, a marked autonomic component is present, which is not present in others.

When the autonomic system is involved, the distal skin often becomes mottled, cyanotic or pink, smooth, and loses hair follicles to become shiny. Capillary refill may become poor. Postural hypotension occurs in advanced stages of autonomic involvement. Treatment of complications of the peripheral nervous system in a patient with diabetes is still poor. Good control of blood glucose levels helps prevent or delay onset of these complications; therefore, young patients today often are put on insulin pumps, and blood glucose is monitored much more closely. The neuropathic pain of diabetes often responds to modern antiepilepsy drugs, as does neuropathic pain of other etiologies. The authors' patients have responded to gabapentin (Neurontin), usually started at very low doses but often increasing to the large doses used in epilepsy. Pain can be well controlled by maintaining trough levels of 6-12 mg/dL.

DIAGNOSIS AND TREATMENT OF INHERITED DISORDERS

Most of the inherited disorders of carbohydrate metabolism fall into a few broad clinical categories.

These are as follows:

- Episodic lactic acidosis in infancy, failure to thrive, and hypotonia with or without features suggesting specific disorders.
- Infantile mental retardation, hypotonia, failure to thrive, and other features
- Mental retardation/developmental delay with features that suggest storage disorders.
- Episodic vomiting in infants or young children.
- Episodic acidosis in childhood or early adult life, often with features that suggest specific disorders.
- Intermittent or episodic ataxia in childhood.
- Cramps and weakness with anaerobic exercise, often with some episodes of muscle breakdown and myoglobinuria, during or after heavy exercise.
- Symmetrical neuromuscular disease with weakness and wasting of proximal muscles.

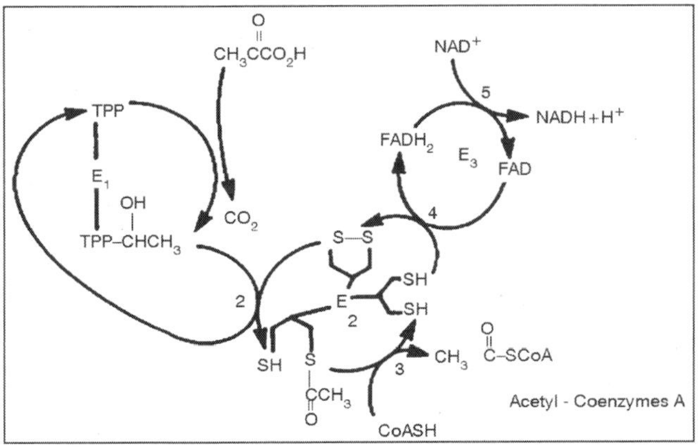

Fig. Pyruvate Dehydrogenase Complex

Episodic lactic acidosis in infancy, failure to thrive, and hypotonia with or without features suggesting specific disorders. In this category, an infant presents with signs of acidosis, tachypnea, marked irritability, and tachycardia. As the acidosis worsens, the infant becomes obtunded or comatose. Muscle tone is very poor; the infant is flaccid and has poor head control. If the infant is more than a few hours or days old, signs of failure to thrive may be seen: weight loss or at least poor gain in weight, microcephaly, and poor linear growth. Prognosis is poor even with current vigourous metabolic management. Without such management, the infant usually dies in a short time. In the absence of other clinical features, this may be the presentation of one of several defects of the PDH multienzyme complex, particularly of its E components.

Scheme of the major reactions of the pyruvate dehydrogenase complex. Initial oxidative decarboxylation of pyruvate is shown as reaction:

- Transfer of the 2-carbon moiety to the lipoyl side chain of lipoyl acetyltransferase (E2) as reaction.
- Formation of acetyl coenzyme A as reaction.
- Oxidation of the lipoyl moiety as reaction.
- Reduction of NAD as reaction.
- The regulatory enzymes acting on pyruvate decarboxylase (E1) are not shown.

The components are called E1 (ie, pyruvate decarboxylase [and the phosphorylase and dephosphorylase that modify the activity of E1]), E2 (ie, lipoyl acetyl transferase), and E3 (ie, lipoamide, also known as dehydrogenase; dihydrolipoamide: NADH oxidoreductase). Several cases of Leigh syndrome appear to have been caused by failure of activation of pyruvate decarboxylase, presumably because of defects in phosphorylation. (Three of the 4 infants described thus far with defects of E3 also had the complex clinical picture described by Leigh.5) In theory, a profound congenital deficiency of thiamine ought to produce episodic lactic acidosis in infants with failure to thrive and hypotonia.

Leigh syndrome (ie, subacute necrotizing encephalo-myelopathy) usually begins in the first or second year of life. Motor and cognitive regression occur along with ataxia, chorea, and tonic spasms. Depending on the age of the patient, history and physical findings include loss of muscle tone and head control, decreased appetite, nausea, vomiting (often with lactic acidosis), irritability to the point of continuous crying, myoclonus, and primarily generalized seizures. Episodes of hyperventilation, apnea and gasps, sobbing without tears, external ophthalmoplegia with pendular or rolling nystagmoid movements of the eyeballs, loss of swallowing, muscle wasting, and loss of reflexes are observed. In cases of postnatal lactic acidosis with elevations in the levels of ammonia, citrulline, and lysine, consider defects of the B-complex of pyruvate carboxylase, the first enzyme in the pathway that generates glucose from fatty acids and amino acids through gluconeogenesis. Death occurs by about 3 months of age.

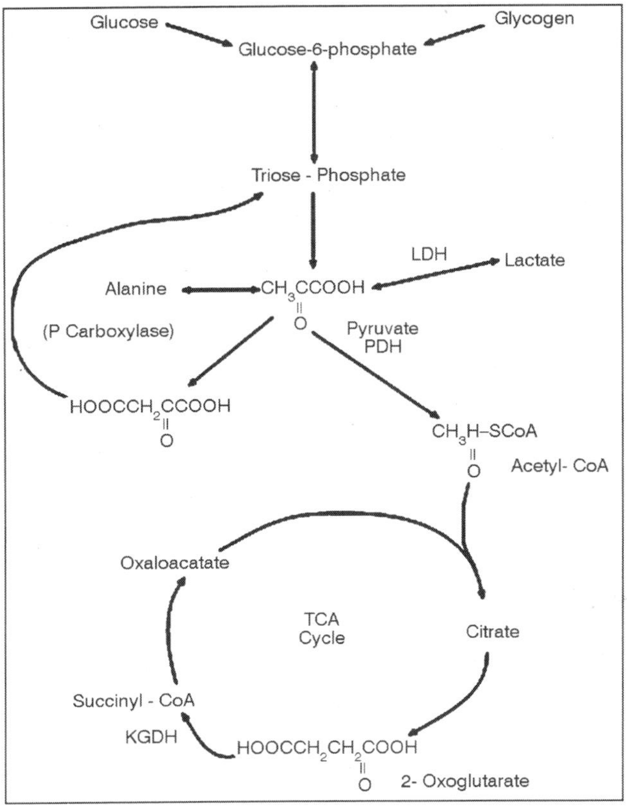

Major metabolic relations of pyruvate. Many of the reactions, such as those between glucose-6-phosphate and pyruvate, pyruvate and alanine, and pyruvate and lactate, are of course reversible. Abbreviations: LDH - lactic dehydrogenase; PDH - pyruvate dehydrogenase complex; p carboxylase - pyruvate carboxylase; TCA cycle - Krebs tricarboxylic acid cycle; KGDH-a-ketoglutarate dehydrogenase complex.

Mental Retardation/Developmental Delay with Features that Suggest Storage Disorders

The features that suggest storage disorders are hepatomegaly, a large forehead with frontal bossing, and macrocephaly. A large number of diseases, including storage disorders caused by inborn errors of metabolism, can cause hepatomegaly in infancy without acidosis or other specific features discussed. Of the disorders of carbohydrate metabolism, these features can be seen with defects of production, storage, or catabolism of glycogen, especially deficiency of the glycogen debrancher enzyme, which leads to hepatomegaly in infancy and early death; defects of 3 of the 4 proteins that make up phosphorylase kinase; and deficiency of glucose-6-phosphatase (ie, von Gierke disease), which is accompanied by gout, lactic acidosis, and elevated lipids.

Differential Diagnosis

Episodic vomiting in infants or young children has a large number of causes, most of which are more common than disorders of carbohydrate metabolism. Nonmetabolic causes include gastroenteritis from viruses, bacteria, and parasites; gastroesophageal reflux disorder; irritable bowel syndrome; small or large bowel obstructions; neurological conditions, such as premigrainous syndromes and brain masses; and abdominal epilepsy. In this clinical setting, inborn errors of the urea cycle and organic acid metabolism need to be considered first, before considering lactose intolerance from inherited disorders of galactose and fructose metabolism.

Workup and Treatment

Hereditary fructose intolerance and defects of fructose 1, 6-diphosphate aldolase are discussed under Episodic lactic acidosis in infancy, failure to thrive, and hypotonia with or without features suggesting specific disorders earlier in this article. Disorders of galactose metabolism are discussed in the subsection Infants with mental retardation, hypotonia, failure to thrive, and other features.

Episodic Acidosis in Childhood or Early Adult Life

This often presents with features of a specific disorder, including many of the differential diagnoses mentioned under Episodic lactic acidosis in infancy, failure to thrive, and hypotonia with or without features suggesting specific disorders. Specific features may enable rapid diagnosis.

Intermittent or Episodic Ataxia

This category is characterized by the sudden, almost abrupt, onset of ataxia, extrapyramidal symptoms in the upper extremities, and dysarthria soon after or in conjunction with an infectious illness or severe emotional upset. In one patient monitored by one of the authors, death of a grandparent and critical school examinations were sufficient stimuli to initiate attacks of ataxia. Physical findings in this patient also included a high arched palate, minimal kyphoscoliosis, minimal pes cavus, and extra rows of lashes on the upper eyelid. When the boy was 8 or 10 years old, a mild attack was precipitated by an environmental challenge with flashing red lights from scintillation counters and Geiger counters for about 30 minutes in a dimly lit room. For this patient and others, the episodes of ataxia usually last about 2 weeks. Although the neurologic symptoms usually are accompanied by increased serum lactic acid, frank acidosis is not always present.

The underlying defect in this patient was shown to be that of the first enzyme of the PDH complex, E1. Eventually, he was placed on a diet high in thiamine and relatively low in carbohydrates; the number of attacks subsided. While this patient graduated from college, others with similar clinical presentations have had at least a mild degree of mental retardation. Work in the 1970s and early 1980s raised the question whether familial cases of a childhood- or adolescent-onset ataxia (one with slower progression than is seen in classical Friedreich ataxia) was associated with moderate defects of the activity of PDH, E3. The association has not been confirmed. Classic Friedreich ataxia appears to be caused by a single genetic defect and does not appear to be associated with measurable abnormalities in the E3 pathway

Differential Diagnosis

Intermittent or episodic ataxia in childhood and adolescence is a rare condition. Porphyria might induce this, and multiple sclerosis, though rare in young people, certainly can present with ataxia; however, multiple sclerosis almost always manifests in other ways. Postinfectious cerebellar dysfunction usually presents as a single episode that clears in weeks or months and does not recur. Griggs and colleagues have described patients with a familial form of intermittent ataxia, unrelated to pyruvate defects, that responds to treatment with acetazolamide.

Workup

E1 is the first enzyme of the PDH multienzyme complex. Phosphorylation inactivates this pyruvate decarboxylase, which activates by removal of the phosphate. The decarboxylase requires thiamine pyrophosphate as a substrate. Patients deficient in this enzyme have elevations of metabolic by-products related to pyruvate degradation during clinical attacks of ataxia (ie, excess lactic acid and alanine in blood or urine, or both). The primary enzyme defect can be demonstrated in cultured fibroblasts and probably can be demonstrated in WBCs or muscle tissue. The substrate, pyruvate, is so labile and the enzyme activity of this and other components of the PDH complex so sensitive to environmental contaminants that carefully controlled assays need to be performed in a research Labouratory.

Treatment

The young man discussed earlier in detail was treated with a low-carbohydrate diet and high doses of thiamine. Some patients with autosomal recessive ataxia with large-Fibre-posterior column sensory loss, areflexia, cardiomyopathy, and bony stigmata have been treated with some success with a modified (titrated) ketogenic diet. Dietary changes were introduced gradually until moderate to high levels of ketones appeared in the urine without producing frank metabolic acidosis.

Black McArdle first described these symptoms of the disease at Guy's Hospital in London.9 (He was so named to distinguish him from his brother"Red," also a distinguished physician there; the pseudonyms referred to the Colour of their hair.) Type V glycogenosis due to muscle phosphorylase deficiency is the clinically mildest of the glycogen storage diseases, and the symptoms may not occur until adolescence or young adulthood. The symptoms tend to be induced during exercise when muscle depends on glycolysis rather than oxidative phosphorylation as a source of fuel. Exercise that includes sprinting or vigourous running most likely will induce symptoms. Often, onset of symptoms comes during what normally would represent the subjective"second wind." Over time, patients may learn to slow down or stop at the first hint of symptoms, and then to begin again, with what has been called a second wind but which is probably a pause in which oxidative metabolism can generate sufficient ATP in muscle so that anaerobic activity can begin again.

If this speculation is true, the"second wind" that is peculiar to myopathies from glycolytic/glycogenolytic defects differs in its nature from the second wind experienced by healthy athletes. The patients therefore can carry on anaerobic exercise in spurts over a long time, presumably with aerobic exercise in between. Affected individuals often can tailor their activities to take a break at the first sign of cramping, and then resume activities shortly thereafter.

Without this pause and resumption in athletes with glycolytic/ glycogenolytic defects, muscles suddenly become so weak that the athlete becomes virtually immobile from painful cramps, and the muscles that were exercised are very sore and tender. This disease and the finding of elevated lactic acid levels in the blood after exercise led to the erroneous notion that muscle cramps in general are caused by lactic acidosis. The muscles in the individuals affected with glycolytic defects continue to be very sore even after the cramps resolve.

If the urine is not frankly red, it must be examined to see if it still contains heme groups. In myopathy due to disorders of glycolysis, the heme pigment proves to be myoglobin, not hemoglobin, when detailed analyses are carried out. With McArdle disease, muscle enzymes and creatine levels usually are elevated in the blood after a symptomatic episode. In this condition, muscle histology demonstrates marked increase in glycogen stores in the myofibrils. Histochemical or immunoassays for specific enzymes can augment Labouratory diagnosis by identifying the precise enzyme defect.

The original enzyme defect described by McArdle9 and then by Schmid and Mahler10 was localized to the glycolysis pathway. The deficiency ultimately was identified in muscle phosphorylase.

Through the past decade, MR spectroscopy of forearm muscles *in vivo* has led to the diagnosis. Deficiencies of several other glycolytic enzymes (eg, muscle phospho-fructokinase, phosphoglycerate kinase, phosphoglycerate mutase, lactate dehydrogenase) produce similar clinical pictures and are indistinguishable from muscle phosphorylase deficiency, unless enzymatic tests are performed.

Muscle breakdown occurs in disorders of glycolysis and glycogenolysis during muscle activity that is largely dependent on glycolysis or anaerobic metabolism. This is in striking contrast to the circumstances of muscle breakdown in disorders related to lipid or ketone metabolism. In these disorders, muscle breaks down, with similar symptoms, under circumstances in which muscles are relying on oxidative metabolism, especially of ketone bodies, as a fuel source. Activity most likely to produce symptoms under these circumstances is prolonged, less vigourous exercise over several hours, which is exacerbated by starvation or high-fat, high-protein meals and occurs more often in a cold or cool environment. Histologically, myofibrils may demonstrate increased lipid droplets, but in contrast to the glycogenoses, glycogen stores are either normal or even slightly depleted.

Differential Diagnosis

The differential diagnosis includes all the acute and symptomatic myopathies, including various forms of polymyositis. Many of these appear to be viral or postinfectious; still others appear to be paraneoplastic or associated with motor neuron degenerations or systemic collagen-vascular diseases but not with problems of intermediary metabolism.

The circumstances under which cramps and weakness occur may provide insight into whether a metabolic disorder is present. If evidence of metabolic etiology appears, differentiating between disorders affecting white muscle (ie, the carbohydrate errors discussed here) and disorders affecting red muscle (ie, errors of lipid and mitochondrial metabolism) is essential, because these disorders can present similar pictures but require very different metabolic treatment strategies. Making the correct diagnosis clearly is important since, for each of these groups of disorders of glycolysis/glycogenolysis, mitochondrial transport, and oxidative phosphorylation, treatment approaches are quite different.

Workup

Prior to the availability of direct enzyme assays, defects of muscle phosphorylase had to be inferred from the clinical features of muscle cramping with vigourous exercise in conjunction with accumulation of lactic acid in the blood and myoglobin in the urine. The forearm ischemia test further substantiated the diagnosis. For this test, a blood pressure cuff is inflated to systolic pressure. With the cuff on, ampules of venous blood are drawn from the antecubital fossa via an indwelling catheter attached to a stopcock before, during, and after 1 minute of forearm exercise (eg, having the patient squeeze a partially inflated blood-pressure cuff repeatedly). In reality, many patients with phosphorylase deficiency cannot continue to exercise for a full minute. Blood samples are assayed for glucose, lactic acid, and sometimes pyruvate. Affected individuals have lower glucose levels and higher lactate and pyruvate levels compared to reference values in blood samples taken 1-2 minutes after exercise.

In the last 25 years, the preferred diagnostic test for McArdle disease has been a phosphorylase assay directly on muscle tissue. Today, histochemical or biochemical techniques can be used to diagnose deficiencies in even the rare enzymes that clinically simulate McArdle disease. Again, these enzymes include muscle phosphofructokinase, phosphoglycerate kinase, phosphoglycerate mutase, and lactate dehydrogenase. In the past 2 decades, diagnosis also has been made by analysis of metabolites *in vivo* on human forearm muscle by means of MR spectroscopy.

Treatment

DiMauro *et al* gave a thorough review of strategies to treat disorders of muscle phosphorylase.11 Pyridoxine, at 50 mg/day, seems to be effective. (Note that at 100 mg/d, this vitamin can produce a painful distal sensory neuropathy.) High-protein diets also were effective, at least for several weeks, but determining the long-term success of dietary treatment will be possible only after adequate clinical trials.

Current therapeutic options include a carbohydrate-rich diet and carbohydrate ingestion shortly before strenuous exercise in combination with medically supervised aerobic training of low to moderate intensity.

Only rarely are cases of progressive neuromuscular diseases caused by defects of glycogen metabolism. When they are, children or young adults generally develop, over months or years, progressive weakness of the limb-girdle muscles and the proximal muscles of the limbs. Patients have no pain under ordinary circumstances, but heavy exercise may induce fatigue and cramps. Symptoms depend on the nature of the underlying metabolic defect. Muscles may or may not be atrophied at presentation but eventually waste. Patients do not complain of muscle tenderness unless cramping occurs. Some patients with debrancher deficiency have evidence of cirrhosis; others become asymptomatic in adult life.

Differential Diagnosis

This condition is essentially indistinguishable from Limb-Girdle Muscular Dystrophy. The symptoms can be demonstrated in a number of acquired diseases, including various forms of polymyositis or inflammatory myopathy, various neuropathic diseases (especially proximal forms of motor neuron disease and the limb-girdle pattern of painless diabetic multiple neuropathy or diabetic plexitis), inherited mitochondrial myopathies (often associated with ragged red Fibres), disorders of lipid metabolism, and/or defects of lipid as carnitine transport at the level of mitochondria. For lipids to enter the mitochondria, they need to be complexed with carnitine derivatives. Primary or secondary carnitine deficiencies as well as deficiencies of enzymes involved in the metabolic pathway can interfere with this process, resulting in a myopathic picture.

Workup

Muscle enzyme levels may be elevated in the serum, but myoglobinuria is an unusual finding. Histochemical analysis of biopsied muscle shows excessive glycogen stores. Partial acid maltase deficiency may result in the adult form of Pompe disease and may be clinically indistinguishable from muscle phosphofructokinase deficiency (reported in at least one case) and glycogen debrancher enzyme deficiency. However, patients with glycogen debrancher enzyme deficiency eventually develop physical, electrical, or radiographic evidence of cardiac ventricle dilatation. Therefore, clinicians are advised to monitor for these changes by ECGs and examinations on an annual basis.

Treatment

Defects of acid maltase in their milder forms are discussed in Mental retardation/developmental delay with features that suggest storage disorders and in the article Lysosomal Storage Disease (ie, Pompe disease). The FDA has recently designated alglucosidase alfa (Myozyme) an orphan drug for use in infants with these disorders. The glycogen debrancher enzyme is discussed above in Mental retardation/developmental delay with features that suggest storage disorders, and muscle phosphofructokinase is discussed above in Cramps and weakness, often with some episodes of muscle breakdown and myoglobinuria, during or after heavy exercise.

FUTURE IN TREATMENT

In many instances, inborn errors of metabolism are difficult or impossible to treat. Novel therapeutic modalities, including gene therapy, need to be explored. Among available gene delivery systems, recombinant adeno-associated virus (AAV) shows promise for the treatment of metabolic disease. Despite the relatively low immunogenicity of AAV vectors, immune responses are also emerging as a factor requiring special attention as efforts accelerate towards human clinical translation. Trials targeting lipoprotein lipase deficiency are showing early evidence of efficacy.12

FERTILITY CONCERNS

Increasing numbers of individuals with inherited metabolic disorders are surviving into adulthood and considering their reproductive options. Lee discusses a practical approach to supporting such individuals, focusing on issues concerning fertility, the impact of pregnancy on metabolism and the metabolic disorder itself on the pregnancy, as well as highlighting the need to pay special attention during the postpartum period.13 He further opines that apart from pregnancies in women with phenylketonuria, data are lacking in this area. Collection of information within registries is needed to aid our understanding of potential problems and counseling of women and their partners.

GLYCOLYSIS

All organisms produce ATP by releasing energy stored in glucose and other sugars.

- Plants make ATP during photosynthesis.
- All other organisms, including plants, must produce ATP by breaking down molecules such as glucose.

AEROBIC RESPIRATION

The process by which a cell uses O2 to"burn" molecules and release energy.

- *The reaction*: $C_6H_{12}O_6 + 6O_2$ >> $6CO_2 + 6H_2O$.
- This reaction is the opposite of photosynthesis

This reaction takes place over the course of three major reaction pathways.

- Glycolysis.
- The Krebs Cycle.
- Electron Transport Phosphorylation (chemiosmosis) Glycolysis (glyco = sugar; lysis = breaking).
- *Goal*: Break glucose down to form two pyruvates.
- *Who*: All life on earth performs glyclolysis
- *Where*: The cytoplasm.
- Glycolysis produces 4 ATP's and 2 NADH, but uses 2 ATP's in the process for a net of 2 ATP and 2 NADH.

This process does not require O_2 and does not yield much energy.

FIRST STAGE OF GLYCOLYSIS

- Glucose (6C) is broken down into 2 PGAL's (Phosphoglyceraldehyde - 3Carbon molecules).
- This requires two ATP's.

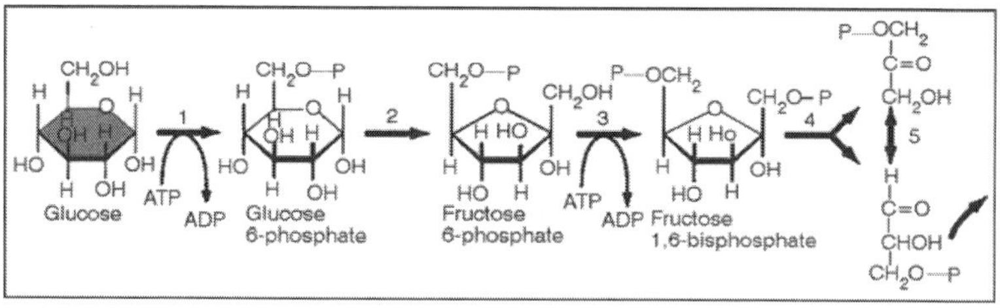

SECOND STAGE OF GLYCOLYSIS

- 2 PGAL's (3C) are converted to 2 pyruvates.
- This creates 4 ATP's and 2 NADH's.
- The net ATP production of Glycolysis is 2 ATP's.

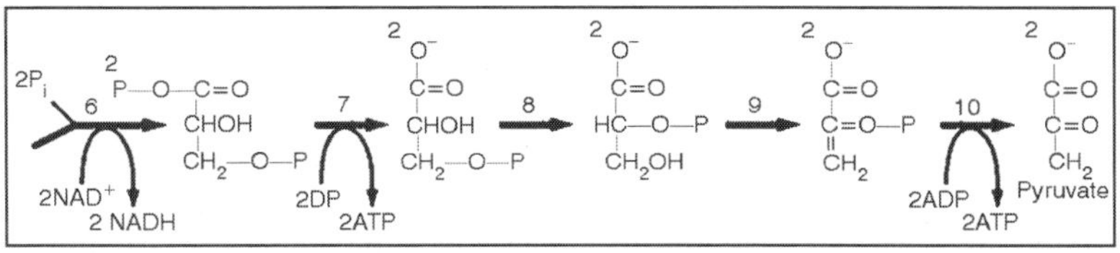

- Goal: take pyruvate and put it into the Krebs cycle, producing NADH and $FADH_2$.
- *Where*: The mitochondria.
- There are two steps.
  - The Conversion of Pyruvate to Acetyl CoA.
  - The Krebs Cycle proper.
- The Krebs cycle and the conversion of pyruvate to Acetyl CoA produce 2 ATP's, 8 NADH's, and 2FADH2's per glucose molecule.

Oxidation of Pyruvate to form Acetyl CoA for Entry Into the Krebs Cycle.

- 2 NADH's are generated (1 per pyruvate).
- 2 CO_2 are released (1 per pyruvate).

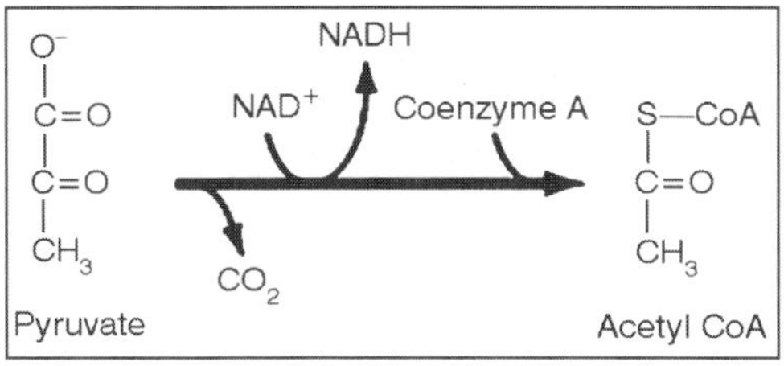

Krebs Cycle

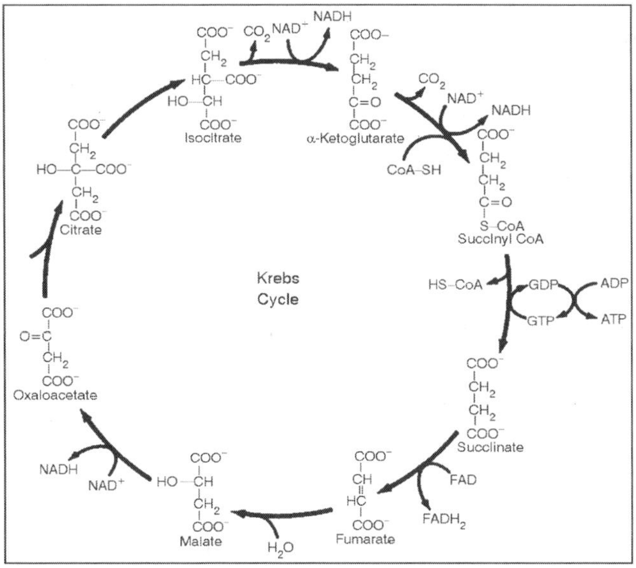

- 6 NADH's are generated (3 per Acetyl CoA that enters).
- 2 $FADH_2$ is generated (1 per Acetyl CoA that enters).
- 2 ATP are generated (1 per Acetyl CoA that enters).
- 4 CO_2's are released (2 per Acetyl CoA that enters).
- Therefore, the total numbers of molecules generated in the oxidation of pyruvate and the Krebs Cycle is:
  - 8 NADH
  - 2 $FADH_2$
  - 2 ATP
  - 6 CO_2

Electron Transport Phosphorylation (Chemiosmosis)

- *Goal*: to break down NADH and $FADH_2$, pumping H+ into the outer compartment of the mitochondria
- *Where*: the mitochondria.
- In this reaction, the ETS creates a gradient which is used to produce ATP, quite like in the chloroplast
- Electron Transport Phosphorylation typically produces 32 ATP's.
- ATP is generated as H+ moves down its concentration gradient through a special enzyme called ATP synthase.

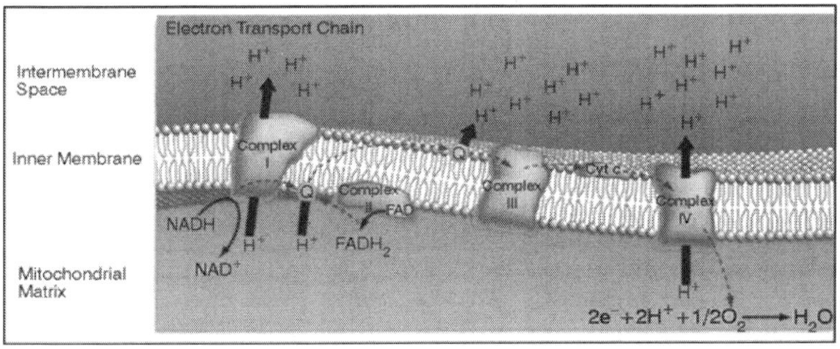

Net Engergy Production from Aerobic Respiration

- *Glycolysis*: 2 ATP.
- *Krebs Cycle*: 2 ATP.
- Electron Transport Phosphorylation: 32 ATP.
  - Each NADH produced in Glycolysis is worth 2 ATP (2 × 2 = 4) – the NADH is worth 3 ATP, but it costs an ATP to transport the NADH into the mitochondria, so there is a net gain of 2 ATP for each NADH produced in gylcolysis.
  - Each NADH produced in the conversion of pyruvate to acetyl COA and Krebs Cycle is worth 3 ATP (8 × 3 = 24).

  - Each FADH2 is worth 2 ATP (2 × 2 = 4).
  - 4 + 24 + 4 = 32.
- Net Energy Production: 36 ATP!

Anaerobic Respiration

- *Goal*: to reduce pyruvate, thus generating NAD+.
- *Where*: the cytoplasm.
- *Why*: in the absence of oxygen, it is the only way to generate NAD+.
- Alcohol Fermentation - occurs in yeasts in many bacteria.
  - The product of fermentation, alcohol, is toxic to the organism.

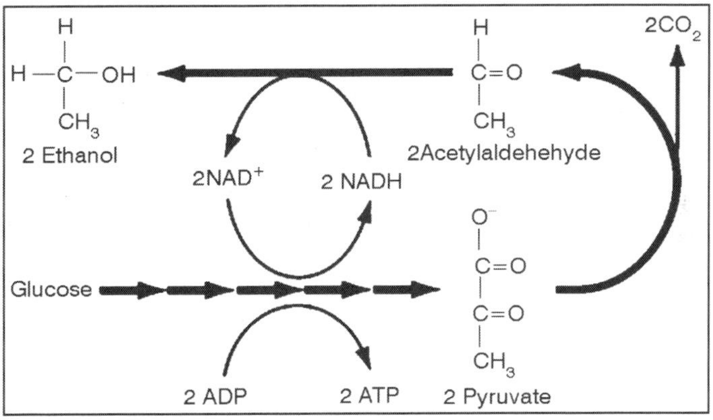

- *Lactic Acid Fermentation*: Occurs in humans and other mammals.
  - The product of Lactic Acid fermentation, lactic acid, is toxic to mammals.
  - This is the"burn" felt when undergoing strenuous activity.

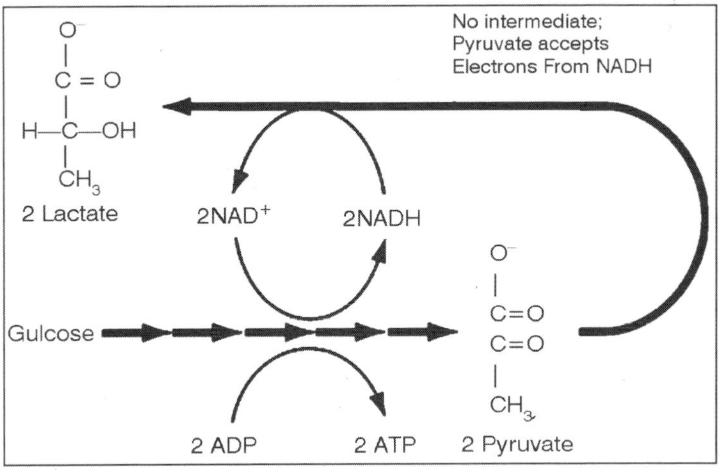

- The only goal of fermentation reactions is to convert NADH to NAD+ (to use in glycolysis).

- No energy is gained.
- Note differences - Anaerobic respiration - 2 ATP's produced (from glycolysis), aerobic respiration - 36 ATP's produced (from glycolysis, Krebs cycle, and Oxidative Phosphorylation).
- Thus, the evolution of an oxygen-rich atmosphere, which facilitated the evolution of aerobic respiration, was crucial in the diversification of life.

Photosynthesis: $6\ CO_2 + 6\ H_2O >> C_6H_{12}O_6 + 6\ O_2$.

Respiration: $C_6H_{12}O_6 + 6\ O_2 >> 6\ CO_2 + 6\ H_2O$.

We see that these reactions are opposites: This is important since the earth is a closed system. All life has a set amount of natural materials to work with, so it is important that they all be cycled through effectively and evenly.

Energy Yields

- *Glucose*: 686 kcal/mol.
- *ATP*: 7.5 kcal/mol.
- 7.5 × 36 = 270 kcal/mol for all ATP's produced.
- 270/ 686 = 39% energy recovered from aerobic respiration

Related Catabolic Processes - Beta Oxidation

- Fats consist of a glycerol backbone with two or three fatty acids connected to it.
- The body absorbs fats and then breaks off the fatty acids from the glycerol.
- Glycerol is converted to glyceraldehyde phosphate, an intermediate of glycolysis.
- The fatty acids are broken down into two-carbon units which are then converted to acetyl CoA.
  - An eight-carbon fatty acid can produce 4 acetyl CoA's.
  - Each acetyl CoA is worth 12 ATP's (3 NADP, 1 $FADH_2$, 1 ATP).
  - Therefore, this short fatty acid is worth 48 ATP's, a fat with three chains of this length would be worth 144 ATP's!
  - This is why fats are such a good source of energy, and are hard to lose if you want to lose weight.

Comparison Between Plants and Animals

- Animal cells and Plant cells contain mitochondria!
  - However, animal cells contain many more mitochondria than plant cells.
- Animal cells get most of their ATP from mitochondria.
- Plant cells get most of their ATP from the chloroplast
  - The ATP generated from the mitochondria is only used when the plant cannot generate ATP directly from the light-dependent reactions.

Molecules used in Glycolysis and Krebs Cycle

- Not all of the molecules that enter Glycolysis and the Krebs Cycle are used for energy.
- Some are used to synthesize fats, nucleotides, amino acids, and other biologically important molecules.

Regulation of Glycolysis and the Krebs Cycle

- *Step 3 of Glycolysis*: The conversion of Fructose 6-phosphate to Fructose 1,6-bisphosphate.
  - Enzyme catalyzing this reaction = Phospho-fructokinase.
  - *"Committing Step"*: Fructose 6-phosphate can be used by the cell for lots of things, but fructose 1,6-bisphosphate has limited use except in glycolysis.
  - Phosphofructokinase inhibited by high levels of ATP.
- ATP is also a substrate - odd, eh?
- Enzyme has two ATP binding sites, one in the active site and one in an allosteric site.
- Low to mid levels of ATP cause ATP to bind to the active site.
- High levels of ATP also enable ATP to bind to allosteric site, causing a conformation change and shutting down the enzyme.
- Conversion of Pyruvate to Acetyl CoA.
  - Enzyme involved in catalyzing this reaction = pyruvate dehydrogenase.
  - High levels of ATP slow down this reaction by phosporylating the enzyme, changing its shape and shutting it down.
- High levels of NADH and Acetyl CoA also inhibit this enzyme.
  - NAD+, CoA, or AMP (an indicator of low ATP) can speed up the reaction.

CARBOHYDRATES MONOSACCHARIDES

- (Monosaccharoses) Dioses ($C_2H_4O_2$) - Glycolose.
- Moses ($C_3H_6O_3$). Aldoses - Glycerose. Ketose - Dioxyacetone.
- Tetroses ($C_4H_8O_4$).
- Aldoses - Erythrose,3 Threose.3 Ketose - Erythrulose.2
- Pentoses ($C_6H_{10}O_5$).
- Aldoses - Arabinose,2 Xylose,2 Ribose,2 Lyxose.8 Ketoses
- - Araboketose3, Xyloketose (ketoxylose).2 [Methyl pentoses ($C_6H_{10}O_5$) - Rhamnose,2 Fucose2].

Some writers use the term polysaccharides to include all carbohydrates other than monosaccharides. Mathews applies it to all carbohydrates more complex than the disaccharides.

Names of a few of the most important carbohydrates are printed in small capitals. Separate mention of the d, l, and dl forms of the various sugars is omitted, since in the study of food and nutrition we are practically concerned only with that one of the three forms which is found in or derived from natural products.

- Hexoses ($C_6H_{12}O_6$).
- Aldoses - Glucose,1 Mannose,2 Galactose,2 Gulose,3 Idose,3Talose,3.
- Allose,2 Altrose.3 Ketoses - Fructose,1 Sorbose,2 Tagatose.3.
- Heptoses ($C_7H_{14}O_7$). Aldose - Mannoheptose.1 Ketose - Sedoheptose.1.

Note: No attempt is here made to summarize the occurrence of any but the tetroses, pentoses, hexoses and heptoses. Glycolose and the trioses if formed in nature are probably too reactive to accumulate sufficiently for identification.

Disaccharides (Disaccharoses)

Dihexoses (Hexobioses) - ($C_{12}H_{22}O_{11}$). Anhydride of glucose + fructose-Sucrose. Anhydrides of glucose + galactose - Lactose, Melibiose. Anhydrides of glucose + glucose- Maltose, Isomaltose, Trehalose, Turanose.

Trisaccharides (Trisaccharoses)

Trihezoses ($C_{18}H_{22}O_{16}$). Anhydride of glucose + galactose +fructose - Raffinose. Anhydride of glucose + glucose + glucose - Melezitose. Anhydride of fructose + fructose + fructose - Secalose.

Tetrasaccharides (Tetrasaccharoses)

Tetrahexoses ($C_{24}H_{42}O_{21}$). Anhydrides of 2 galactose + glucose + fructose - Stachyose, Lupeose.

Polysaccharides (Polysaccharoses)

Pentosans (chief constituents of gums and mucilages). Anhydrides of xylose - Xylans. Anhydrides of arabinose - Arabans.

Hexosans

Anhydrides of glucose-Starch, Cellulose, Glycogen, Dextrin (and other"dextrans"). Anhydrides of mannose - Mannans. Anhydrides of galactose-Galactans (pectins). Anhydrides of fructose - Inulin (and other"levulans").

- Occurs free in nature.
- Not yet found free in nature (or only in small amounts) but obtained by hydrolysis or fermentation of natural product.
- Known only (with certainty) as a Labouratory product.

SIMPLE BIOCHEMICAL TESTS

Materials

- Benedict's solution (not"quantitative")
- Iodine solution (I/KI)

- Waterbaths set at about 100°C
- Boiling tubes Racks
- 1% glucose soln
- 1% fructose soln
- % sucrose soln
- other sugars?
- 1% starch sol solution X

HCl, NaOH, litmus/universal indicator paper filter funnels, papers, flasks? measuring cylinders (10, 25, 50, 100 ml) plastic pipettes glass pipettes (1, 5, 10 ml?), pipette pumps

Objectives

Familiarisation with Benedict's test - for reducing sugars modification of Benedict's test for non-reducing sugars iodine test dilution techniques sensitivity of above tests.

Procedure

See also Roberts Students' manual p. 42 You will be expected to write up and hand in a concise account of your techniques and the results. There is no need to use large amounts of test substances or reagents.

SAFETY

Irritant Solutions Wear Safety Glasses And Lab Coats Hot Water And Steam Take Care: Keep Working Area Clear:

- Carry out Benedict's test on the named sugar solutions and starch as provided.
- Use fixed (noted!) quantities of test substance and reagents.
- Do not contaminate stock solutions shared by the group.
- Note results - colour and intensity - separately from conclusions.
  - *Refinement*: Filter the contents of the tube in which you have been testing glucose (using paper previously labelled with pencil). Consider the colour of the filtrate, and accordingly modify the quantities of reagents used. Repeat filtration process. Explain your conclusions in your write-up.
- Dilute the sugar solutions (say ×10, ×100 etc), and carry out Benedict's test on suitable samples (i.e. amounts same as before).
- Explain the dilution technique you used in your write-up.
- What does this tell you about the sensitivity of the test?
- Using sucrose, carry out the test for non reducing sugars. Having obtained a negative result from ordinary benedict's test, a second test is performed after hydrolysing it to reducing sugars (acid hydrolysis, followed by neutralisation), then performing Benedict's test:-

Add... drops/ml HCl - leave for... minutes at room temperature (... °C) or... mins at 100 °C - add... drops/ml NaOH (same strength) - check neutrality with litmus (colour......) or universal indicator paper (colour.....).

- Test solution X using a combination of the above tests.
- Test starch solution using iodine solution (I/KI). You could use drops in a recessed tile, but for sensitivity purposes it is probably best to use boiling tubes and drops of iodine.
  - Dilute starch × 10, × 100, etc. and repeat (standardised) test.
  - Warm the last tubes - i.e. the least blue coloured (100 °C waterbath)), and observe colour.
  - Cool (the lower end of) the tube under the tap.
  - Conclusion?

GLYCOGEN STORAGE DISEASES OF CARBOHYDRATE DISORDERS

Glycogen storage diseases are caused by deficiencies of enzymes involved in glycogen synthesis or breakdown; the deficiencies may occur in the liver or muscles and cause hypoglycemia or deposition of abnormal amounts or types of glycogen (or its intermediate metabolites) in tissues.

Inheritance for glycogen storage diseases (GSDs) is autosomal recessive except for GSD type VIII/IX, which is X-linked. Incidence is estimated at about 1/25,000 births, which may be an underestimate because milder subclinical forms may be undiagnosed. For a more complete listing of glycogen storage diseases, see Table Glycogen Storage Diseases and Disorders of Gluconeogenesis.

Age of onset, clinical manifestations, and severity vary by type, but symptoms and signs are most commonly those of hypoglycemia and myopathy. Diagnosis is suspected by history, examination, and detection of glycogen and intermediate metabolites in tissues by MRI or biopsy.

Defects in glycolysis (rare) may cause syndromes similar to GSDs. Deficiencies of phosphoglycerate kinase, phosphoglycerate mutase, and lactate dehydrogenase mimic the myopathies of GSD types V and VII; deficiencies of glucose transport protein 2 (Fanconi-Bickel syndrome) mimic the hepatopathy of other GSD types (eg, I, III, IV, VI).

GALACTOSEMIA

Galactosemia is caused by inherited deficiencies in enzymes that convert galactose to glucose. Symptoms and signs include hepatic and renal dysfunction, cognitive deficits, cataracts, and premature ovarian failure. Diagnosis is by enzyme analysis of RBCs. Treatment is dietary elimination of galactose. Physical prognosis is good with treatment, but cognitive and performance parametres are often subnormal.

Galactose is found in dairy products, fruits, and vegetables; autosomal recessive enzyme deficiencies cause 3 clinical syndromes.

Galactose-1-phosphate Uridyl Transferase Deficiency

This deficiency causes classic galactosemia. Incidence is 1/62,000 births; carrier frequency is 1/125. Infants become anorectic and jaundiced within a few days or weeks of consuming breast milk or lactose-containing formula. Vomiting, hepatomegaly, poor growth, lethargy, diarrhea, and septicemia (usually Escherichia coli) develop, as does renal dysfunction (eg, proteinuria, aminoaciduria, Fanconi syndrome), leading to metabolic acidosis and edema. Hemolytic anemia may also occur. Without treatment, children remain short and develop cognitive, speech, gait, and balance deficits in their teenage years; many also have cataracts, osteomalacia (caused by hypercalciuria), and premature ovarian failure. Patients with the Duarte variant have a much milder phenotype.

Galactokinase Deficiency

Patients develop cataracts from production of galactitol, which osmotically damages lens Fibres; idiopathic intracranial hypertension (pseudotumor cerebri) is rare. Incidence is 1/40,000 births.

Uridine Diphosphate Galactose 4-epimerase Deficiency

There are benign and severe phenotypes. Incidence of the benign form is 1/23,000 births in Japan; no incidence data are available for the more severe form. The benign form is restricted to RBCs and WBCs and causes no clinical abnormalities. The severe form causes a syndrome indistinguishable from classic galactosemia, although sometimes with hearing loss.

DIAGNOSIS AND TREATMENT

Diagnosis is suggested clinically and supported by elevated galactose levels and the presence of reducing substances other than glucose (eg, galactose, galactose 1-phosphate) in the urine; it is confirmed by enzyme analysis of RBCs, hepatic tissue, or both. Most states require that neonates be screened for galactose-1-phosphate uridyl transferase deficiency. Treatment is elimination of all sources of galactose in the diet, most notably lactose, which is a source of galactose present in all dairy products, including milk-based infant formulas and a sweetener used in many foods. A lactose-free diet prevents acute toxicity and reverses some manifestations (eg, cataracts) but may not prevent neurocognitive deficits. Many patients require supplemental Ca and vitamins. For patients with epimerase deficiency, some galactose intake is critical to ensure a supply of uridine-5?-diphosphate-galactose (UDP-galactose) for various metabolic processes.

DISORDERS OF FRUCTOSE METABOLISM

Deficiency of enzymes that metabolize fructose may be asymptomatic or cause hypoglycemia. Fructose is a monosaccharide that is present in high concentrations in fruit and honey and is a constituent of sucrose and sorbitol.

Fructose 1-phosphate Aldolase Deficiency

This deficiency causes the clinical syndrome of hereditary fructose intolerance. Inheritance is autosomal recessive; incidence is estimated at 1/20,000 births. Infants are healthy until they ingest fructose; fructose 1-phosphate then accumulates, causing hypoglycemia, nausea and vomiting, abdominal pain, sweating, tremors, confusion, lethargy, seizures, and coma. Prolonged ingestion may cause cirrhosis, mental deterioration, and proximal renal tubular acidosis with urinary loss of phosphate and glucose.

Diagnosis is suggested by symptoms in relation to recent fructose intake and is confirmed by enzyme analysis of liver biopsy tissue or by induction of hypoglycemia by fructose infusion, 200 mg/kg IV. Diagnosis and identification of heterozygous carriers of the mutated gene can also be made by direct DNA analysis. Short-term treatment is glucose for hypoglycemia; long-term treatment is exclusion of dietary fructose, sucrose, and sorbitol. Many patients develop a natural aversion to fructose-containing food. Prognosis is excellent with treatment.

Fructokinase Deficiency

This deficiency causes benign elevation of blood and urine fructose levels (benign fructosuria). Inheritance is autosomal recessive; incidence is about 1/130,000 births. The condition is asymptomatic and diagnosed accidentally when a non-glucose reducing substance is detected in urine.

Deficiency of Fructose-1,6-bisphosphatase

This deficiency compromises gluconeogenesis and results in fasting hypoglycemia, ketosis, and acidosis. This deficiency can be fatal in neonates. Inheritance is autosomal recessive; incidence is unknown. Febrile illness can trigger episodes. Acute treatment is oral or IV glucose. Tolerance to fasting generally improves with age.

DISORDERS OF PYRUVATE METABOLISM

Inability to metabolize pyruvate causes lactic acidosis and a variety of CNS abnormalities. Pyruvate is an important substrate in carbohydrate metabolism.

Pyruvate Dehydro®genase Deficiency

Pyruvate dehydrogenase is a multi-enzyme complex responsible for the generation of acetyl CoA from pyruvate for the Krebs cycle. Deficiency results in elevation of pyruvate and thus elevation of lactic acid levels. Inheritance is X-linked or autosomal recessive.

Clinical manifestations vary in severity but include lactic acidosis and CNS malformations and other postnatal changes, including cystic lesions of the cerebral cortex, brain stem, and basal ganglia; ataxia; and psychomotor retardation. Diagnosis is confirmed by enzyme analysis of skin fibroblasts, DNA testing, or both. There is no clearly effective treatment, although a low-carbohydrate or ketogenic diet and dietary thiamin supplementation have been beneficial for some patients.

Pyruvate Carboxylase Deficiency

Pyruvate carboxylase is an enzyme important for gluconeogenesis from pyruvate and alanine generated in muscle. Deficiency may be primary, or secondary to deficiency of holocarboxylase synthetase, biotin, or biotinidase; inheritance for both is autosomal recessive and both result in lactic acidosis.

Primary deficiency incidence is < 1/250,000 births but may be higher in certain American Indian populations. Psychomotor retardation with seizures and spasticity are the major clinical manifestations. Labouratory abnormalities include hyperammonemia; lactic acidosis; ketoacidosis; elevated levels of plasma lysine, citrulline, alanine, and proline; and increased excretion of ?-ketoglutarate. Diagnosis is confirmed by enzyme analysis of cultured skin fibroblasts. Secondary deficiency is clinically similar, with failure to thrive, seizures, and other organic aciduria. There is no effective treatment, but selected patients with primary deficiency and all those with secondary deficiencies should be given biotin supplementation 5 to 20 mg po once/day.

OTHER DISORDERS

Phosphoenolpyruvate carboxykinase deficiency impairs gluconeogenesis and results in symptoms and signs similar to the hepatic forms of GSD but without hepatic glycogen accumulation. Other deficiencies include those of glycolytic enzymes or enzymes in the pentose phosphate pathway. Common examples are pyruvate kinase deficiency (Anemias Caused by Hemolysis: Embden-Meyerhof Pathway Defects) and glucose-6-phosphate dehydrogenase (G6PD) deficiency (see Anemias Caused by Hemolysis: Glucose-6-Phosphate Dehydrogenase (G6PD) Deficiency), both of which may result in hemolytic anemia. Wernicke-Korsakoff syndrome (Vitamin Deficiency, Dependency, and Toxicity: Symptoms and Signs) is caused by a partial deficiency of transketolase, which is an enzyme for the pentose phosphate pathway that requires thiamin as a cofactor.

DISORDERS OF CARBOHYDRATE

Disorders of carbohydrate metabolism occur in many forms. The most common disorders are acquired. Acquired or secondary derangements in carbohydrate metabolism, such as diabetic ketoacidosis, hyperosmolar coma, and hypoglycemia, all affect the central nervous system. Many forms and variants of peripheral nerve disease also are seen in diabetes. The remaining disorders of carbohydrate metabolism are the rare inborn errors of metabolism (ie, genetic defects).

The inherited defects affecting carbohydrate metabolism that have been discovered so far are inherited as autosomal recessive traits. Although multiple affected siblings may be identified in a kindred, most affected individuals are the first identified in a family. Most of the known defects of carbohydrate metabolism appear to be due to point mutations. Whether this will still hold true after the discoveries that the Human Genome Project will make possible is not known.

Most of the inherited disorders of carbohydrate metabolism fall into a few broad clinical syndromes, which are classified by age of onset as follows:

- Infants and early childhood.
  - Episodic lactic acidosis from early infancy, failure to thrive, and hypotonia with or without features that may suggest specific defects.
  - Infantile or early childhood mental retardation, hypotonia, failure to thrive, and other features.
  - Mental retardation/developmental delay with features that suggest storage disorders.
  - Episodic vomiting in infants or young children.
- Childhood or adolescence.
  - Episodic acidosis in childhood or early adulthood, often with features of a specific disorder.
  - Intermittent or episodic ataxia in childhood and adolescence
- In adults and, more rarely, in adolescents or older children.
  - Cramps and weakness, often with some episodes of muscle breakdown and myoglobinuria, during or after heavy anaerobic exercise.
  - Symmetrical neuromuscular disease with weakness and wasting of proximal muscles.

PATHOPHYSIOLOGICAL DISORDERS

The brain, heart, skeletal muscle, and liver depend on ketone bodies or on glucose for energy and for carbon chains to synthesize cellular proteins For infants as well as individuals with decreased sugar intake because of starvation or dietary choices, ketones are the preferred fuel and source of carbon chains. Nervous and muscle tissues preferentially oxidize ketones over fatty acids, glucose, or amino acids under normal conditions.

Structure	Metabolite	Enzyme
$CH_3–CH(OH)–CH_2–C(=O)–OH$	Beta-hydroxy-butyrate (3 hyroxy- butyrate)	
		3Hyroxy- Butyrate Dehydrogenase
$CH_3–C(=O)–OH_2–C(=O)–OH$	Acetoacetate	
		Acetoaetate-CoA Synthetase
$CH_3–C(=O)–CH_2–C(=O)\sim CoA$	Acetoacetyl ~ CoA	
		Thiolase
$CH_3C(=O)\sim CoA$ $CH_3C(=O)\sim CoA$	2Molecules of Actyl ~/CoA	

Fig. Ketone Metabolism

KETONE METABOLISM

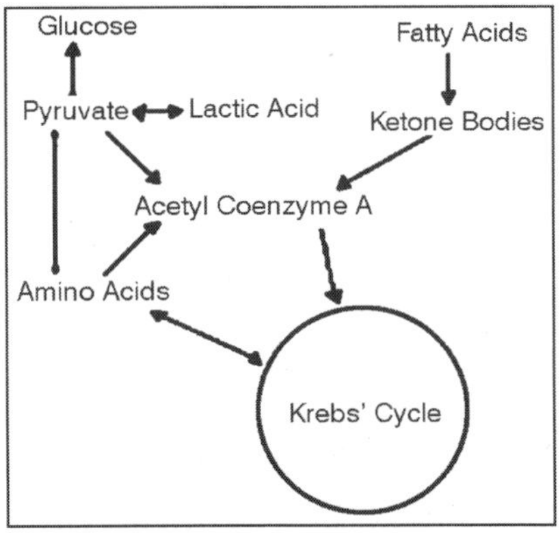

Krebs Cycle

Ketone bodies are largely beta-hydroxybutyrate and acetoacetate. These are formed in the liver and kidneys from the breakdown of fatty acids and a few amino acids. The brains of adults who live on a high-carbohydrate diet lose the ability to synthesize the key enzymes needed to metabolize ketones. These include beta-hydroxybutyrate dehydrogenase and acetoacetyl coenzyme A (CoA) synthetase. For the brain to resynthesize these enzymes to effective levels, starvation for 3-7 days or a high-fat diet is needed. Neural tissue has far more metabolic activity than other tissues of the body. The brain comprises only a small Percentage of total body weight-approximately 1.4 kg in an average 70-kg person-but uses 20% of the body's oxygen at rest. In addition, neurons (and glia) are constantly synthesizing a variety of neurotransmitters, proteins for axonal flow, and proteins and lipids for regeneration of synaptic vesicles and other components of membranes. Many of these chemicals are synthesized in the brain, in part or wholly from glucose or ketone bodies.

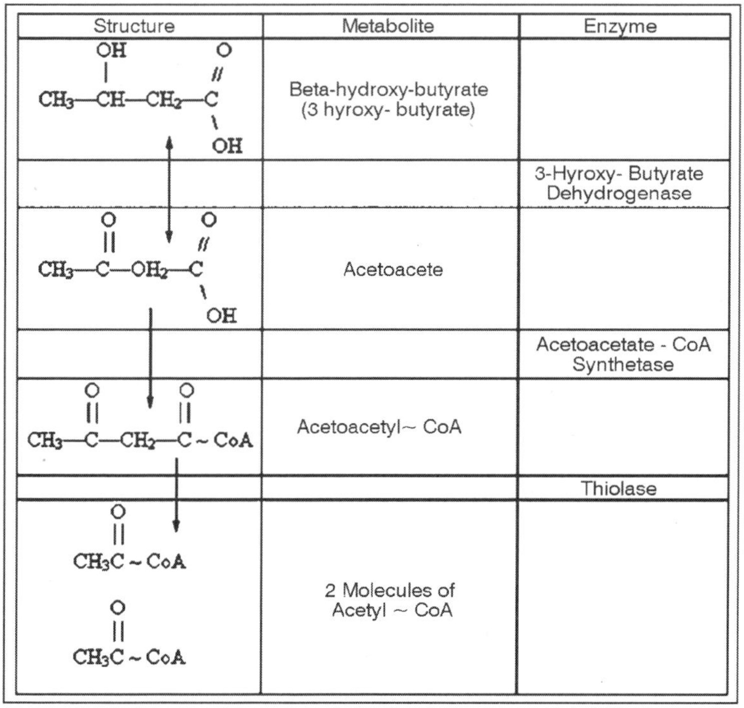

Fig. Ketone Metabolism

Nonglucose carbohydrates (eg, galactose, mannose, inositol) are clinically important in fetal and neonatal nutrition; however, little is known about their metabolism in the neonate. Glucose and galactose increase postprandially and several other carbohydrates contained in milk do not. Brown *et al*, in their attempt to determine whether postprandial changes in plasma carbohydrate and sugar alcohol concentrations are affected by clinical variables such as postnatal age, milk type, feeding volume, or feeding duration in term newborns, found that galactose is almost cleared completely by the neonatal liver. This provides novel evidence of a marked difference in plasma galactose concentrations [Gal] between relatively younger and older neonates.1

Acute hypoglycemia, as in an alcoholic or a diabetic who has taken too much insulin, causes the brain to be deprived rapidly and abruptly of blood glucose after it has depended for a long time on glucose rather than ketones as its fuel. Beta-hydroxybutyrate dehydrogenase and acetoacetyl CoA synthetase activity in the brain is insufficient to enable the blood ketones to sustain the brain's metabolism. Without fuel to burn, the neural tissue cannot carry out the high level of metabolism needed to pump ions, fire messages, or transmit intercellular chemical messages, let alone repair itself. The groggy state of sudden memory loss (ie, disorientation) and other intellectual deficits that internists tend to call confusion sets in rapidly with acute hypoglycemia, followed just as rapidly by a downhill course into obtundation, coma, and death. If the average adult brain can withstand only 4-6 minutes of oxygen deprivation, it can withstand only about 40-45 minutes without fuel.

Most physicians, as medical students or interns, have had experience treating diabetic ketoacidosis. Although the experience tends to cast ketone bodies as toxins, they are not. The acidosis produced by ketone bodies is toxic and is the direct cause of the Kussmaul breathing typical of patients with diabetic ketoacidosis. Patients with diabetes in this state usually have had normal to elevated blood glucose for the previous days or weeks. Enough glucose enters the brain cells during this time to inhibit synthesis of beta-hydroxybutyrate dehydrogenase and acetoacetyl CoA synthetase.

Again, the ketone-utilizing enzymes in the brain are not working at effective levels. By the time frank ketoacidosis sets in, too little insulin is left for the blood glucose to be able to sustain the high metabolic activity of the brain. The ketone bodies accumulate but cannot provide energy or carbon for the brain. The hydrogen ion associated with the ketones also accumulates, thereby stimulating (or overstimulating) acid receptors in the floor of the fourth ventricle and interfering with reuptake of potassium in the renal tubules. The coma of hyperosmolar diabetic disorder seems to be related directly to the addition of large concentrations of glucose in the serum, the cerebrospinal fluid (CSF), the extracellular fluid of the brain and, perhaps to some degree, the intracellular fluid as well. The cause of high concentrations of glucose in these patients but not of ketone bodies remains unknown. The lack of ketones probably explains the lack of acidosis. CSF concentrations of glucose mirror those of the serum with a delay of approximately 2 hours. Patients do not regain consciousness immediately when the blood glucose is lowered.

PATHOPHYSIOLOGY OF INHERITED DISORDERS

Many of the clinical features of the inherited disorders of carbohydrate metabolism are caused by the following:

- Lack of glucose for the metabolism of brain, muscle, liver, or kidney (in circumstances in which ketone bodies cannot be used)
- Inability to break down glucose to pyruvate
- Inability to oxidize pyruvate fully in the Krebs cycle

Some features of the inherited diseases are due to excessive storage of abnormal substrates or to normal substrates that cells cannot degrade normally because they lack a specific enzymatic activity. In some diseases, abnormal metabolites (intermediates of pathways that use carbohydrates) block the normal function of the pathway or of related pathways. Defects of the enzymes of the pentose shunt interfere with the normal production of nucleic acids, which are needed by cells as second messengers and as coenzymes of intermediary metabolism, as well as components of RNA and DNA.

Galactose accumulates in the lens of the eye in galactose kinase deficiency Metabolism of galactose. The galactose ring differs from glucose in that the hydroxyl group (OH) at carbon-3 is up and not down. To convert galactose to glucose, the ring is first bound to uridine diphosphate (UDP). In this form, the sugar moiety can be epimerized, putting the OH down and forming glucose. The UDP-glucose then is converted to glucose-1-phosphate, then to glucose-6-phosphate, and subsequently metabolized further.

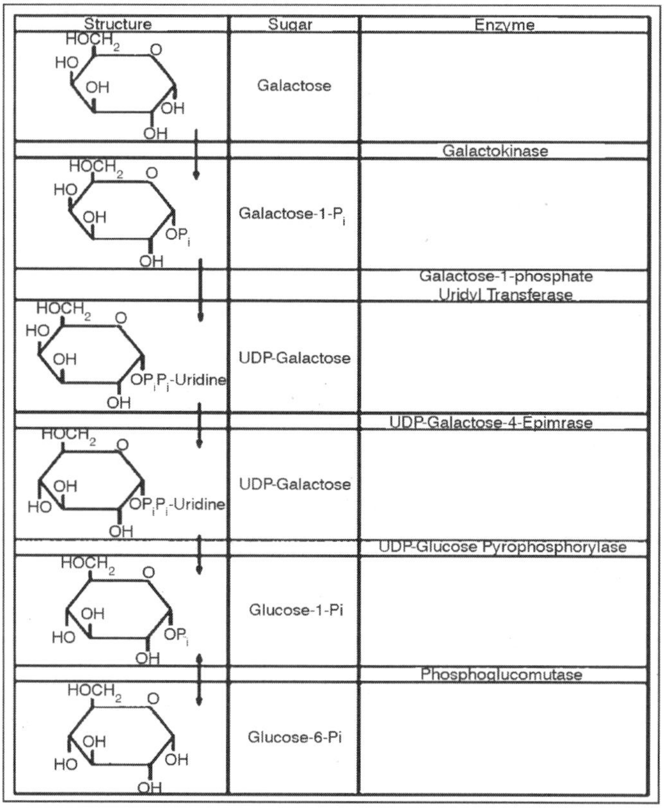

Fig. Metabolism of Galactose

The osmotic pressure generated causes water to enter the lens; this disruption leads to cataract formation. A similar mechanism may play a role in the pathophysiology of diabetic peripheral neuropathy, which was at one time thought to be due, in part, to the accumulation of sorbitol, a relatively inert but osmotically active metabolite of glucose, and hence to excess water in the cells of peripheral nerves.

Media files 1-8 for the major biochemical pathways involved in carbohydrate metabolism. The expression of a disease often depends on other factors in addition to the causative agent. In inherited disease, variations in the other 30,000 genes, varying diets, infectious disease, access to medical care, and even upbringing and cultural effects can influence the clinical expression of a disease. Intermittent ataxia of childhood from pyruvate decarboxylase deficiency tends to appear in conjunction with certain infections or emotional stress, just as is the case with the neurological manifestations of phytanic acid storage disease

PATTERNS OF OCCURRENCE

Frequency

The acquired disorders of carbohydrate metabolism are fairly common, both in the United States and internationally. Hypoglycemia is a common cause of neurological disease, especially acute mental deterioration, memory loss, disorientation, obtundation, and coma, among both alcoholics and patients with diabetes who are treated with insulin. Hyperinsulinemia from other causes is rare, but pancreatic tumors could be the cause.

Diabetes, with its various neurological complications, is among the most common disorders treated in adult patients. Diabetic ketoacidosis still occurs, though education and close medical follow-up make it less common than it was several decades ago. Hyperosmolar coma is also less a problem than when it was first brought to the attention of internists by Plum and Posner's classic monograph Diagnosis of Stupor and Coma.2 Hyperosmolar coma still occurs and needs to be kept in mind while evaluating an obtunded patient. The inherited disorders of carbohydrate metabolism are rare. Severe defects of the pyruvate dehydrogenase (PDH) complex and the benign chemical anomaly called pentosuria have been reported in very few (2-6) patients. Patients with inherited disorders of carbohydrate metabolism tend to be seen at tertiary Centres, where specialist teams are available.

The team may comprise pediatricians, pediatric neurologists, biochemical geneticists, psychiatrists, neurologists with a special interest in mental retardation, and neurologists interested in neuromuscular diseases and the more unusual ataxias. Neurologists in general practice see few, if any, cases of inherited disorders of carbohydrate metabolism in their careers. This depends partly on whether the practice is in an area in which the carrier state for the inherited disorders is more common and partly on whether the neurologist is on the alert for these disorders. As in every area of medicine, inherited carbohydrate disorders that are looked for tend to be found and those that are not looked for tend to be overlooked. Although each of these disorders was described between 1930 and 1990, certainly they were present in the populations of medically sophisticated countries decades, perhaps centuries, earlier.

Glass *et al* performed a retrospective chart review of new pediatric patients seen during 1998 by specialists of the Division of Clinical and Metabolic Genetics of the Hospital for Sick Children in Toronto, the largest such referral Centre in the country. They found that 81% of specific genetic metabolic diagnoses were made within 1 month of being seen in consultation. Four years after initial consultation, a specific diagnosis was not reached in 5% of patients. The authors concluded that the specific diagnosis of inborn errors of metabolism at a major medical genetic referral Centre tended to be made quickly, or never. Some of the causes of delays in diagnosis include the lack of ready access to existing diagnostic Labouratory testing, technical barriers to the identification of specific metabolic or genetic defects, and incomplete knowledge of genetic defects causing inherited metabolic diseases.

Mortality

Hypoglycemia, diabetic ketoacidosis, and hyperosmolar coma are all potentially fatal but potentially curable conditions. Mortality depends on the time taken by the patient or patient's family to present medical attention, the time taken by the doctor to consider the diagnosis and begin correct treatment, the thoroughness of the treatment, and the time taken to diagnose and thoroughly treat other associated medical problems. The patient with acidosis benefits very little if the ketosis is corrected but an underlying cause is missed, for example, by a delayed spinal tap if meningitis or encephalitis happens to be the illness that has tipped the metabolic balance.

Some of the inherited disorders of carbohydrate metabolism are fatal. For some of these conditions, the more profoundly the mutation interferes with enzyme activity, the more likely the disorder will be fatal. Other inherited disorders usually are not fatal if treated early and thoughtfully. Pentosuria is not even a disease in the strict sense; it is more a metabolic curiosity without significant clinical consequences. In general, patients with inherited carbohydrate disorders that primarily affect the brain or heart tend to die early; those whose disorders affect the liver and kidney die later on; and those whose disorders affect only muscle have lower mortality rates, unless things go very wrong. Some concern exists among physicians who study heat illness in military recruits that subclinical inherited predispositions to disorders of muscle and of RBCs might be among the factors that predispose to heat illness. As heat illness is the major cause of death of military recruits in boot camp, this is an issue that deserves careful study.

Race

No racial predilection has been reported in most inherited disorders of carbohydrate metabolism. One well-known exception is pentosuria, a very rare genetic occurrence of no clinical significance, which is so far confined to Ashkenazi Jews. The pentose excreted in this condition is xylulose.

Sex

No sex predilection has been reported; the known inherited disorders of carbohydrate metabolism are all autosomal recessive.

Age

A general rule exists in inherited diseases. If the degree of deficiency of an enzyme varies from family to family, symptoms are likely to appear earliest in families in which the mutation virtually inactivates the enzyme activity and latest in families in which the activity is only just below a certain minimum symptomatic threshold. Profound defects of activity of the PDH complex are associated with lactic acidosis, hypotonia, profound mental retardation, and often death in the first months or first year of life. Somewhat milder defects and vigourous treatment can be compatible with survival into early to mid childhood. Moderate defects of the PDH complex are associated with ataxia and other symptoms and signs presenting in childhood and compatible with graduation from college and survival at least into the adult years. Data from studies in the 1970s and 1980s raised the question whether mild defects of this complex might be associated with brain diseases whose onset is in adolescence or early adult life and that are compatible with long survival. These studies have not yet been refuted or disproved.

5

The Comparative Biochemistry of Myoglobin and Hemoglobin

A comparison of the properties of hemoglobin and myoglobin offers insights into allosteric phenomena, even though these proteins are notenzymes. Hemoglobin displays sigmoid-shaped O2-binding curves (Figure). The unusual shape of these curves was once a great enigma in biochemistry. Such curves closely resemble allosteric enzyme:substrate saturation graphs. In contrast, myoglobin's interaction with oxygen obeys classical Michaelis-Menten-type substrate saturation behaviour.

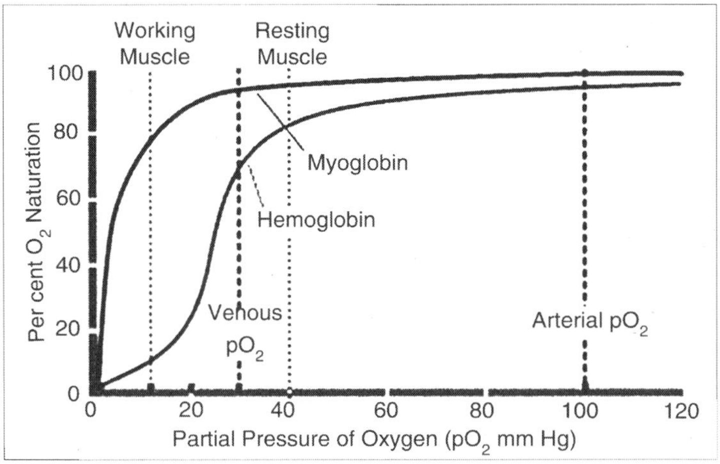

Fig. O_2-binding Curves for Hemoglobin and Myoglobin.

Before examining myoglobin and hemoglobin in detail, let us first encapsulate the lesson: Myoglobin is a compact globular protein composed of a single polypeptide chain 153 amino acids in length; its molecular mass is 17.2 kD (Figure). It contains heme, a porphyrin ring system complexing an iron ion, as its prosthetic group. Oxygen binds to Mb via its heme. Hemoglobin (Hb) is also a compact globular protein, but Hb is a tetramer. It consists of four polypeptide chains, each of which is very similar structurally to the myoglobin polypeptide chain, and each bears a heme group. Thus, a hemoglobin molecule can bind four O_2 molecules. In adult human Hb, there are two identical chains of 141 amino acids, the α-chains, and two identical β-chains, each of 146 residues. The human Hb molecule is an $\alpha_2\beta_2$-type tetramer of molecular mass 64.45 kD. The tetrameric nature of Hb is crucial to its biological function: When a molecule of O_2binds to a heme in Hb, the heme Fe ion is drawn into the plane of the porphyrin ring. This slight movement sets off a chain of conformational events that are transmitted to adjacent subunits, dramatically enhancing the affinity of their heme groups for O_2. That is, the binding of O_2 to one heme of Hb makes it easier for the Hb molecule to bind additional equivalents of O_2. Hemoglobin is a marvelously constructed molecular machine. Let us dissect its mechanism, beginning with its monomeric counterpart, the myoglobin molecule.

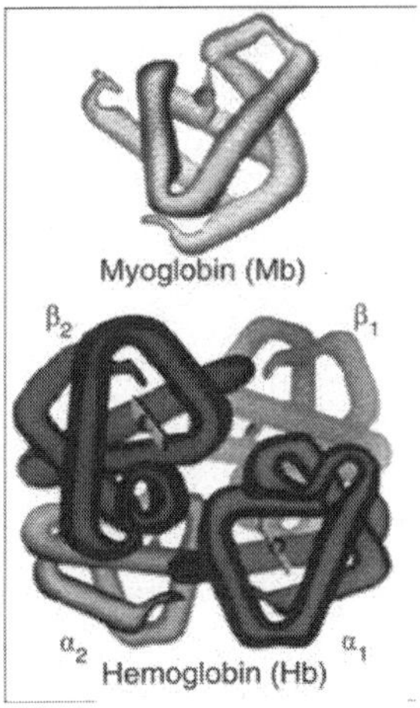

Fig. The Myoglobin and Hemoglobin Molecules.

Myoglobin (sperm whale): one polypeptide chain of 153 aa residues (mass 5 17.2 kD) has one heme (mass = 652 D) and binds one O_2. Hemoglobin (human): four polypeptide chains, two of 141 aa residues (a) and two of 146 residues (β); mass = 64.45 kD. Each polypeptide has a heme; the Hb tetramer binds four O_2. (Irving Geis)

MYOGLOBIN

Myoglobin is the oxygen-storage protein of muscle. The muscles of diving mammals such as seals and whales are especially rich in this protein, which serves as a store for O2 during the animal's prolonged periods underwater. Myoglobin is abundant in skeletal and cardiac muscle of nondiving animals as well. Myoglobin is the cause of the characteristic red colour of muscle.

THE MB POLYPEPTIDE CRADLES THE HEME GROUP

The myoglobin polypeptide chain is folded to form a cradle (4.4 × 4.4 ×

2.5 nm) that nestles the heme prosthetic group (Figure 15.24). O_2 binding depends on the heme's oxidation state. The iron ion in the heme of myoglobin is in the +2 oxidation state, that is, the ferrous form.

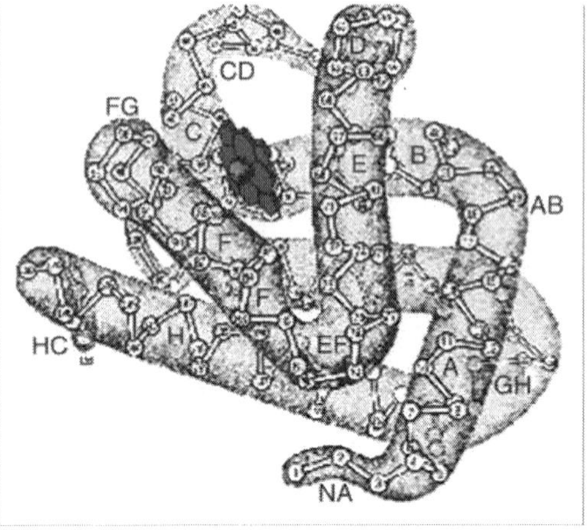

Fig. Detailed Structure of the Myoglobin Molecule.

This is the form that binds O_2. Oxidation of the ferrous form to the +3 ferric form yieldsmetmyoglobin, which will not bind O_2. It is interesting to note that free heme in solution will readily interact with O_2 also, but the oxygen quickly oxidizes the iron atom to the ferric state. Fe3+:protoporphyrin IX is referred to as hematin. Thus, the polypeptide of myoglobin may be viewed as serving three critical functions: it cradles the heme group, it protects the heme iron atom from oxidation, and it provides a pocket into which the O_2 can fit. The myoglobin polypeptide chain consists of eight helical segments, designated by the letters A through H, counting from the N-terminus. These helices, ranging in length from 7 to 26 residues, are linked by short, unordered regions that are named for the helices they connect, as in the AB region or the EF region. The individual amino acids in the polypeptide are indicated according to their position within the various segments, as in His F8, the eighth residue in helix F, or Phe CD1, the first amino acid in the interhelical CD region. Occasionally, amino acids are specified in the conventional way, that is, by the relative position in the chain, as in Gly153. The heme group is cradled within the folded polypeptide chain. (Irving Geis)

O_2 BINDING TO MB

Iron ions, whether ferrous or ferric, prefer to interact with six ligands, four of which share a common plane. The fifth and sixth ligands lie above and below this plane (see Figure). In heme, four of the ligands are provided by the nitrogen atoms of the four pyrroles. A fifth ligand is donated by the imidazole side chain of amino acid residue His F8. When myoglobin binds O_2 to become oxymyoglobin, the O_2 molecule adds to the heme iron ion as the sixth ligand (Figure). O_2 adds end on to the heme iron, but it is not oriented perpendicular to the plane of the heme. Rather, it is tilted about 60° with respect to the perpendicular.

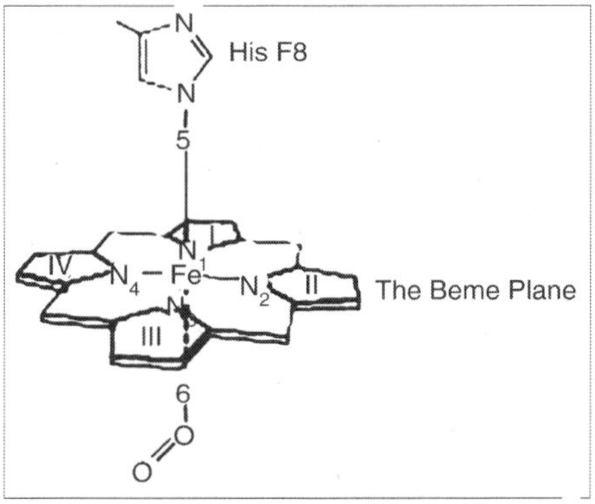

Fig. The Six Liganding Positions of An Iron Ion.

Four ligands lie in the same plane; the remaining two are, respectively, above and below this plane. In myoglobin, His F8 is the fifth ligand; in oxymyoglobin, O_2 becomes the sixth. In deoxymyoglobin, the sixth ligand position is vacant, and in metmyoglobin, a water molecule fills the O_2 site and becomes the sixth ligand for the ferric atom. On the oxygen-binding side of the heme lies another histidine residue, His E7. While its imidazole function lies too far away to interact with the Fe atom, it is close enough to contact the O_2. Therefore, the O_2-binding site is a sterically hindered region. Biologically important properties stem from this hindrance.

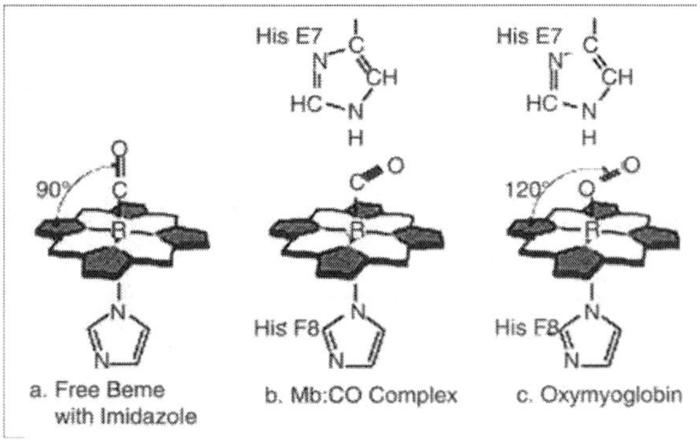

Fig. Oxygen and Carbon Monoxide Binding to the Heme Group of Myoglobin.

For example, the affinity of free heme in solution for carbon monoxide (CO) is 25,000 times greater than its affinity for O_2. But CO only binds 250 times more tightly than O_2 to the heme of myoglobin, because His E7 forces the CO molecule to tilt away from a preferred perpendicular alignment with the plane of the heme (Figure 15.26). This diminished affinity of myoglobin for CO guards against the possibility that traces of CO produced during metabolism might occupy all of the heme sites, effectively preventing O_2 from binding. Nevertheless, CO is a potent poison and can cause death by asphyxiation.

O_2 BINDING ALTERS MB CONFORMATION

What happens when the heme group of myoglobin binds oxygen? X-ray crystallography has revealed that a crucial change occurs in the position of the iron atom relative to the plane of the heme. In deoxymyoglobin, the ferrous ion has but five ligands, and it lies 0.055 nm above the plane of the heme, in the direction of His F8. The iron:porphyrin complex is therefore dome-shaped.

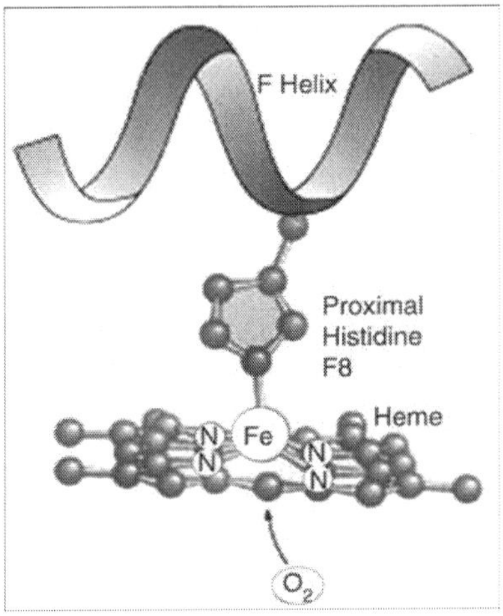

Fig. The Displacement of the Fe ion of the Heme of Deoxymyoglobin from the Plane of the Porphyrin Ring System by the Pull of His F8. In Oxymyoglobin, the Bound O^2 Counteracts this Effect.

When O_2 binds, the iron atom is pulled back towards the porphyrin plane and is now displaced from it by only 0.026 nm (Figure). The consequences of this small motion are trivial as far as the biological role of myoglobin is concerned. However, as we shall soon see, this slight movement profoundly affects the properties of hemoglobin. Its action on His F8 is magnified through changes in polypeptide conformation that alter subunit interactions in the Hb tetramer. These changes in subunit relationships are the fundamental cause of the allosteric properties of hemoglobin.

THE PHYSIOLOGICAL SIGNIFICANCE OF COOPERATIVE

The relative oxygen affinities of hemoglobin and myoglobin reflect their respective physiological roles. Myoglobin, as an oxygen storage protein, has a greater affinity for O2 than hemoglobin at all oxygen pressures. Hemoglobin, as the oxygen carrier, becomes saturated with O_2 in the lungs, where the partial pressure of O_2 (pO_2) is about 100 torr.1 In the capillaries of tissues, pO_2 is typically 40 torr, and oxygen is released from Hb. In muscle, some of it can be bound by myoglobin, to be stored for use in times of severe oxygen deprivation, such as during strenuous exercise.

HEMOGLOBIN AND MYOGLOBIN

Ancient life forms evolved in the absence of oxygen and were capable only of anaerobic metabolism. As the earth's atmosphere changed over time, so too did living things. Indeed, the production of O_2 by photosynthesis was a major factor in altering the atmosphere. Evolution to an oxygen-based metabolism was highly beneficial. Aerobic metabolism of sugars, for example, yields far more energy than corresponding anaerobic processes. Two important oxygen-binding proteins appeared in the course of evolution so that aerobic metabolic processes were no longer limited by the solubility of O_2 in water. These proteins are represented in animals as hemoglobin (Hb) in blood and myoglobin (Mb) in muscle. Because hemoglobin and myoglobin are two of the most-studied proteins in nature, they have become paradigms of protein structure and function. Moreover, hemoglobin is a model for protein quaternary structure and allosteric function. The binding of O_2 by hemoglobin, and its modulation by effectors such as protons, CO_2, and 2,3 - bisphosphoglycerate, depend on interactions between subunits in the Hb tetramer. Subunit-subunit interactions in Hb reveal much about the functional significance of quaternary associations and allosteric regulation.

MODEL FOR THE ALLOSTERIC BEHAVIOUR OF HEMOGLOBIN

A model for the allosteric behaviour of hemoglobin is based on recent observations that oxygen is accessible only to the heme groups of the a -chains when hemoglobin is in the T conformational state. Perutz has pointed out that the heme environment of β-chains in the T state is virtually inaccessible because of steric hindrance by amino acid residues in the E helix. This hindrance disappears when the hemoglobin molecule undergoes transition to the R conformational state. Binding of O_2 to the β-chains is thus dependent on the T to R conformational shift, and this shift is triggered by the subtle changes that occur when O_2 binds to the a -chain heme groups.

Changes in the Heme Iron upon O_2 Binding

In deoxyhemoglobin, the six d electrons of the heme Fe21 exist as four unpaired electrons and one electron pair, and five ligands can be accommodated: the four N-atoms of the porphyrin ring system and histidine F8. In this electronic configuration, the iron atom is paramagnetic and in the high-spin state. When the heme binds O_2 as a sixth ligand, these electrons are rearranged into three e_2 pairs and the iron changes to the low-spin state and is diamagnetic. This change in spin state allows the bond between the Fe21 ion and histidine F8 to become perpendicular to the heme plane and to shorten.

In addition, interactions between the porphyrin N-atoms and the iron strengthen. Also, high-spin Fe21 has a greater atomic volume than low-spin Fe21 because its four unpaired e2 occupy four orbitals rather than two when the electrons are paired in low-spin Fe21. So, low-spin iron is less sterically hindered and able to move nearer to the porphyrin plane.

H^+ PROMOTES THE DISSOCIATION OF OXYGEN FROM HEMOGLOBIN

Protons, carbon dioxide, and chloride ions, as well as the metabolite 2,3 -bisphosphoglycerate (or BPG), all affect the binding of O_2 by hemoglobin. Their effects have interesting ramifications, which we shall see as we discuss them in turn. Deoxyhemoglobin has a higher affinity for protons than oxyhemoglobin. Thus, as the pH decreases, dissociation of O_2 from hemoglobin is enhanced. In simple symbolism, ignoring the stoichiometry of O_2 or H+ involved:

$$HbO_2 + H^+ \rightleftharpoons HbH^+ + O_2$$

Expressed another way, H+ is an antagonist of oxygen binding by Hb, and the saturation curve of Hb for O_2 is displaced to the right as acidity increases (Figure 15.34). This phenomenon is called the Bohr effect, after its discoverer, the Danish physiologist Christian Bohr (the father of Niels Bohr, the atomic physicist).

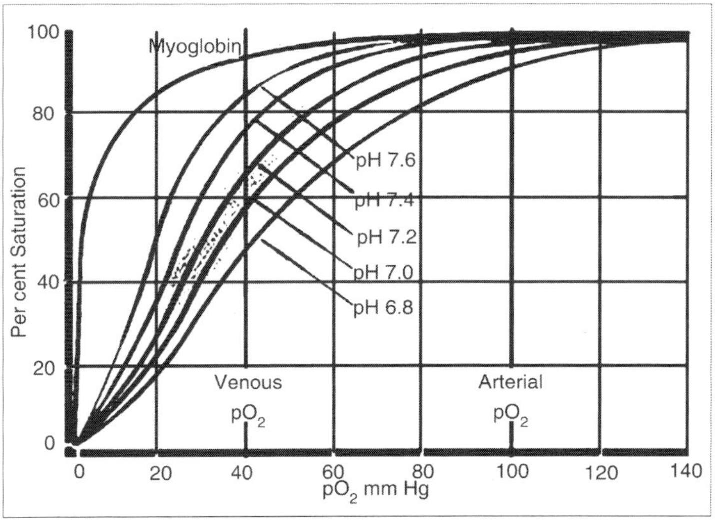

Fig. The Oxygen Saturation Curves for Myoglobin and for Hemoglobin at Five Different pH Values: 7.6, 7.4, 7.2, 7.0, and 6.8.

The effect has important physiological significance because actively metabolizing tissues produce acid, promoting O_2 release where it is most needed. About two protons are taken up by deoxyhemoglobin. The N-termini of the two a -chains and the His β146 residues have been implicated as the major players in the Bohr effect. (The pKa of a free amino terminus in a protein is about 8.0, but the pKa of a protein histidine imidazole is around 6.5.) Neighbouring carboxylate groups of Asp β94 residues help to stabilize the protonated state of the His β146 imidazoles that occur in deoxyhemoglobin. However, when Hb binds O_2, changes in the conformation of β-chains upon Hb oxygenation move the negative Asp function away, and dissociation of the imidazole protons is favoured.

CO_2 ALSO PROMOTES THE DISSOCIATION OF O_2 FROM HEMOGLOBIN

Carbon dioxide has an effect on O_2 binding by Hb that is similar to that of H^+, partly because it produces H+ when it dissolves in the blood:

The enzyme carbonic anhydrase promotes the hydration of CO_2. Many of the protons formed upon ionization of carbonic acid are picked up by Hb as O_2 dissociates. The bicarbonate ions are transported with the blood back to the lungs. When Hb becomes oxygenated again in the lungs, H1 is released and reacts with HCO_3- to re-form H_2CO_3, from which CO_2 is liberated. The CO_2 is then exhaled as a gas.

In addition, some CO_2 is directly transported by hemoglobin in the form of carbamate (—NHCOO-). Free a -amino groups of Hb react with CO2 reversibly:

$$R{-}NH_2 + CO_2 \rightarrow R{-}NH{-}COO^- + H+$$

This reaction is driven to the right in tissues by the high CO_2 concentration; the equilibrium shifts the other way in the lungs where $[CO_2]$ is low. Thus, carbamylation of the N-termini converts them to anionic functions, which then form salt links with the cationic side chains of Arg a 141 that stabilize the deoxy or T state of hemoglobin.

In addition to CO_2, Cl- and BPG also bind better to deoxyhemoglobin than to oxyhemoglobin, causing a shift in equilibrium in favour of O2 release. These various effects are demonstrated by the shift in the oxygen saturation curves for Hb in the presence of one or more of these substances (Figure). Note that the O2-binding curve for Hb + BPG + CO2 fits that of whole blood very well.

Allosteric Effector for Hemoglobin

The binding of 2,3 -bisphosphoglycerate (BPG) to Hb promotes the release of O_2 (Figure). Erythrocytes (red blood cells) normally contain about 4.5 mM BPG, a concentration equivalent to that of tetrameric hemoglobin molecules. Interestingly, this equivalence is maintained in the Hb:BPG binding stoichiometry because the tetrameric Hb molecule has but one binding site for BPG. This site is situated within the central cavity formed by the association of the four subunits. The strongly negative BPG molecule (Figure) is electrostatically bound via interactions with the positively charged functional groups of each Lys β82, His $β_2$, His β143, and the NH_3^+-terminal group of each β-chain.

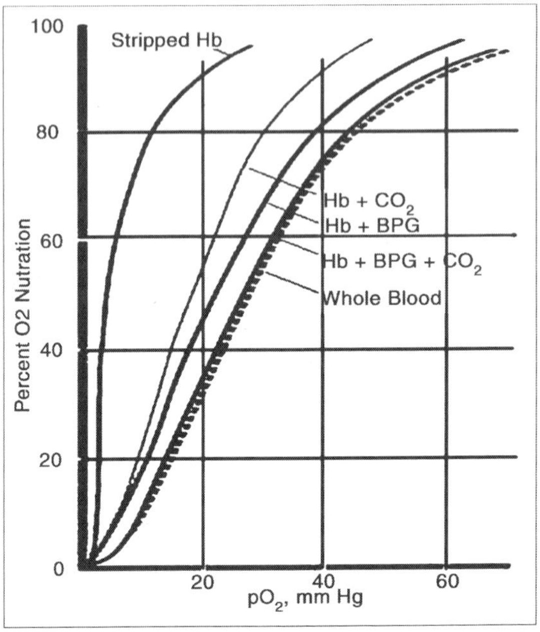

Fig. Oxygen-binding Curves of Blood and of Hemoglobin in the Absence and Presence of CO_2 and BPG. From left to right: Stripped Hb, Hb + CO_2, Hb + BPG, Hb + BPG + CO_2, and whole Blood.

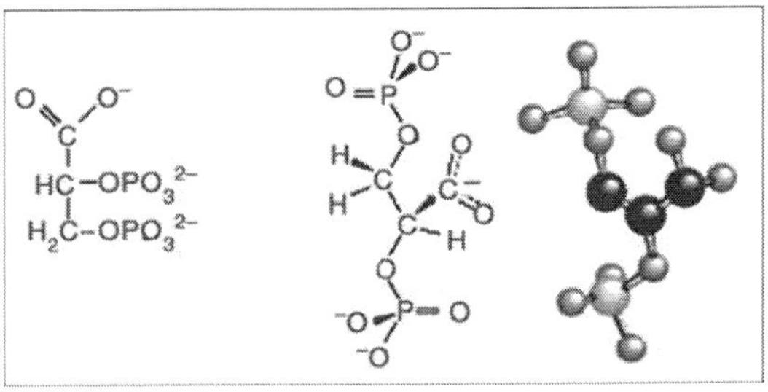

Fig. The Structure, in Ionic form, of BPG or 2,3 -bisphosphoglycerate, an Important Allosteric Effector for Hemoglobin.

These positively charged residues are arranged to form an electrostatic pocket complementary to the conformation and charge distribution of BPG (Figure). In effect, BPG cross-links the two β-subunits. The ionic bonds between BPG and the two β-chains aid in stabilizing the conformation of Hb in its deoxy form, thereby favouring the dissociation of oxygen. In oxyhemoglobin, this central cavity is too small for BPG to fit. Or, to put it another way, the conformational changes in the Hb molecule that accompany O_2-binding perturb the BPG-binding site so that BPG can no longer be accommodated. Thus, BPG and O_2 are mutually exclusive allosteric effectors for Hb, even though their binding sites are physically distinct.

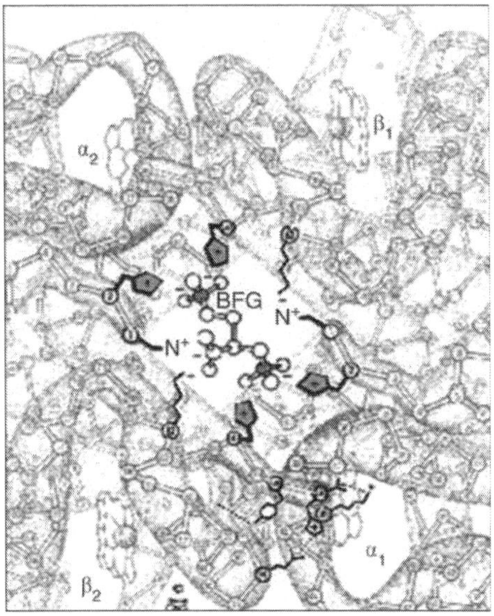

Fig. The Ionic Binding of BPG to the two β-subunits of Hb. (Irving Geis)

PHYSIOLOGICAL SIGNIFICANCE OF BPG BINDING

The importance of the BPG effect is evident in Figure. Hemoglobin stripped of BPG is virtually saturated with O2 at a pO_2 of only 20 torr, and it cannot release its oxygen within tissues, where the pO_2 is typically 40 torr. BPG shifts the oxygen saturation curve of Hb to the right, making the Hb an O_2 delivery system eminently suited to the needs of the organism.

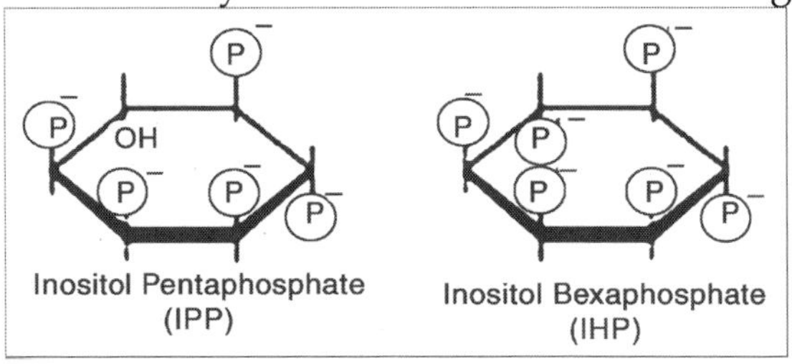

Fig. The Structures of Inositol Pentaphosphate and Inositol Hexaphosphate, the Functional Analogs of BPG in Birds and Reptiles.

BPG serves this vital function in humans, most primates, and a number of other mammals. However, the hemoglobins of cattle, sheep, goats, deer, and other animals have an intrinsically lower affinity for O_2, and these Hbs are relatively unaffected by BPG. In fish, whose erythrocytes contain mitochondria, the regulatory role of BPG is filled by ATP or GTP. In reptiles and birds, a different organophosphate serves, namely inositol pentaphosphate (IPP) or inositol hexaphosphate (IHP).

Fetal Hemoglobin Has a Higher Affinity for O_2

The fetus depends on its mother for an adequate supply of oxygen, but its circulatory system is entirely independent. Gas exchange takes place across the placenta. Ideally then, fetal Hb should be able to absorb O_2 better than maternal Hb so that an effective transfer of oxygen can occur. Fetal Hb differs from adult Hb in that the β-chains are replaced by very similar, but not identical, 146-residue subunits called g -chains (gamma chains). Fetal Hb is thus a_2 g_2.

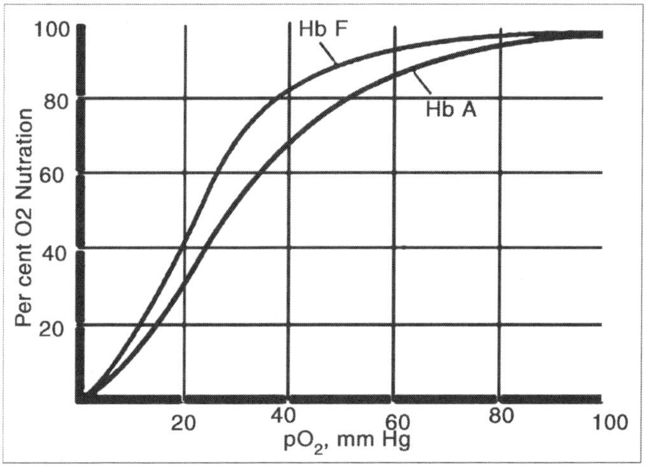

Fig. Comparison of the Oxygen Saturation Curves of Hb A and Hb F Under Similar Conditions of pH and [BPG].

Recall that BPG functions through its interaction with the β-chains. BPG binds less effectively with the g -chains of fetal Hb (also called Hb F). (Fetal g -chains have Ser instead of His at position 143, and thus lack two of the positive charges in the central BPG-binding cavity.) Figure compares the relative affinities of adult Hb (also known as Hb A) and Hb F for O_2 under similar conditions of pH and [BPG]. Note that Hb F binds O_2 at pO_2 values where most of the oxygen has dissociated from Hb A. Much of the difference can be attributed to the diminished capacity of Hb F to bind BPG; Hb F thus has an intrinsically greater affinity for O_2, and oxygen transfer from mother to fetus is ensured.

SICKLE-CELL ANEMIA

In 1904, a Chicago physician treated a 20-year-old black college student complaining of headache, weakness, and dizziness. The blood of this patient revealed serious anemia — only half the normal number of red cells were present. Many of these cells were abnormally shaped; in fact, instead of the characteristic disc shape, these erythrocytes were elongate and crescentlike in form, a feature that eventually gave name to the disease sickle-cell anemia. These sickle cells pass less freely through the capillaries, impairing circulation and causing tissue damage. Further, these cells are more fragile and rupture more easily than normal red cells, leading to anemia.

SICKLE-CELL ANEMIA IS A MOLECULAR DISEASE

A single amino acid substitution in the β-chains of Hb causes sickle-cell anemia. Replacement of the glutamate residue at position 6 in the β-chain by a valine residue marks the only chemical difference between Hb A and sickle-cell hemoglobin, Hb S. The amino acid residues at position β6 lie at the surface of the hemoglobin molecule. In Hb A, the ionic R groups of the Glu residues fit this environment. In contrast, the aliphatic side chains of the Val residues in Hb S create hydrophobic protrusions where none existed before. To the detriment of individuals who carry this trait, a hydrophobic pocket forms in the EF corner of each β-chain of Hb when it is in the deoxy state, and this pocket nicely accommodates the Val side chain of a neighbouring Hb S molecule (Figure). This interaction leads to the aggregation of Hb S molecules into long, chainlike polymeric structures. The obvious consequence is that deoxyHb S is less soluble than deoxyHb A. The concentration of hemoglobin in red blood cells is high (about 150 mg/mL), so that even in normal circumstances it is on the verge of crystallization. The formation of insoluble deoxyHb S fibres distorts the red cell into the elongated sickle shape characteristic of the disease. In certain regions of Africa, the sickle-cell trait is found in 20% of the people. Why does such a deleterious heritable condition persist in the population? For reasons as yet unkonw, individuals with the trait are less susceptible to the most virulent form of malaria. The geographic distribution of malaria and the sickle-cell trait are positively correlated.

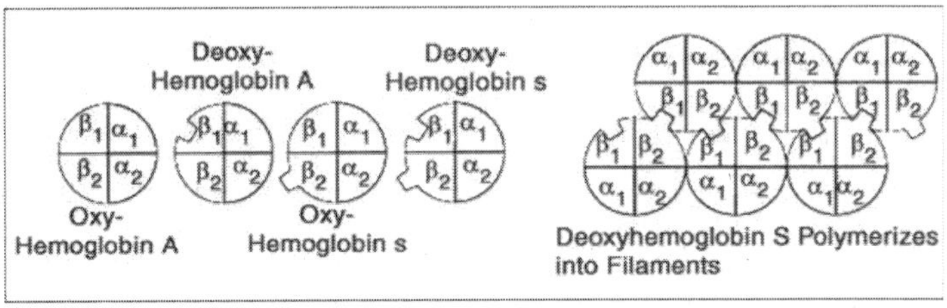

Fig. The Polymerization of Hb S via the Interactions between the Hydrophobic Val Side Chains at Position β6 and the Hydrophobic Pockets in the EF corners of β-chains in Neighbouring Hb Molecules.

The protruding "block" on Oxy S represents the Val hydrophobic protrusion. The complementary hydrophobic pocket in the EF corner of the β-chains is represented by a square-shaped indentation. (This indentation is probably present in Hb A also.) Only the β_2 Val protrusions and the β_1 EF pockets are shown. (The β_1 Val protrusions and the β_2 EF pockets are not involved, although they are present.)

Hemoglobin and Nitric Oxide

Nitric oxide (NO) is a simple gaseous molecule whose many remarkable physiological functions are still being discovered. For example, NO is known to act as a neurotransmitter and as a second messenger in signal transduction. Further,endothelial relaxing factor (ERF, also known as endothelium-derived relaxing factor, or EDRF), an elusive hormonelike agent that acts to relax the musculature of the walls (endothelium) of blood vessels and lower blood pressure, has been identified as NO. It has long been known that NO is a high-affinity ligand for Hb, binding to its heme-Fe21 atom with an affinity 10,000 times greater than that of O_2. An enigma thus arises: Why is NO not instantaneously bound by Hb within human erythrocytes and prevented from exerting its vasodilation properties? The reason that Hb doesn't block the action of NO is due to a unique interaction between Cys93β of Hb and NO recently described by Li Jia, Celia and Joseph Bonaventura, and Johnathan Stamler at Duke University. Nitric oxide reacts with the sulfhydryl group of Cys93β, forming an S-nitroso derivative:

$$-CH_2 - S - N = 0$$

This S-nitroso group is in equilibrium with other S-nitroso compounds formed by reaction of NO? with small-molecule thiols such as free cysteine or glutathione (an isoglutamylcysteinylglycine tripeptide):

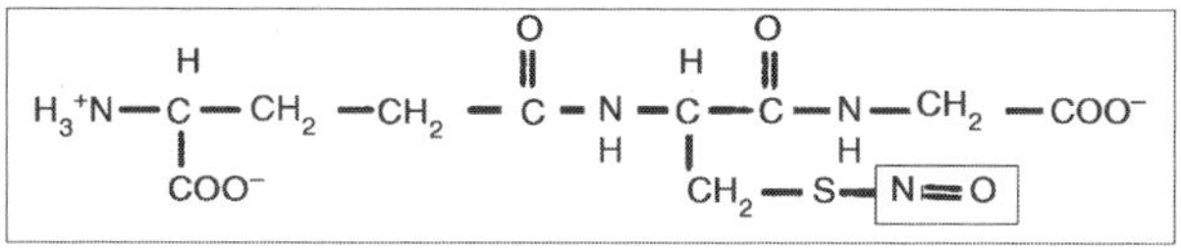

These small-molecule thiols serve to transfer NO from erythrocytes to endothelial receptors, where it acts to relax vascular tension. NO itself is a reactive free-radical compound whose biological half-life is very short (1 – 5 sec). S-nitrosoglutathione has a half-life of several hours.

The reactions between Hb and NO are complex. NO forms a ligand with the heme-Fe_{2+} that is quite stable in the absence of O_2. However, in the presence of O_2, NO is oxidized to NO32 and the heme-Fe_{2+} of Hb is oxidized to Fe_{3+}, forming methemoglobin. Fortunately, the interaction of Hb with NO is controlled by the allosteric transition between R-state Hb (oxyHb) and T-state Hb (deoxyHb). Cys93β is more exposed and reactive in R-state Hb than in T-state Hb, and binding of NO to Cys93βprecludes reaction of NO with heme iron. Upon release of O2 from Hb in tissues, Hb shifts conformation from R state to T state, and binding of NO at Cys93β is no longer favoured. Consequently, NO is released from Cys93β and transferred to small-molecule thiols for delivery to endothelial receptors, causing capillary vasodilation. This mechanism also explains the puzzling observation that free Hb produced by recombinant DNA methodology for use as a whole blood substitute causes a transient rise of 10 to 12 mm Hg in diastolic blood pressure in experimental clinical trials. (Conventional whole blood transfusion has no such effect.) It is now apparent that the "synthetic" Hb, which has no bound NO, is binding NO in the blood and preventing its vasoregulatory function.

THE STRUCTURE OF THE HEMOGLOBIN MOLECULE

The hemoglobin is an $a_2\beta_2$ tetramer. Each of the four subunits has a conformation virtually identical to that of myoglobin. Two different types of subunits, a and β, are necessary to achieve cooperative O_2-binding by Hb.

Fig. Conformational Drawings of the a - and β-chains of Hb and the Myoglobin Chain. (Irving Geis)

The β-chain at 146 amino acid residues is shorter than the myoglobin chain (153 residues), mainly because its final helical segment (the H helix) is shorter. The a -chain (141 residues) also has a shortened H helix and lacks the D helix as well. Max Perutz, who has devoted his life to elucidating the atomic structure of Hb, noted very early in his studies that the molecule was highly symmetrical. The actual arrangement of the four subunits with respect to one another is shown in Figure for horse methemoglobin. All vertebrate hemoglobins show a three-dimensional structure essentially the same as this. The subunits pack in a tetrahedral array, creating a roughly spherical molecule 6.4 × 5.5 × 5.0 nm.

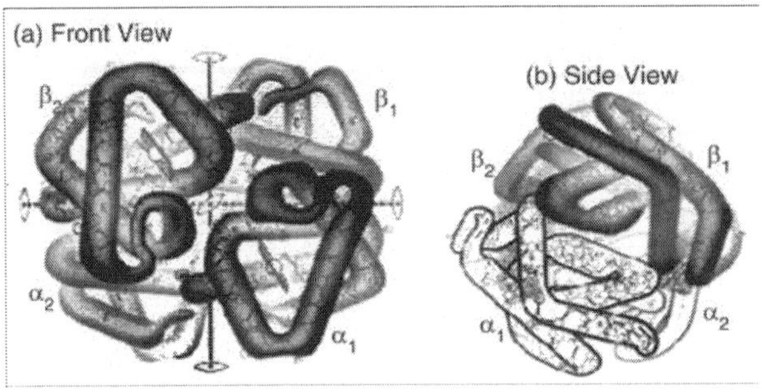

Fig. The Arrangement of Subunits in Horse Methemoglobin, the First Hemoglobin whose Structure was Determined by X-ray Diffraction. The Iron Atoms on metHb are in the Oxidized, Ferric (Fe^{3+}) State. (Irving Geis)

The four heme groups, nestled within the easily recognizable cleft formed between the E and F helices of each polypeptide, are exposed at the surface of the molecule. The heme groups are quite far apart; 2.5 nm separates the closest iron ions, those of hemes a_1 and β_2, and those of hemes a_2 and β_1. The subunit interactions are mostly between dissimilar chains: each of the a -chains is in contact with both β-chains, but there are few a-a or α-α interactions.

The Physiological Significance of the Hb:O_2 Interaction

We can determine quantitatively the physiological significance of the sigmoid nature of the hemoglobin oxygen-binding curve, or, in other words, the biological importance of cooperativity.

$$\frac{Y}{(1-Y)} = \left(\frac{pO_2}{p_{50}}\right)^n$$

The equation describes the relationship between pO_2, the affinity of hemoglobin for O_2 (defined as P_{50}, the partial pressure of O_2 giving half-maximal saturation of Hb with O_2), and the fraction of hemoglobin with O_2 bound, Y, versus the fraction of Hb with no O_2 bound, (1 – Y). The coefficient n is the Hill coefficient, an index of the cooperativity (sigmoidicity) of the hemoglobin oxygen-binding curve. Taking p O_2 in the lungs as 100 torr, P_{50} as 26 torr, and n as 2.8, Y, the fractional saturation of the hemoglobin heme groups with O2, is 0.98.

If p O_2 were to fall to 10 torr within the capillaries of an exercising muscle, Y would drop to 0.06. The oxygen delivered under these conditions would be proportional to the difference, Ylungs-Ymuscle, which is 0.92. That is, virtually all the oxygen carried by Hb would be released. Suppose instead that hemoglobin binding of O_2 were not cooperative; in that case, the hemoglobin oxygen-binding curve would be hyperbolic, and n=1.0. Then Y in the lungs would be 0.79 and Y in the capillaries, 0.28, and the difference in Y values would be 0.51. Thus, under these conditions, the cooperativity of oxygen binding by Hb means that 0.92/0.51 or 1.8 times as much O2can be delivered.

OXYGENATION MARKEDLY ALTERS THE QUATERNARY STRUCTURE OF HB

Crystals of deoxyhemoglobin shatter when exposed to O_2. Further, X-ray crystallographic analysis reveals that oxy- and deoxyhemoglobin differ markedly in quaternary structure. In particular, specific a β-subunit interactions change. The ab contacts are of two kinds. The a_1 $\beta_1$and a_2 β_2 contacts involve helices B, G, and H and the GH corner. These contacts are extensive and important to subunit packing; they remain unchanged when hemoglobin goes from its deoxy to its oxy form. The a_1 β_2 and a_2 β_1 contacts are called sliding contacts. They principally involve helices C and G and the FG corner (Figure).

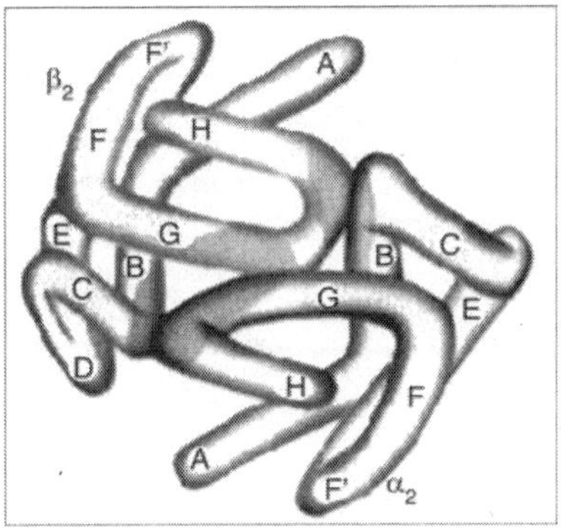

Fig. Side view of One of the Two a β dimers in Hb, with Packing Contacts Indicated in Blue. The Sliding Contacts made with the Other Dimer are Shown.

When hemoglobin undergoes a conformational change as a result of ligand binding to the heme, these contacts are altered. Hemoglobin, as a conformationally dynamic molecule, consists of two dimeric halves, an a_1 β_1-subunit pair and an a_2 β_2-subunit pair. Each aβ dimer moves as a rigid body, and the two halves of the molecule slide past each other upon oxygenation of the heme. The two halves rotate some 15° about an imaginary pivot passing through the a β-subunits; some atoms at the interface between a β dimers are relocated by as much as 0.6 nm.

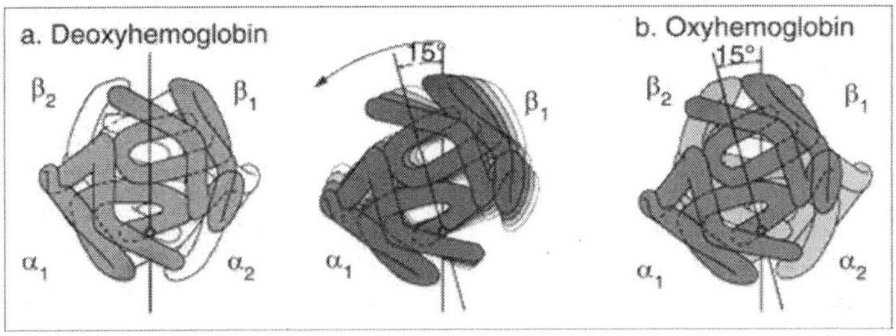

Fig. Subunit Motion in Hemoglobin when the Molecule goes from the (a) Deoxy to the (b) Oxy form. (Irving Geis)

Movement of the Heme Iron by Less Than 0.04 nm Induces the Conformational Change in Hemoglobin

In deoxyhemoglobin, histidine F8 is liganded to the heme iron ion, but steric constraints force the Fe^{2+}:His -N bond to be tilted about 8° from the perpendicular to the plane of the heme. Steric repulsion between histidine F8 and the nitrogen atoms of the porphyrin ring system, combined with electrostatic repulsions between the electrons of Fe^{2+} and the porphyrin π-electrons, forces the iron atom to lie out of the porphyrin plane by about 0.06 nm. Changes in electronic and steric factors upon heme oxygenation allow the Fe^{2+} atom to move about 0.039 nm closer to the plane of the porphyrin, so now it is displaced only 0.021 nm above the plane. It is as if the O_2 were drawing the heme Fe^{2+} into the porphyrin plane (Figure). This modest displacement of 0.039 nm seems a trivial distance, but its biological consequences are far-reaching. As the iron atom moves, it drags histidine F8 along with it, causing helix F, the EF corner, and the FG corner to follow. These shifts are transmitted to the subunit interfaces, where they trigger conformational readjustments that lead to the rupture of interchain salt links.

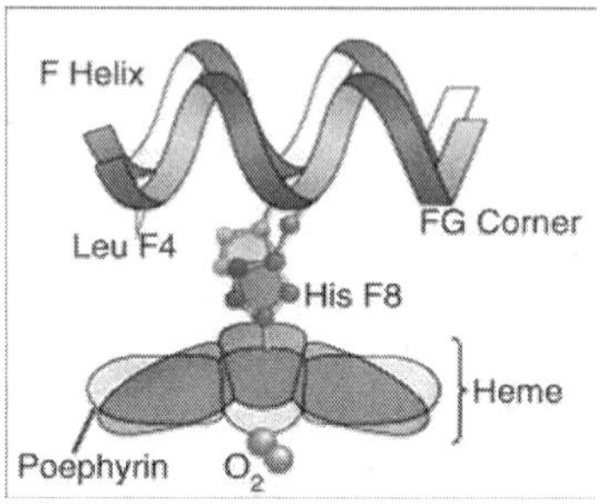

Fig. Changes in the Position of the Heme iron Atom upon Oxygenation Lead to Conformational Changes in the Hemoglobin Molecule.

THE OXY AND DEOXY FORMS OF HEMOGLOBIN REPRESENT

Hemoglobin resists oxygenation because the deoxy form is stabilized by specific hydrogen bonds and salt bridges (ion-pair bonds). All of these interactions are broken in oxyhemoglobin, as the molecule stabilizes into a new conformation. A crucial H bond in this transition involves a particular tyrosine residue. Both a- and β-subunits have Tyr as the penultimate C-terminal residue (Tyr a 140=Tyr HC2; Tyr β145 =Tyr HC2, respectively). The phenolic -OH groups of these Tyr residues form intrachain H bonds to the peptide C=O function contributed by Val FG5 in deoxyhemoglobin. (Val FG5 is a 93 and β98, respectively.) The shift in helix F upon oxygenation leads to rupture of this Tyr HC2:Val FG5 hydrogen bond.

Further, eight salt bridges linking the polypeptide chains are broken as hemoglobin goes from the deoxy to the oxy form. Six of these salt links are between different subunits. Four of these six involve either carboxyl-terminal or amino-terminal amino acids in the chains; two are between the amino termini and the carboxyl termini of the a-chains, and two join the carboxyl termini of the β-chains to the e-NH31 groups of the two Lys a 140 residues. The other two interchain electrostatic bonds link Arg and Asp residues in the two a -chains. In addition, ionic interactions between Asp β94 and His β146 form an intrachain salt bridge in each β-subunit. In deoxy-hemoglobin, with all of these interactions intact, the C-termini of the four subunits are restrained, and this conformational state is termed T, the tense or taut form. In oxyhemoglobin, these C-termini have almost complete freedom of rotation, and the molecule is now in its R, or relaxed, form.

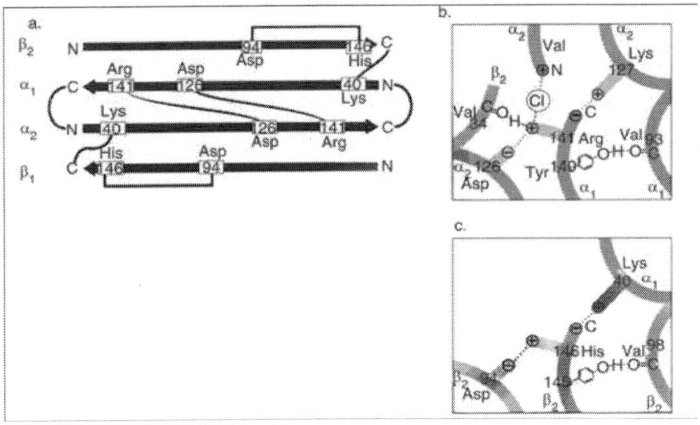

Fig. Salt Bridges between Different Subunits in Hemoglobin.

These noncovalent, electrostatic interactions are disrupted upon oxygenation. Arga^{141} and Hisβ^{146} are the C-termini of the a-and β-polypeptide chains.

(a) The various intrachain and interchain salt links formed among the a-and b-chains of deoxy-hemoglobin.

(b) A focus on those salt bridges and hydrogen bonds involving interactions between N-terminal and C-terminal residues in the a -chains. Note the Cl- ion, which bridges ionic interactions between the N-terminus of a 2 and the R group of Arga141.

(c) A focus on the salt bridges and hydrogen bonds in which the residues located at the C-termini of b-chains are involved. All of these links are abolished in the deoxy to oxy transition. (Irving Geis)

6

Enzymes

As major gene products, enzymes in the CNS are likely candidates for heritable variation. The range of enzymes active in neural tissue is wide and provides a wealth of material for pharmacological study. At a time when research on Behavioural control by various neurotransmitter systems is so predominant, it is perhaps easy to overlook other kinds of enzyme systems that could markedly influence drug response. Not the least of these are those biochemical pathways involved in energy metabolism.

Small perturbations in cerebral energy metabolism can have potentially major effects on a wide variety of Behaviours. The example provided by hypoglycemia and its effects on cognitive and affective characters is clear. The brain is a high energy utilization organ. All specialized functions of neural tissue depend on the more basic biochemistry of energy sources. Although we are not aware of genetic studies of enzymes involved in such energy metabolism and pharmacological alterations of activity, the caveat is relevant in the context of the distinction between direct and indirect drug effects on Behaviour. A drug such as alcohol may very well modify neurotransmitter activity and some related Behaviour by actions on synthesis, degradation, receptor binding characteristics, or channel alteration. Nu merous other "side" effects might also occur. For example, ethanol consumption produces a marked alteration in general metabolism, producing among other things a general acidosis and increasing blood a-hydroxybutyrate levels. As â hydroxybutyrate is a usable energy source for brain tissue, marked changes in cerebral energy metabolism are likely to follow.

Such "indirect" effects on neural function, and perhaps Behaviour, would certainly appear subject to the same sort of heritable variation as other biochemical systems. Thus the degree of Behavioural modification by indirect as well as direct action of the pharmacological agent might also be influenced by genetic factors. While hoping not to be redundant, we believe the conclusion is obivous. That is, gene action may affect pharamcological responsiveness in a variety of ways other than via the specific neural system controlling the phenotype of interest. Keeping the foregoing caution in mind, numerous informative avenues of research on neuro-transmitter system regulation by genetic variables are available.

The most well researched of these is probably the area of catecholamine neurotransmitters. Norepinephrine and dopamine systems are subject to functional alteration by numerous pharmacological agents; many of these agents are of use in a clinical setting precisely because of their Behavioural effects. These functional alterations can occur as a result of inhibition or induction of the enzymes responsible for synthesis and degradation of these catecholamine transmitters. Early work by Ciaranello, Barchas, Kessler, and Barchas first established the fact of strain differences in brain activity of tyrosine hydroxylase (TH), the rate limiting enzyme in catecholamine biosynthesis.

The same report also demonstrated marked strain differences in adrenal phenylethanolamine-N methyl transferase (PNMT), the enzyme responsible for epinephrine production from norepinephrine. We here describe further examination of these mouse strain differences in TH activity in a context that serves as an example for pharmacogenetic study as well as neural development. In midbrain of CBA/J mice, the activity of TH is 20per cent less than in BALB/cJ mice. A comparable difference in immunocyto chemically detectable dopamine neurons in midbrain suggests that this difference in TH activity is due to more cells with the enzyme, rather than to greater specific activity per cell. A correspondingly more dense innervation of the striatum, by what are presumably the axons of these midbrain nigral cells, occurs in the BALB/c mice. It is likely that this difference is related in a general way to the fact that the BALB/c brain is simply a larger one. Nonetheless, this represents a clear example of genetic control over neuronal organization and neurochemical characteristics of a Behaviourally important system. The differences in density of striatal innervation apparently predispose a differential response to two agents with dopaminergic activity, amphetamine and apomorphine.

BALB/c mice are more responsive to the locomotor stimulant effects of amphetamine and less sensitive to the stereotypy producing effects of apomorphine, a dopamine receptor agonist. Interestingly, these charac teristics follow a developmental time course that lags slightly behind the appearance of the TH activity difference. Apparently both pre and postsynaptic functions of the dopamine system are different between the two strains as reflected in the actions of the two drugs. A more detailed understanding of the genetic control over dopamine cell proliferation, growth, and effects on target tissue in this model system will answer many questions not only of pharmacogenetic interest, but also of basic importance to developmental neurobiology.

In this example of TH activity differences, the enzyme activity itself is probably not the focal point of either gene or pharmacological action. As noted, a more basic developmental issue is central. Yet the enzyme is a clear marker for a dramatic genetic influence. In other work, TH is seen to be more directly involved.

Gamma-butyrolactone (GBL), through its metabolite gamma-hydroxy butyric acid (GHB), produces an inhibition of impulse transmission in dopamine neurons. A compensatory activation of TH occurs in these cells and an accumulation of dopamine occurs. In the LS and SS mice, GLB has hypnotic effects that differ in the same direction as ethanol effects. Of interest here is the fact that dopamine accumulation is 50per cent greater in the LS mice after GBL treatment. Presumably the activation of TH was greater in LS mice. This activation occurs via kinetic rather than molecular genetic induction and apparently reflects a differential effectiveness of GHB at a subcellular site in LS and SS dopamine neurons. Although several studies of neurotransmitter systems' enzymes exist, there is yet to be a complete characterization of all biosynthetic and degradatory enzymes, transmitter levels and turnover rate, and receptor kinetics for any single system where genetic variation is also of interest. The difficulties of such a massive undertaking are obvious, but the benefits accruing to all three fields, neuropharmacology of drug action, neurochemical organization and dynamics, and mechanisms of genetic action are equally impressive. The field must surely progress along these lines.

ALCOHOL

The genetic effects on a given Behavioural response do not necessarily imply similar genetic effects on other phenotypes. The degree to which a similar genetic pattern occurs for more than one phenotype is a measure of the genetic correlation of these phenotypes. Yet taskor situation specific genetic effects are not rare. The LS and SS mice are usually very different in their response to ethanol. Yet in one study, across a wide range of doses, a particular Behaviour was not differentially affected. When ethanol was used as a poison (UCS) in a conditioned taste aversion experiment, the magnitude of aversion produced was similar in the two lines. Thus, whatever systems underlie the loss of righting relfex differences do not control response to ethanol in the taste aversion experiment. A second example may be drawn from the literature on strain differences in ethanol response. Although BALB/c mice lose the righting reflex longer than C_{57}BL/6 mice, they show a greater locomootor stimulation following low dose ethanol administration. It may be of interest that, in these two strains, amphetamine shows exactly opposite effects on locomotor activity. BALB/c mice are not activated, whereas C_{57}BL/10 mice are thus a negative genetic correlation is suggested for these effects of amphetamine and ethanol.

MEASURES OF RESPONSES TO OPIATES

It is becoming increasingly apparent that strain differences are readily obtained using the standard measures of opiate induced analgesia and locomotor activity. However, it is equally apparent that both the magnitude and the direction of these differences change with the particular measures and procedures employed. For example, C_{57}BL mice have been characterized as being sensitive to morphine induced hyperactivity in open field tests and toggle boxes. However, Eidelberg, Erspamer, Kreinick, and Harris have reported that C_{57}BL mice respond to several doses of morphine with decreased activity when the measure is recorded on an activity wheel. Similarly, measures of analgesia, such as the hot plate technique, tail flick procedures, and abdominal writhing induced by agents such as acetic acid often yield very different strain distribution patterns. The exact nature of these differences among measures has been reviewed elsewhere.

For the present, the disparate results of these procedures lead us to argue for careful descriptions of the methods employed when general phenomena associated with opiate responsivity. However, we share the optimism of Crabbe and Belknap that consistent patterns among several dimensions of opiate responsiveness may yet emerge as the results of further investigations are known.

GENETIC CORRELATIONS

The basis of many inferences in pharmacogenetic research is the comparison of effects of more than one drug on the same genetic material. This is implicitly an example of genetic correlation. Both correlated drug effects and correlated effects of a single drug on several Behaviours are means for analyzing the domain of effects of that pharmacological agent, or the common physiological substrate of a variety of agents. The quantitative genetic methods for deriving genetic correlations are readily available and are not presented in detail here. Instead, the major experimental approaches to genetic correlation methods and the general usefulness of such correlations.

One of the most utilized approaches to genetic correlations has been the examination of correlated characters in selected lines. A generalized example would find a set of lines selected for response to drug X, characterized at a variety of physiological levels, and perhaps other types of responses to the same drug being assessed as well. The usual argument is that care must be taken in interpretation of any correlated characters especially if replicated lines and unselected control lines are not available. The basis of this assertion is that when only two lines are sampled and two characters are measured on those genotypes, then the degrees of freedom for assessing the correlation are $N - 2 = 0$. Two points are worthwhile in counter. A set of once replicated lines would produce a situation with $df = 2$, hardly a powerful statistical test. Only when the numbers of genotypes exceeds 20 will actual correlations below.40 be detectable. In addition, correlational degrees of freedom are based on the assumption that both variables generating the covariance are free to vary. This is clearly not the case after a selection has been accomplished.

The selected phenotype is at that point fixed. Only the as yet unmeasured phenotype is "free" to vary. Perhaps a more Bayesian approach is desirable. The most rational alternative to numerous replicates is a simultaneous measurement of several phenotypes during the course of selection. As selection progresses, those characters sharing a true genetic covariance with the selected phenotype will diverge in somewhat parallel fashion. As this approach forces one to identify relevant variables even before the selection begins, it is a less than perfect one.

Obviously a more practical approach is required. We would advocate a rational approach that makes use of a priori knowledge of the pharmacological system being examined and physiological intervention if the question being asked is one of mechanism of drug action. For example, in the LS and SS mice it is not surprising that alcohols other than ethanol have a differential effect.

However, if a correlated character is, for example, an altered preference for saccharin sweetened water, then one must wonder if drift is responsible. In sum, no single approach seems completely suitable. Therefore, only careful and objective scrutiny can avoid erroneous conclusions. An additional tool in genetic correlation analysis is the diallel cross and also the triple test cross. The inclusion of non homozygous animals in the diallel cross analysis is probably an advantage, but again, unless the number of genotypes is large, statistical power is low. The use of genetic correlations to assess commonalities of drug action is a powerful approach. Only the use of cross tolerance studies can so clearly demon strate common substrates of action as can the presence of a genetic correlation. The aforementioned effects of higher order alcohols on LS and SS mice led to the conclusion that considerably different mechanisms of action underlie the effects of alcohols and other sedative hypnotics that did not differentially affect the two lines.

It is conceivable that, in a genetic system with high heritabilities, an experimenter could test a very large range of drug classes and attempt to classify them via one of a number of classification techniques on the basis of the genetic correlations. The alternate approach examines the effects of a single drug on numerous Behaviours. Examination of the kinds of genotype patterns of Behaviours affected similarly by a given drug should indicate similar substrates for those Behaviours. It would also be of interest to examine the pattern of genetic correlations in drugged and undrugged mice separately. Changes in those correlations would suggest an ability of the drug to dissociate genetically based covariation of Behaviours. Thus even at a strictly Behavioural level of analysis, the genetic correlation could be a very useful tool.

DEVELOPMENTAL PHENOMENA

As already mentioned, the examination of drug responses in the developing organism and concomitant study of neural development is a powerful approach to the study of drug action and neural development. The addition of the genetic variable as a unique tool to tease apart neuropharmacological mechanisms is a potentially fruitful tactic. Correlated biochemical and pharmacological changes are more easily linked if the correlation holds in genotypes that show different time courses of developmental events.

An example from the literature on alcohol preference in mice is instructive. Wood reviews this literature on age related differences in response to and preference for alcohol by labouratory rodents. Two studies in particular are instructive both for purposes of the genetic variable and its interaction with an experiential variable. BALB/c mice are, as adults, avoiders of ethanol. Yet, from 4-9 weeks of age, they will voluntarily consume a fair amount of ethanol. Consumption diminished to near zero by 9-10 weeks unless a choice between alcohol and water was provided during the 4 to 8-week period. Interestingly, mice forced to drink alcohol during this period would typically not consume the alcohol as adults. Thus the fact of choice in young mice appears to predispose alcohol consumption in adults. This plasticity of phenotype or lack of phenostability makes clear the complex nature of the neural and Behavioural systems on which the genes can act only very indirectly.

A more general view of pharmacogenetic research must realise that a most critical variable in drug effectiveness is the functional organization of the complex target organ called the nervous system. It is ultimately the genetic influence on the assemblage of the brain and other physiological systems that provides the substrate for individual differences produced in drug responsivity. A given drug acting on similar receptors in two brains that differ in organization of the neural network in which these receptors reside might produce different Behavioural outcomes. The mechanism of such gene effects on neural development may be as simple as cell proliferation and migration rates or as complex as regulation of the timing of critical and sensitive periods for Behavioural and emotional development. The study of such biochemical characters as those already enumerated may yield important findings. Ultimately, however, a concern with developmental issues in pharmacogenetic work will yield much more information about the mechanisms of gene and drug action.

GENERAL CHARACTERISTICS OF ENZYMES

Enzymes are usually proteins of high molecular weight that acts as catalysts in biological systems. They accelerate hundreds of reactions taking place simultaneously in the cell, without undergoing any permanent chemical change and without affecting the reaction equilibrium; by lowering the activation energy of the reaction via binding the substrate and forming an enzymesubstrate complex. Enzymes are specific, versatile, and very effective biological catalysts, resulting in much higher reaction rates as compared to chemically catalyzed reactions under ambient conditions. Enzyme specificity is thought to be a consequence of its elaborate three-dimensional conformation which allows formation of the active site responsible for the catalytic ability of the enzyme. Enzymes catalyze only the reactions of very narrow ranges of reactants, which may consist of a small number of closely related classes of compounds, a single class of compounds, or a single compound. This means that the chosen reaction can be catalyzed to the exclusion of side-reactions, eliminating undesirable by-products. Thus, higher productivities may be achieved, reducing material costs.

Another distinguishing characteristic of enzymes is their frequent need for cofactors. A cofactor is a nonprotein compound which combines with an otherwise inactive protein to give a catalytically active complex. The simplest cofactors are metal ions like Ca^{2+}, Zn^{2+}, Co^{2+}, etc. Each enzyme has a specific temperature and pH range where it functions to its optimal capacity, the optima for these proteins usually lie between 37-47°C, and pH optima range from acidic, *i.e.*, 1.0, to alkaline, *i.e.*, 10.5. Biologically active enzymes may be extracted from any living organism. A very wide range of sources is used for commercial enzyme production- from *Actinoplanes* to *Zymomonas*, from spinach to snake venome. Of the hundred or so enzymes being used industrially, over half are from fungi and yeast and over a third are from bacteria, with the remainder divided between animal (%8) and plant (%4) sources. It is a remarkable fact that enzymes with the same name but obtained from different organisms often have different amino acid sequences and hence different properties and catalytic activities.

Enzymes can be isolated and are active outside the living cell. They are such efficient catalysts that they accelerate chemical reactions measurably, even at concentrations so low that they cannot be detected by most chemical tests for protein. Like other chemical reactions, enzyme-catalyzed reactions proceed only when accompanied by a decrease in free energy; at equilibrium the concentrations of reactants and products are the same in the presence of an enzyme as in its absence. An enzyme can catalyze an indefinite amount of chemical change without itself being diminished or altered by the reaction. However, because most isolated enzymes are relatively unstable, they often gradually lose activity under the conditions employed for their study.

Table. Some Representative Enzymes, their Sources, and Reaction Specificities

Enzyme	Some sources	Reaction catalyzed
Pepsin	Gastric juice peptides and amino acids	Hydrolysis of proteins to
Urease	Jackbean, bacteria ammonia and carbon	Hydrolysis of urea to dioxide
Amylase	Saliva, pancreatic juice	Hydrolysis of starch to maltose

Phosphorylase	Muscle, liver, plants of starch or glycogen to	Reversible phosphorolysis glucose- 1-
phosphate		
Transaminases plant tissues	Many animal and from an amino acid to a	Transfer of an amino group keto acid
Phosphohexose isomerase	Muscle, yeast	Interconversion of glucose-6- phosphate and fructose-
	6- phosphate	
Pyruvic carboxylase	Yeast, bacteria, plants	Decarboxylation of pyruvate to acetaldehyde and carbon dioxide
Catalase	Erythrocytes, liver	Decomposition of hydrogen peroxide to oxygen and water
Alcohol dehydrogenase	Liver	Oxidation of ethanol to acetaldehyde
Xanthine oxidase	Milk, liver	Oxidation of xanthine and hypoxanthine to uric acid

Chemical Nature

All enzymes are proteins. Their molecular weights range from about 10,000 to more than 1,000,000. Like other proteins, enzymes consist of chains of amino acids linked together by peptide bonds. An enzyme molecule may contain one or more of these polypeptide chains. The sequence of amino acids within the polypeptide chains is characteristic for each enzyme and is believed to determine the unique three-dimensional conformation in which the chains are folded. This conformation, which is necessary for the activity of the enzyme, is stabilized by interactions of amino acids in different parts of the peptide chains with each other and with the surrounding medium.

These interactions are relatively weak and may be disrupted readily by high temperatures, acid or alkaline conditions, or changes in the polarity of the medium. Such changes lead to an unfolding of the peptide chains (denaturation) and a concomitant loss of enzymatic activity, solubility, and other properties characteristic of the native enzyme. Enzyme denaturation is sometimes reversible. Many enzymes contain an additional, nonprotein component, termed a coenzyme or prosthetic group.

This may be an organic molecule, often a vitamin derivative, or a metal ion. The coenzyme, in most instances, participates directly in the catalytic reaction. For example, it may serve as an intermediate carrier of a group being transferred from one substrate to another. Some enzymes have coenzymes that are tightly bound to the protein and difficult to remove, while others have coenzymes that dissociate readily. When the protein moiety (the apoenzyme) and the coenzyme are separated from each other, neither possesses the catalytic properties of the original conjugated protein (the holoenzyme). By simply mixing the apoenzyme and the coenzyme together, the fully active holoenzyme can often be reconstituted. The same coenzyme may be associated with many enzymes which catalyze different reactions. It is thus primarily the nature of the apoenzyme rather than that of the coenzyme which determines the specificity of the reaction. The complete amino acid sequence of several enzymes has been determined by chemical methods. By x-ray crystallographic methods even the exact three-dimensional molecular structure of a few enzymes has been deduced.

Classification and Nomenclature

Enzymes are usually classified and named according to the reaction they catalyze. The principal classes are as follows. Oxidoreductases catalyze reactions involving electron transfer, and play an important role in cellular respiration and energy production. Some of them participate in the process of oxidative phosphorylation, whereby the energy released by the oxidation of carbohydrates and fats is utilized for the synthesis of adenosine triphosphate (ATP) and thus made directly available for energy-requiring reactions. Transferases catalyze the transfer of a particular chemical group from one substance to another. Thus, transaminases transfer amino groups, transmethylases transfer methyl groups, and so on. An important subclass of this group are the kinases, which catalyze the phosphorylation of their substrates by transferring a phosphate group, usually from ATP, thereby activating an otherwise metabolically inert compound for further transformations.

Hydrolases catalyze the hydrolysis of proteins (proteinases and peptidases), nucleic acids (nucleases), starch (amylases), fats (lipases), phosphate esters (phosphatases), and other substances. Many hydrolases are secreted by the stomach, pancreas, and intestine and are responsible for the digestion of foods. Others participate in more specialized cellular functions. For example, cholinesterase, which catalyzes the hydrolysis of acetylcholine, plays an important role in the transmission of nervous impulses. Lyases catalyze the nonhydrolytic cleavage of their substrate with the formation of a double bond. Examples are decarboxylases, which remove carboxyl groups as carbon dioxide, and dehydrases, which remove a molecule of water. The reverse reactions are catalyzed by the same enzymes. Isomerases catalyze the interconversion of isomeric compounds. Ligases, or synthetases, catalyze endergonic syntheses coupled with the exergonic hydrolysis of ATP. They allow the chemical energy stored in ATP to be utilized for driving reactions uphill.

Specificity

The majority of enzymes catalyze only one type of reaction and act on only one compound or on a group of closely related compounds. There must exist between an enzyme and its substrate a close fit, or complementarity. In many cases, a small structural change, even in a part of the molecule remote from that altered by the enzymatic reaction, abolishes the ability of a compound to serve as a substrate. An example of an enzyme highly specific for a single substrate is urease, which catalyzes the hydrolysis of urea to carbon dioxide and ammonia. On the other hand, some enzymes exhibit a less restricted specificity and act on a number of different compounds that possess a particular chemical group. This is termed group specificity.

A remarkable property of many enzymes is their high degree of stereospecificity, that is, their ability to discriminate between asymmetric molecules of the right-handed and left-handed configurations. An example of a stereospecific enzyme is L -amino acid oxidase. This enzyme catalyzes the oxidation of a variety of amino acids of the type $R{-}CH(NH_2)COOH$. The rate of oxidation varies greatly, depending on the nature of the R group, but only amino acids of the L configuration react.

ENZYMES AND LIFE PROCESSES

The living cell is the site of tremendous biochemical activity called metabolism. This is the process of chemical and physical change which goes on continually in the living organism. Build-up of new tissue, replacement of old tissue, conversion of food to energy, disposal of waste materials, reproduction - all the activities that we characterize as "life." This building up and tearing down takes place in the face of an apparent paradox. The greatest majority of these biochemical reactions do not take place spontaneously. The phenomenon of catalysis makes possible biochemical reactions necessary for all life processes. Catalysis is defined as the acceleration of a chemical reaction by some substance which itself undergoes no permanent chemical change. The catalysts of biochemical reactions are enzymes and are responsible for bringing about almost all of the chemical reactions in living organisms. Without enzymes, these reactions take place at a rate far too slow for the pace of metabolism.

The oxidation of a fatty acid to carbon dioxide and water is not a gentle process in a test tube - extremes of pH, high temperatures and corrosive chemicals are required. Yet in the body, such a reaction takes place smoothly and rapidly within a narrow range of pH and temperature. In the laboratory, the average protein must be boiled for about 24 hours in a 20% HCl solution to achieve a complete breakdown. In the body, the breakdown takes place in four hours or less under conditions of mild physiological temperature and pH.

CHEMICAL NATURE OF ENZYMES

All known enzymes are proteins. They are high molecular weight compounds made up principally of chains of amino acids linked together by peptide bonds.

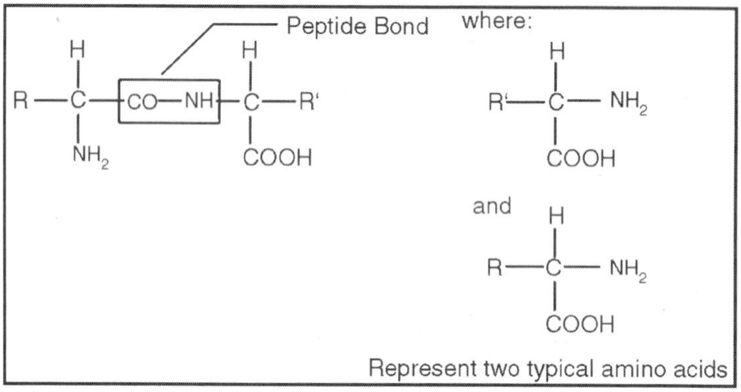

Fig. Typical Protein Structure - Two Amino Acids Joined by a Peptide Bond.

Enzymes can be denatured and precipitated with salts, solvents and other reagents. They have molecular weights ranging from 10,000 to 2,000,000. Many enzymes require the presence of other compounds - cofactors - before their catalytic activity can be exerted. This entire active complex is referred to as the holoenzyme; *i.e.*, apoenzyme (protein portion) plus the cofactor (coenzyme, prosthetic group or metal-ion-activator) is called the holoenzyme.

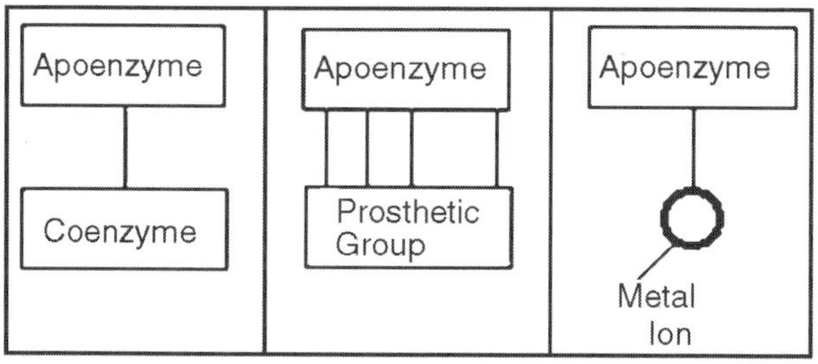

Fig. Holoenzymes- Apoenzymes Plus Various Types of Cofactors.

Apoenzyme + Cofactor = Holoenzyme

According to Holum, the cofactor may be:

- *A coenzyme*: A non-protein organic substance which is dialyzable, thermostable and loosely attached to the protein part.
- *A prosthetic group*: An organic substance which is dialyzable and thermostable which is firmly attached to the protein or apoenzyme portion.
- *A metal-ion-activator*: These include K^+, Fe^{++}, Fe^{+++}, Cu^{++}, Co^{++}, Zn^{++}, Mn^{++}, Mg^{++}, Ca^{++}, and Mo^{+++}.

Specificity of Enzymes

One of the properties of enzymes that makes them so important as diagnostic and research tools is the specificity they exhibit relative to the reactions they catalyze. A few enzymes exhibit absolute specificity; that is, they will catalyze only one particular reaction. Other enzymes will be specific for a particular type of chemical bond or functional group.

In general, there are four distinct types of specificity:

1. *Absolute specificity*: The enzyme will catalyze only one reaction.
2. *Group specificity*: The enzyme will act only on molecules that have specific functional groups, such as amino, phosphate and methyl groups.
3. *Linkage specificity*: The enzyme will act on a particular type of chemical bond regardless of the rest of the molecular structure.
4. *Stereochemical specificity*: The enzyme will act on a particular steric or optical isomer.

Though enzymes exhibit great degrees of specificity, cofactors may serve many apoenzymes. For example, nicotinamide adenine dinucleotide (NAD) is a coenzyme for a great number of dehydrogenase reactions in which it acts as a hydrogen acceptor. Among them are the alcohol dehydrogenase, malate dehydrogenase and lactate dehydrogenase reactions.

Naming and Classification

Except for some of the original enzymes such as pepsin, rennin, and trypsin, most enzyme names end in "ase". The International Union of Biochemistry (I.U.B.) initiated standards of enzyme nomenclature which recommend that enzyme names indicate both the substrate acted upon and the type of reaction catalyzed. Under this system, the enzyme uricase is called urate: O_2 oxidoreductase, while the enzyme glutamic oxaloacetic transaminase (GOT) is called L-aspartate: 2-oxoglutarate aminotransferase.

Enzymes can be classified by the kind of chemical reaction catalyzed:

- Addition or removal of water
  - *Hydrolases*: These include esterases, carbohydrases, nucleases, deaminases, amidases, and proteases
  - Hydrases such as fumarase, enolase, aconitase and carbonic anhydrase
- Transfer of electrons
  - Oxidases
  - Dehydrogenases
- Transfer of a radical
  - *Transglycosidases*: Of monosaccharides
  - *Transphosphorylases and phosphomutases*: Of a phosphate group
  - *Transaminases*: Of amino group
  - *Transmethylases*: Of a methyl group
  - *Transacetylases*: Of an acetyl group
- Splitting or forming a C-C bond
  - Desmolases
- Changing geometry or structure of a molecule
  - Isomerases
- Joining two molecules through hydrolysis of pyrophosphate bond in ATP or other triphosphate
  - Ligases

ENZYME CLASSIFICATIONS

Traditionally, enzymes were simply assigned names by the investigator who discovered the enzyme. As knowledge expanded, systems of enzyme classification became more comprehensive and complex. Currently enzymes are grouped into six functional classes by the International Union of Biochemists (I.U.B.).

Number	Classification	Biochemical Properties
1	Oxidoreductases	Act on many chemical groupings to add or remove hydrogen atoms.
2	Transferases	Transfer functional groups between donor and acceptor molecules. Kinases are specialized transferases that regulate metabolism by transferring phosphate from ATP to other molecules.
3	Hydrolases	Add water across a bond, hydrolyzing it.
4.	Lyases	Add water, ammonia or carbon dioxide across double bonds, or remove these elements to produce double bonds.
5 D	Isomerases	Carry out many kinds of isomerization: L to isomerizations, mutase reactions (shifts of chemical groups) and others.
6	Ligases	Catalyze reactions in which two chemical groups are joined (or ligated) with the use of energy from ATP.

These rules give each enzyme a unique number. The I.U.B. system also specifies a textual name for each enzyme. The enzyme's name is comprised of the names of the substrate(s), the product(s) and the enzyme's functional class. Because many enzymes, such as alcohol dehydrogenase, are widely known in the scientific community by their common names, the change to I.U.B.-approved nomenclature has been slow. In everyday usage, most enzymes are still called by their common name.

Enzymes are also classified on the basis of their composition. Enzymes composed wholly of protein are known as simple enzymes in contrast to complex enzymes, which are composed of protein plus a relatively small organic molecule. Complex enzymes are also known as holoenzymes. In this terminology the protein component is known as the apoenzyme, while the non-protein component is known as the coenzyme or prosthetic group where prosthetic complex in which the small organic molecule is bound to the apoenzyme by covalent bonds; when the binding between the apoenzyme and non-protein components is non-covalent, the small organic molecule is called a coenzyme. Many prosthetic groups and coenzymes are water-soluble derivatives of vitamins. It should be noted that the main clinical symptoms of dietary vitamin insufficiency generally arise from the malfunction of enzymes, which lack sufficient cofactors derived from vitamins to maintain homeostasis.

The non-protein component of an enzyme may be as simple as a metal ion or as complex as a small non-protein organic molecule. Enzymes that require a metal in their composition are known as metalloenzymes if they bind and retain their metal atom(s) under all conditions with very high affinity. Those which have a lower affinity for metal ion, but still require the metal ion for activity, are known as metal-activated enzymes.

ENZYMES IN THE DIAGNOSIS OF PATHOLOGY

The measurement of the serum levels of numerous enzymes has been shown to be of diagnostic significance. This is because the presence of these enzymes in the serum indicates that tissue or cellular damage has occurred resulting in the release of intracellular components into the blood. Hence, when a physician indicates that he/she is going to assay for liver enzymes, the purpose is to ascertain the potential for liver cell damage. Commonly assayed enzymes are the amino transferases: alanine transaminase, ALT (sometimes still referred to as serum glutamate-pyruvate aminotransferase, SGPT) and aspartate aminotransferase, AST (also referred to as serum glutamate-oxaloacetate aminotransferase, SGOT); lactate dehydrogenase, LDH; creatine kinase, CK (also called creatine phosphokinase, CPK); gamma-glutamyl transpeptidase, GGT. Other enzymes are assayed under a variety of different clinical situations but they will not be covered here.

The typical liver enzymes measured are AST and ALT. ALT is particularly diagnostic of liver involvement as this enzyme is found predominantly in hepatocytes. When assaying for both ALT and AST the ratio of the level of these two enzymes can also be diagnostic. Normally in liver disease or damage that is not of viral origin the ratio of ALT/AST is less than 1. However, with viral hepatitis the ALT/AST ratio will be greater than 1. Measurement of AST is useful not only for liver involvement but also for heart disease or damage. The level of AST elevation in the serum is directly proportional to the number of cells involved as well as on the time following injury that the AST assay was performed. Following injury, levels of AST rise within 8 hours and peak 24-36 hours later. Within 3-7 days the level of AST should return to pre-injury levels, provided a continuous insult is not present or further injury occurs. Although measurement of AST is not, in and of itself, diagnostic for myocardial infarction, taken together with LDH and CK measurements the level of AST is useful for timing of the infarct.

The measurement of LDH is especially diagnostic for myocardial infarction because this enzyme exist in 5 closely related, but slightly different forms (isozymes).

The 5 types and their normal distribution and levels in non-disease/injury are listed below:

1. *LDH* 1: Found in heart and red-blood cells and is 17% - 27% of the normal serum total.
2. *LDH* 2: Found in heart and red-blood cells and is 27% - 37% of the normal serum total.
3. *LDH* 3: Found in a variety of organs and is 18% - 25% of the normal serum total.
4. *LDH* 4: Found in a variety of organs and is 3% - 8% of the normal serum total.
5. *LDH* 5: Found in liver and skeletal muscle and is 0% - 5% of the normal serum total.

Following a myocardial infarct the serum levels of LDH rise within 24-48 hours reaching a peak by 2-3 days and return to normal in 5-10 days. Especially diagnostic is a comparison of the LDH-1/LDH-2 ratio. Normally, this ration is less than 1. A reversal of this ration is referred to as a "flipped LDH". Following an acute myocardial infarct the flipped LDH ratio will appear in 12-24 hours and is definitely present by 48 hours in over 80% of patients. Also important is the fact that persons suffering chest pain due to angina only will not likely have altered LDH levels.

CPK is found primarily in heart and skeletal muscle as well as the brain. Therefore, measurement of serum CPK levels is a good diagnostic for injury to these tissues. The levels of CPK will rise within 6 hours of injury and peak by around 18 hours. If the injury is not persistent the level of CK returns to normal within 2-3 days.

Like LDH, there are tissue-specific isozymes of CPK and there designations:

- CPK3 (CPK-MM) is the predominant isozyme in muscle and is 100% of the normal serum total.
- CPK2 (CPK-MB) accounts for about 35% of the CPK activity in cardiac muscle, but less than 5% in skeletal muscle and is 0% of the normal serum total.
- CPK1 (CPK-BB) is the characteristic isozyme in brain and is in significant amounts in smooth muscle and is 0% of the normal serum total.

Since most of the released CPK after a myocardial infarction is CPK-MB, an increased ratio of CPK-MB to total CPK may help in diagnosis of an acute infarction, but an increase of total CPK in itself may not. CPK-MB levels rise 3–6 hours after a myocardial infarct and peak 12–24 hours later if no further damage occurs and returns to normal 12–48 hours after the infarct.

REGULATION OF ENZYME ACTIVITY

While it is clear that enzymes are responsible for the catalysis of almost all biochemical reactions, it is important to also recognize that rarely, if ever, do enzymatic reactions proceed in isolation. The most common scenario is that enzymes catalyze individual steps of multi-step metabolic pathways, as is the case with glycolysis, gluconeogenesis or the synthesis of fatty acids. As a consequence of these lock- step sequences of reactions, any given enzyme is dependent on the activity of preceding reaction steps for its substrate. In humans, substrate concentration is dependent on food supply and is not usually a physiologically important mechanism for the routine regulation of enzyme activity. Enzyme concentration, by contrast, is continually modulated in response to physiological needs.

Three principal mechanisms are known to regulate the concentration of active enzyme in tissues:

1. Regulation of gene expression controls the quantity and rate of enzyme synthesis.
2. Proteolytic enzyme activity determines the rate of enzyme degradation.
3. Covalent modification of preexisting pools of inactive proenzymes produces active enzymes.

Enzyme synthesis and proteolytic degradation are comparatively slow mechanisms for regulating enzyme concentration, with response times of hours, days or even weeks. Proenzyme activation is a more rapid method of increasing enzyme activity but, as a regulatory mechanism, it has the disadvantage of not being a reversible process. Proenzymes are generally synthesized in abundance, stored in secretory granules and covalently activated upon release from their storage sites. Examples of important proenzymes include pepsinogen, trypsinogen and chymotrypsinogen, which give rise to the proteolytic digestive enzymes. Likewise, many of the proteins involved in the cascade of chemical reactions responsible for blood clotting are synthesized as proenzymes. Other important proteins, such as peptide hormones and collagen, are also derived by covalent modification of precursors. Another mechanism of regulating enzyme activity is to sequester enzymes in compartments where access to their substrates is limited. For example, the proteolysis of cell proteins and glycolipids by enzymes responsible for their degradation is controlled by sequestering these enzymes within the lysosome.

In contrast to regulatory mechanisms that alter enzyme concentration, there is an important group of regulatory mechanisms that do not affect enzyme concentration, are reversible and rapid in action, and actually carry out most of the moment to moment physiological regulation of enzyme activity. These mechanisms include allosteric regulation, regulation by reversible covalent modification and regulation by control proteins such as calmodulin. Reversible covalent modification is a major mechanism for the rapid and transient regulation of enzyme activity. The best examples, again, come from studies on the regulation of glycogen metabolism where phosphorylation of glycogen synthase and glycogen phosphorylase kinase results in the stimulation of glycogen degradation while glycogen synthesis is coordinately inhibited. Numerous other enzymes of intermediary metabolism are affected by phosphorylation, either positively or negatively.

These covalent phosphorylations can be reversed by a separate sub-subclass of enzymes known as phosphatases. Recent research has indicated that the aberrant phosphorylation of growth factor and hormone receptors, as well as of proteins that regulate cell division, often leads to unregulated cell growth or cancer. The usual sites for phosphate addition to proteins are the serine, threonine and tyrosine R group hydroxyl residues.

ENZYMES AS BIOLOGICAL CATALYSTS

In cells and organisms most reactions are catalyzed by enzymes, which are regenerated during the course of a reaction. These biological catalysts are physiologically important because they speed up the rates of reactions that would otherwise be too slow to support life. Enzymes increase reaction rates, sometimes by as much as one million-fold, but more typically by about one thousand fold. Catalysts speed up the forward and reverse reactions proportionately so that, although the magnitude of the rate constants of the forward and reverse reactions is are increased, the ratio of the rate constants remains the same in the presence or absence of enzyme. Since the equilibrium constant is equal to a ratio of rate constants, it is apparent that enzymes and other catalysts have no effect on the equilibrium constant of the reactions they catalyze.

Enzymes increase reaction rates by decreasing the amount of energy required to form a complex of reactants that is competent to produce reaction products. This complex is known as the activated state or transition state complex for the reaction. Enzymes and other catalysts accelerate reactions by lowering the energy of the transition state. The free energy required to form an activated complex is much lower in the catalyzed reaction. The amount of energy required to achieve the transition state is lowered; consequently, at any instant a greater proportion of the molecules in the population can achieve the transition state. The result is that the reaction rate is increased.

Michaelis-Menten Kinetics

In typical enzyme-catalyzed reactions, reactant and product concentrations are usually hundreds or thousands of times greater than the enzyme concentration. Consequently, each enzyme molecule catalyzes the conversion to product of many reactant molecules. In biochemical reactions, reactants are commonly known as substrates. The catalytic event that converts substrate to product involves the formation of a transition state, and it occurs most easily at a specific binding site on the enzyme. This site, called the catalytic site of the enzyme, has been evolutionarily structured to provide specific, high-affinity binding of substrate(s) and to provide an environment that favours the catalytic events. The complex that forms, when substrate(s) and enzyme combine, is called the enzyme substrate (ES) complex. Reaction products arise when the ES complex breaks down releasing free enzyme.

Between the binding of substrate to enzyme, and the reappearance of free enzyme and product, a series of complex events must take place. At a minimum an ES complex must be formed; this complex must pass to the transition state (ES*); and the transition state complex must advance to an enzyme product complex (EP). The latter is finally competent to dissociate to product and free enzyme.

The series of events can be shown thus:

$$E + S \leftrightarrow ES \leftrightarrow ES^* \leftrightarrow EP \leftrightarrow E + P$$

The kinetics of simple reactions like that above were first characterized by biochemists Michaelis and Menten. The concepts underlying their analysis of enzyme kinetics continue to provide the cornerstone for understanding metabolism today, and for the development and clinical use of drugs aimed at selectively altering rate constants and interfering with the progress of disease states. The Michaelis-Menten equation is a quantitative description of the relationship among the rate of an enzyme- catalyzed reaction [v_1], the concentration of substrate [S] and two constants, V_{max} and K_m (which are set by the particular equation). The symbols used in the Michaelis-Menten equation refer to the reaction rate [v_1], maximum reaction rate (V_{max}), substrate concentration [S] and the Michaelis-Menten constant (K_m).

$$V_1 = \frac{V_{max}[s]}{\{K_m + [S]\}}$$

The Michaelis-Menten equation can be used to demonstrate that at the substrate concentration that produces exactly half of the maximum reaction rate, *i.e.* ½ V_{max}, the substrate concentration is numerically equal to K_m. This fact provides a simple yet powerful bioanalytical tool that has been used to characterize both normal and altered enzymes, such as those that produce the symptoms of genetic diseases. Rearranging the Michaelis-Menten equation leads to:

$$K_m = [S]\left\{\left[\frac{V_{max}}{V_1}\right] - 1\right\}$$

From this equation it should be apparent that when the substrate concentration is half that required to support the maximum rate of reaction, the observed rate, v_1, will, be equal to V_{max} divided by 2; in other words, v_1 = [V_{max}/2]. At this substrate concentration V_{max}/v_1 will be exactly equal to 2, with the result that:

$$[S](1) = K_m$$

The latter is an algebraic statement of the fact that, for enzymes of the Michaelis-Menten type, when the observed reaction rate is half of the maximum possible reaction rate, the substrate concentration is numerically equal to the Michaelis-Menten constant. In this derivation, the units of K_m are those used to specify the concentration of S, usually Molarity.

The Michaelis-Menten equation has the same form as the equation for a rectangular hyperbola; graphical analysis of reaction rate (v) versus substrate concentration [S] produces a hyperbolic rate plot.

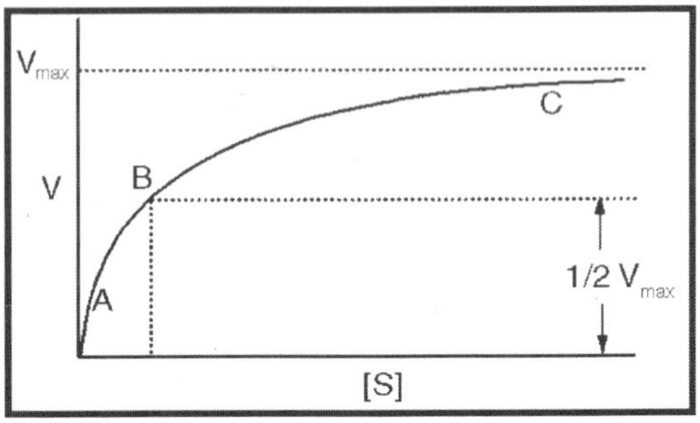

Fig. Plot of Substrate Concentration Versus Reaction Velocity

The key features of the plot are marked by points A, B and C. At high substrate concentrations the rate represented by point C the rate of the reaction is almost equal to V_{max}, and the difference in rate at nearby concentrations of substrate is almost negligible. If the Michaelis-Menten plot is extrapolated to infinitely high substrate concentrations, the extrapolated rate is equal to V_{max}. When the reaction rate becomes independent of substrate concentration, or nearly so, the rate is said to be zero order. (Note that the reaction is zero order only with respect to this substrate. If the reaction has two substrates, it may or may not be zero order with respect to the second substrate).

The very small differences in reaction velocity at substrate concentrations around point C (near V_{max}) reflect the fact that at these concentrations almost all of the enzyme molecules are bound to substrate and the rate is virtually independent of substrate, hence zero order. At lower substrate concentrations, such as at points A and B, the lower reaction velocities indicate that at any moment only a portion of the enzyme molecules are bound to the substrate. In fact, at the substrate concentration denoted by point B, exactly half the enzyme molecules are in an ES complex at any instant and the rate is exactly one half of V_{max}. At substrate concentrations near point A the rate appears to be directly proportional to substrate concentration, and the reaction rate is said to be first order.

Inhibition of Enzyme Catalyzed Reactions

To avoid dealing with curvilinear plots of enzyme catalyzed reactions, biochemists Lineweaver and Burk introduced an analysis of enzyme kinetics based on the following rearrangement of the Michaelis-Menten equation:

$$\frac{1}{V} = \left[\frac{K_m(1)}{V_{max}(S)} + \frac{1}{V_{max}}\right]$$

Plots of 1/v versus 1/[S] yield straight lines having a slope of K_m/V_{max} and an intercept on the ordinate at $1/V_{max}$.

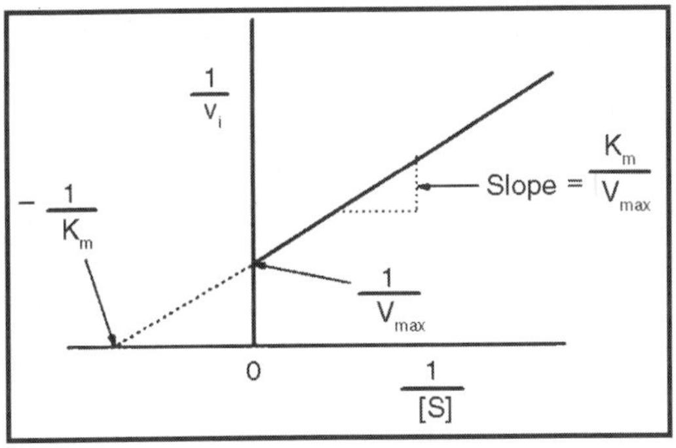

Fig. A Lineweaver-Burk Plot

An alternative linear transformation of the Michaelis-Menten equation is the Eadie-Hofstee transformation:

$$\frac{v}{[S]} = -v\left[\frac{1}{K_m} +\right] + \left[\frac{V_{max}}{K_m}\right]$$

and when v/[S] is plotted on the y-axis versus v on the x-axis, the result is a linear plot with a slope of $-1/K_m$ and the value V_{max}/K_m as the intercept on the y-axis and V_{max} as the intercept on the x-axis.

Both the Lineweaver-Burk and Eadie-Hofstee transformation of the Michaelis-Menton equation are useful in the analysis of enzyme inhibition. Well- known examples of such therapy include the use of methotrexate in cancer chemotherapy to semi-selectively inhibit DNA synthesis of malignant cells, the use of aspirin to inhibit the synthesis of prostaglandins which are at least partly responsible for the aches and pains of arthritis, and the use of sulfa drugs to inhibit the folic acid synthesis that is essential for the metabolism and growth of disease-causing bacteria. In addition, many poisons, such as cyanide, carbon monoxide and polychlorinated biphenols (PCBs) produce their life-threatening effects by means of enzyme inhibition.

Enzyme inhibitors fall into two broad classes: those causing irreversible inactivation of enzymes and those whose inhibitory effects can be reversed. Inhibitors of the first class usually cause an inactivating, covalent modification of enzyme structure. Cyanide is a classic example of an irreversible enzyme inhibitor: by covalently binding mitochondrial cytochrome oxidase, it inhibits all the reactions associated with electron transport.

The kinetic effect of irreversible inhibitors is to decrease the concentration of active enzyme, thus decreasing the maximum possible concentration of ES complex. Since the limiting enzyme reaction rate is often k_2[ES], it is clear that under these circumstances the reduction of enzyme concentration will lead to decreased reaction rates. Note that when enzymes in cells are only partially inhibited by irreversible inhibitors, the remaining unmodified enzyme molecules are not distinguishable from those in untreated cells; in particular, they have the same turnover number and the same K_m. Turnover number, related to V_{max}, is defined as the maximum number of moles of substrate that can be converted to product per mole of catalytic site per second. Irreversible inhibitors are usually considered to be poisons and are generally unsuitable for therapeutic purposes.

Reversible inhibitors can be divided into two main categories; competitive inhibitors and noncompetitive inhibitors, with a third category, uncompetitive inhibitors, rarely encountered.

Inhibitor Type	Binding Site on Enzyme	Kinetic effect
Competitive Inhibitor	Specifically at the catalytic site, where it competes with substrate for binding in a dynamic equilibrium-like process. Inhibition is reversible by substrate.	V_{max} is unchanged; K_m, as defined by [S] required for ½ maximal activity, is increased.
Noncompetitive Inhibitor	Binds E or ES complex other than at the catalytic site. Substrate binding unaltered, but ESI complex cannot form products. Inhibition cannot be reversed by substrate. concentration.	K_m appears unaltered; V_{max} is decreased proportionately to inhibitor
Uncompetitive Inhibitor	Binds only to ES complexes at locations other than the catalytic site. Substrate binding modifies enzyme structure, making inhibitor- binding site available. Inhibition cannot be reversed by substrate.	Apparent V_{max} decreased; K_m, as defined by [S] required for ½ maximal activity, is decreased.

The hallmark of all the reversible inhibitors is that when the inhibitor concentration drops, enzyme activity is regenerated. Usually these inhibitors bind to enzymes by non-covalent forces and the inhibitor maintains a reversible equilibrium with the enzyme. The equilibrium constant for the dissociation of enzyme inhibitor complexes is known as K_i:

$$K_i = \frac{[E][I]}{[EI]}$$

The importance of K_I is that in all enzyme reactions where substrate, inhibitor and enzyme interact, the normal K_m and or V_{max} for substrate enzyme interaction appear to be altered. These changes are a consequence of the influence of K_i on the overall rate equation for the reaction. The effects of K_i are best observed in Lineweaver-Burk plots.

Probably the best known reversible inhibitors are competitive inhibitors, which always bind at the catalytic or active site of the enzyme. Most drugs that alter enzyme activity are of this type. Competitive inhibitors are especially attractive as clinical modulators of enzyme activity because they offer two routes for the reversal of enzyme inhibition, while other reversible inhibitors offer only one. First, as with all kinds of reversible inhibitors, a decreasing concentration of the inhibitor reverses the equilibrium regenerating active free enzyme. Second, since substrate and competitive inhibitors both bind at the same site they compete with one another for binding

Raising the concentration of substrate (S), while holding the concentration of inhibitor constant, provides the second route for reversal of competitive inhibition. The greater the proportion of substrate, the greater the proportion of enzyme present in competent ES complexes. As noted earlier, high concentrations of substrate can displace virtually all competitive inhibitor bound to active sites. Thus, it is apparent that V_{max} should be unchanged by competitive inhibitors. This characteristic of competitive inhibitors is reflected in the identical vertical-axis intercepts of Lineweaver-Burk plots, with and without inhibitor.

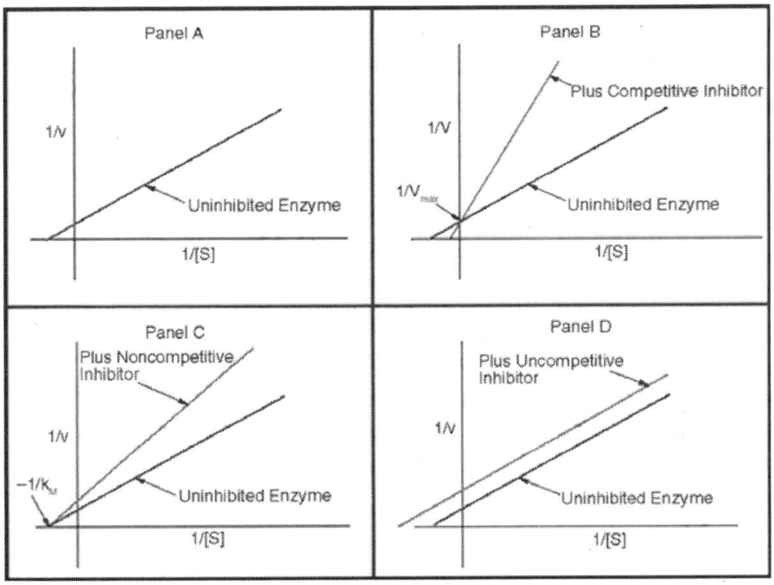

Fig. Lineweaver-Burk Plots of Inhibited Enzymes

Since attaining V_{max} requires appreciably higher substrate concentrations in the presence of competitive inhibitor, K_m (the substrate concentration at half maximal velocity) is also higher, as demonstrated by the differing negative intercepts on the horizontal axis in panel B. Analogously, panel C illustrates that noncompetitive inhibitors appear to have no effect on the intercept at the x-axis implying that noncompetitive inhibitors have no effect on the K_m of the enzymes they inhibit. Since noncompetitive inhibitors do not interfere in the equilibration of enzyme, substrate and ES complexes, the K_m's of Michaelis-Menten type enzymes are not expected to be affected by noncompetitive inhibitors, as demonstrated by x-axis intercepts in panel C. However, because complexes that contain inhibitor (ESI) are incapable of progressing to reaction products, the effect of a noncompetitive inhibitor is to reduce the concentration of ES complexes that can advance to product. Since $V_{max} = k_2[E_{total}]$, and the concentration of competent E_{total} is diminished by the amount of ESI formed, noncompetitive inhibitors are expected to decrease V_{max}, as illustrated by the y-axis intercepts in panel C. A corresponding analysis of uncompetitive inhibition leads to the expectation that these inhibitors should change the apparent values of K_m as well as V_{max}. Changing both constants leads to double reciprocal plots, in which intercepts on the x and y axes are proportionately changed; this leads to the production of parallel lines in inhibited and uninhibited reactions.

ENZYMES IN MEDICINE

Since the mid-1950s there has been a considerable increase in both measurement of enzyme activities and the use of purified enzymes in clinical practice. In recent years, many enzymes have been isolated and purified, and this made it possible to use enzymes to determine the concentration of substrates and products of clinical importance. A further development, arising from the increased availability of purified enzymes, has been targeted to enzyme therapy.

Clinical Diagnosis

The measurement of enzyme activities in serum is of major importance as an aid in diagnosis, being used as means of monitoring progress after therapy, recovery after surgery, and detection of transplant rejection. Urine can be also analysed for determination of enzyme activities since the detection of certain enzymes in urine may indicate kidney damage or failure. On the other hand, the concentration of certain metabolites in serum or urine may be determined by enzymatic methods. The method consists of using an enzyme to transform a metabolite into its product and then estimate the amount of transformed substrate. The use of enzymatic methods presents several advantages, such as the high specificity of the enzyme to estimate the concentration of the metabolite in the presence of other substances, avoiding the need of purification steps prior to chemical analysis. In addition, enzymatic reactions are performed at mild conditions, allowing the analysis of labile compounds that would be degraded by harsher chemical methods. The cost of purified enzymes may be, however, too high to support routine analysis, but the use of immobilized enzymes allows for enzyme reuse and the application of immobilized enzymes for diagnostic assays and as biosensors. The determination of serum metabolites in serum by enzymatic methods includes a wide range of substances such as glucose, uric acid, urea, cholesterol, cholesterol esters, triglycerides, and creatine, among others.

Enzyme Therapy

Many inborn metabolic disorders are associated with the absence of activity of one particular enzyme normally found in the body. Of the 1250 autosomal recessive human genetic diseases, over 200 involve errors in metabolism that result from specific known enzyme deficiencies. The initial identification of the disease may be difficult to determine and normally requires a tissue biopsy. For some genetic diseases, DNA probes are now available that can be used on small amounts of blood, cells, or amniotic fluid. Albinism, for example, is often caused by the absence of tyrosinase, an enzyme essential for the production of cellular pigments. Tyrosinase is a coppercontaining enzyme that catalyzes the first two rate-limiting steps in the melanin biosynthetic pathway, the oxidation of tyrosine to dopa and the subsequent dehydrogenation of dopa to dopaquinone.

The human tyrosinase gene, encoding 529 amino acids, consists of five exons spanning more than 50 Kb of DNA in chromosomes. When homozygous mutations of the tyrosine gene result in complete absence of melanogenic activity, such a patient, categorized as tyrosinasenegative oculocutaneous albinism, will never develop any melanin pigment in the skin, hair, and eyes throughout his or her life. Some inborn errors in metabolism are relatively harmless, *e.g.*, albinism or alkaptonuria, but others must be detected early if the defect is to be circumvented. This is the case of phenylketonuria, where the enzyme phenylalanine 4-monooxygenase (enzyme that converts phenylalanine into tyrosine) is missing.

Phenylketonuria results in the accumulation of phenylalanine, which may cause mental retardation. Patients with phenylalanine 4-monooxygenase deficiency must follow a phenylalanine-free diet in order to avoid the accumulation of deleterious effects. Phenylalanine is, however, an amino acid essential to maintain growth and protein turnover, and thus it must be supplied in a minimal amount required to maintain normal metabolism. Such a dietary scheme is normally carried out by lowering the amount of protein consumed, but this may cause a deficiency in other essential amino acids. A possible alternative to this therapy is to replace the missing enzyme. It may be difficult to find, however, an enzyme with the same function from a human source since the direct administration of enzymes from other sources into the body would cause an adverse immunological response. A possible approach to circumvent this problem involves the isolation of the enzyme within a microcapsule, fibre, or gel, which will protect the enzyme from proteolysis and avoid the undesirable immunological response.

The pharmacological properties of enzymes have been employed to replace enzymes that are missing or defective as a consequence of an inherited disease or malfunction of an organ where they are normally synthesized or to accomplish a certain biological effect that is dependent on the catalytic activity of the enzyme.

Therefore, depending on the treatment, the administration of enzymes as therapeutic agents can be subdivided into two categories:

- The topical application of an enzyme as an extracellular agent
- The intracellular applications of enzymes to treat metabolic deficiency and related disease.

The main areas where enzyme therapy has been applied are the degradation of necrotic tissue by the use of proteolytic enzymes, removal of toxic compounds from the blood, treatment of genetic deficiency diseases and cancer, and treatment of pancreatic insufficiency. Enzymes may be administered either intra- or extracorporeally, depending on the objective. If the enzyme is to be used for the removal or transformation of a substance present in the blood (*e.g.*, toxic metabolite or a blood clot), then it is only necessary for the enzyme to be present in the blood and not necessary for the enzyme to enter the intracellular compartments.

This type of application may be either intra- or extracorporeal using a bypass as in kidney dialysis. For intracellular therapy, it is necessary for the enzyme to be taken up by the appropriate target cells. Although the attempts made with enzyme therapy in clinical trials have so far had limited success, it is reasonable to assume that the delivery of enzymes would constitute a feasible approach for the treatment of certain diseases in the near future.

ENZYME IMMOBILIZATION TECHNOLOGY

The term *immobilized enzyme* was adopted in 1971 at the first Enzyme Engineering Conference. It describes enzymes physically confined at or localized in a certain region of space with retention of catalytic activity and which can be used repeatedly and continuously. The immobilization of biocatalysts (not only enzymes but also other bioactive molecules such as growth factors and hormones, cellular organelles, microbial cells, and plant and animal cells) is attracting worldwide attention in biotechnology applications. In general, immobilized biocatalysts are more stable and easier to handle compared with their free counterparts.

At present, applications of immobilized biocatalysts include the production of useful compounds by stereospecific or regiospecific bioconversion, the production of energy by biological processes, the selective treatment of specific pollutants to solve environmental problems, continuous analyses of compounds with a high sensitivity and specificity, and medical uses such as new types of drugs for enzyme therapy or artificial organs. Immobilized enzymes are already being used in medical applications for clinical diagnosis and also for intra- and extracorporeal enzyme therapy. Applications in clinical analysis are mainly related to biosensors, which have been used to detect the presence of various organic compounds for many years. For example, glucose oxidase and catalase have been used to measure blood glucose concentration, and cholesterol oxidase and cholesterol esterase to determine cholesterol levels. In addition, enzymes can be immobilized on different prosthetic devices or used extracorporeally (*e.g.*, artificial heart, artificial lung, artificial kidney, equipment for hemodialysis and specific blood purification) as surface modifiers in order to increase the biocompatibility of these devices and to prevent blood clotting.

Methods for Immobilizing Enzymes in Polymeric Carriers

Various methods have been developed for the immobilization of biocatalysts, which are being used extensively today. A wide range of support materials has also been employed for enzyme immobilization. The support type can be classified according to their chemical composition, such as organic or inorganic supports, and the former can be further classified into natural or synthetic matrices. Immobilization techniques can be divided into different categories: physical, chemical, enzymatic, and genetic engineering methods.

Adsorption

The adsorption of an enzyme onto a support or film material is the simplest method of obtaining an immobilized enzyme. Basically, the enzyme is attached to the support material by noncovalent linkages and does not require any preactivation step of the support. The interactions formed between the enzyme and the support material will be dependent on the existing surface chemistry of the support and on the type of amino acids exposed at the surface of the enzyme molecule. Enzyme immobilization by adsorption involves, normally, weak interactions between the support and the enzyme such as ionic or hydrophobic interactions, hydrogen bonding, and van der Waals forces. Most of the support materials available have sufficient surface-charge properties suitable for immobilization by adsorption. They include inorganic carriers (ceramic, alumina, activated carbon, kaolinite, bentonite, porous glass), organic synthetic carriers (nylon, polystyrene), and natural organic carriers (chitosan, dextran, gelatin, cellulose, starch).

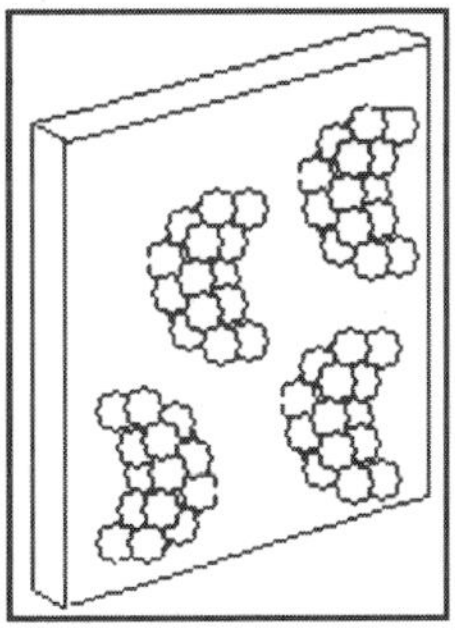

Fig. Biocatalysts Bound to a Carrier by Adsorption.

The method consists of simply mixing an aqueous solution of enzyme with the support material for a period of time, after which the excess enzyme is washed away from the immobilized enzyme on the support. The procedure requires strict control of the pH and ionic strength, because these can alter the charges of the enzyme and the support and therefore affect the level of adsorption. A simple shift in pH can cancel ionic interactions and promote the release of the enzyme from the support.

The main advantages of adsorption are the method simplicity, the little effect on the conformation/activity of the biocatalyst, and the possibility of regenerating inactive enzyme by addition of fresh enzyme. The main disadvantage is the desorption of the biocatalyst from the support due to the weak interactions established. The enzyme desorption can easily occur by changes in the environment medium such as pH, temperature, solvent, and ionic strength or in the case of extended reactions.

Ionic Binding

Immobilization via ionic binding is based, mainly, on ionic binding of enzyme molecules or active molecule to solid supports containing ionic charges. In this method, the amount of enzyme bound to the carrier and the activity after immobilization depends on the nature of the carrier. How the enzyme is bound to the carrier. In some cases, physical adsorption may also take place. The main difference between ionic binding and physical adsorption is the strength of the interaction, which is much stronger for ionic binding, although less strong than covalent binding. The preparation of immobilized enzymes using ionic binding is based on the same procedure as described for physical adsorption.

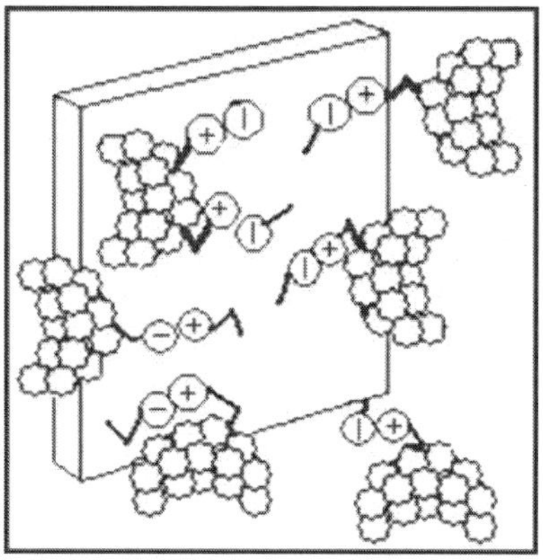

Fig. Biocatalysts Bound to a Carrier by Ionic Binding.

The ionic nature of the binding forces between the enzyme and the support also depends on pH variations, support charge, enzyme concentrations, and temperature. The supports used for ionic binding may be based on polysaccharide derivatives (*e.g.*, diethylamino-ethylcellulose, dextran, chitosan, carboxymethyl-cellulose), synthetic polymers (*e.g.*, polystyrene derivatives, polyethylene vinylalcohol), and inorganic materials (*e.g.*, ambertite, alumina, silicates, bentonite, sepiolite, silica gel, etc.). The immobilization by ionic binding has the advantage that changes in the enzyme conformation only occur in a small extent, resulting in immobilized enzymes with high enzymatic activities. The main disadvantage is the possible interference of other ions, and special attention should be paid in maintaining the correct ionic strength and pH conditions in order to prevent their easy detachment.

Covalent Binding by Chemical Coupling

The covalent binding method is based on the binding of enzymes, or other active molecules, to a support or matrix by means of covalent bonds. The bond is normally formed between a functional group present on the support surface and amino acid residues on the surface of the enzyme. Those which are most often involved in covalent binding are the amino (NH_2) group of lysine or arginine, carboxyl (CO_2H) group of aspartic acid or glutamic acid, hydroxyl (OH) group of serine or threonine, and sulphydryl (SH) group of cysteine. There are many reaction procedures for joining an enzyme to a material with a covalent bond (diazotation, amino bond, Schiff's base formation, amidation reactions, thiol-disulfide, peptide bond, and alkylation reactions). The connection between the support and the biocatalyst can be achieved either by direct linkage between the components or via an intercalated link of different length, the so-called spacer or harm. The advantage of using a spacer molecule is that it gives a greater degree of mobility to the coupled biocatalysts so that its activity can, under certain circumstances, be higher than if it is bound directly to the support. It is important to choose a method that will not involve the reaction with the amino acids present in the active site, since this could inactivate the enzyme. Basically, two steps are involved in the covalent binding of enzymes to a support material.

First, functional groups on the support material are activated by specific reagents (*e.g.*, cyanogen bromide, carbodiimide, aminoalkylethoxysilane, isothiocyanate, and epichlorohydrin, etc.). A large range of support materials is available for covalent binding, and this extensive range reflects the fact that no ideal matrix exists. Therefore, the advantages and disadvantages of a given matrix must be taken into account when considering the appropriate procedure for a given enzyme immobilization. Immobilization of enzymes through covalent attachment has also been demonstrated to induce higher resistance to temperature, denaturants, and organic solvents in several cases.

The extent of these improvements may depend on other conditions of the system, *e.g.*, the nature of the enzyme, type of support, and the method of immobilization. Many factors may influence the selection of the support, and some of the more important are its cost and availability, the binding capacity (amount of enzyme bound per given weight of matrix), hydrophilicity (the ability to incorporate water into the matrix and stability of matrix), structural rigidity, and durability during applications.

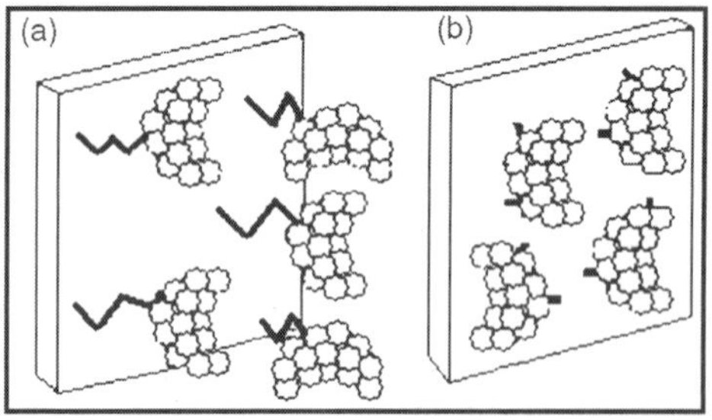

Fig. Covalent Bond Between the Biocatalysts and a Carrier with (a) and without Spacer (b).

Natural polymers, which are very hydrophilic, are popular support materials for enzyme immobilization since the residues in these polymers contain hydroxyl groups, which are ideal functional groups for participating in covalent bonds. A frequently encountered disadvantage of immobilization by covalent binding is that it places great stress on the enzyme. The necessary harshness of the immobilization procedure nearly always leads to considerable changes in conformation and a resultant loss of catalytic activity.

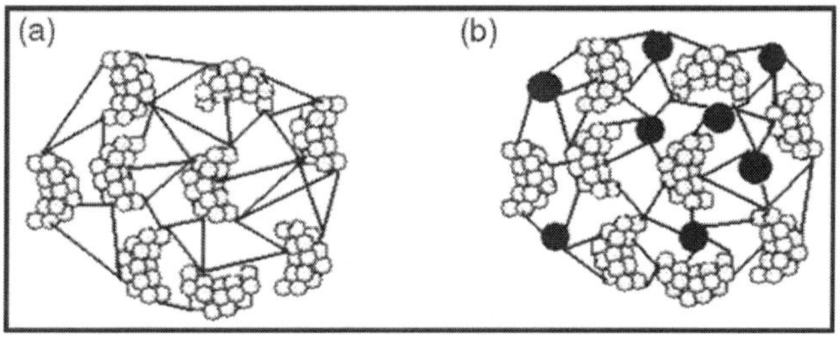

Fig. Biocatalysts Immobilized by Means of Crosslinking (a) and Co-Crosslinking with Inert Molecules Incorporated (b).

Crosslinking

The crosslinking method is based on the formation of covalent bonds between the enzyme or active molecules, by means of bi- or multifunctional reagents. The individual biocatalytic units (enzymes, organelles, whole cells) are joined to one another with the help of bi- or multifunctional reagents (*e.g.*, glutardialdehyde, glutaraldehyde, glyoxal, diisocyanates, hexamethylene diisocyanate, toluene diisocyanate, etc.). Enzyme crosslinking involves normally the amino groups of the lysine but, in occasional cases, the sulfhydryl groups of cysteine, phenolic OH groups of tyrosine, or the imidazol group of histidine can also be used for binding. How the biocatalysts can be linked by a simple crosslinking process and also by co-crosslinking, in which inert molecules are incorporated in the high-polymer network in order to improve the mechanical and enzymatic immobilized preparation. The advantages and disadvantages of a given matrix must be taken into account when considering the appropriate procedure for a given enzyme immobilization. One advantage is the simplicity of the process.

The main disadvantages are the fragility of the particles produced in some cases and diffusion limitations. Since crosslinking and co-crosslinking usually involve covalent bonds, immobilized biocatalysts in this way frequently undergo changes in the conformation with a resultant loss of activity. The isomerization of glucose process is a very important example of the industrial application using biocatalysts crosslinked with glutaraldehyde. Some of the immobilized preparations used in these large-scale processes are produced simply by glutaraldehyde treatment of bacterial cell masses that have formed fine particles.

BIOREACTORS FOR EXTRACORPOREAL ENZYME THERAPY

Extracorporeal shunts have been proposed for the treatment of several clinical conditions. The most likely applications for enzymatic treatment are the removal of urea during kidney failure, removal of toxins (*e.g.*, paracetamol) during liver failure, or the reduction of key metabolites from the circulation to treat cancer. Urease is one of the most important enzymes in biomedical applications. Urease is an enzyme that catalyzes the hydrolysis of urea to form ammonia and carbon dioxide. Urea is one of the main metabolic end products, and the removal of its excess has been a major problem for patients suffering from renal failure. Hence, its immobilization by entrapment has been investigated by many workers for applications in biosensors and as artificial kidneys.

The most attention has been given to the development of enzyme reactors, where the urea would be removed and the dialysis fluid prepared for further use. The use of this enzyme is often limited due to its high cost, availability in small amounts, instability, and the limited possibility of feasible recovery of these biocatalysts from a reaction mixture. Numerous synthetic and natural polymeric supports have been used for urease immobilization, and their uses in medical and technical fields are well reported. The covalent bond of urease in different supports has been reported in many studies. Some commonly used supports are chitosan-poly(glycidil methacrylate), carboxy-methylcellulose, polyurethane, sepharose-2B, polyacrylamide, ion exchange resins, copolymers of polyglycidylmethacrylate, calcium alginate beads, poly(vinyl alcohol) (PVA), hydroxyapatite, 2-dimethylaminoethylmethacrylate, poly(ethylene glycol dimethacry-late/2-hydroxy ethylene methacrylate) microbeads, poly(caprolactone)/starch, and poly(orthoesters).

L-asparaginase has been used for treating leukemias and disseminating cancers that require asparagines for growth, but this treatment presents several serious side effects. To overcome these problems, immobilized enzyme derivatives have been prepared on various supports for extracorporeal treatment. Blood can be passed over the immobilized enzyme, thus depleting the asparagine supply needed by the cancer cells. The enzyme does not come into direct contact with the organs to which it is toxic, and hypersensitivity reactions do not occur. With this type of treatment, however, the blood plasma must be first separated from the cells to minimize cell damage and then passed through a separate column containing the immobilized enzyme.

This process requires that the blood remain outside the body for relatively long periods of time, resulting in the denaturation and depletion of many plasma proteins. Some techniques have been developed to minimize this problem, such as the use of a porous hollow-fibre plasmapheresis device. With this system, the plasma can be separated from the whole blood and contact with immobilized enzyme in one passage, thus minimizing its time outside the body and then reducing the damage to the plasma proteins. Adsorption techniques have been used to immobilize asparaginase onto hollow fibres after first coating the fibres with albumin and then crosslinking the enzyme with glutaraldehyde. Maciel and Minim also reported that the use of *L*-asparaginase covalently attached to nylon tubing may constitute a useful system to be used in clinical applications. Bilirubin oxidase is also an example of enzyme used in extracorporeal applications. All human newborns accumulate bilirubin to levels greater that those in adults, and 20% accumulate enough to stain their skin, resulting in jaundice. Bilirubin binds to cellular and mitochondrial membranes, causing cell death in a variety of tissues. Clinically, bilirubin toxicity may lead to mental retardation, cerebral palsy, deafness, seizures, or death. The most common treatments for jaundiced infants are phototherapy and exchange transfusion.

This technique presents serious problems such as hypoglycemia, hypocalcemia, acidosis, transmission of infectious, etc. Lavin *et al.* reported the use of a highly specific enzyme to remove bilirubin from the bloodstream using a small reactor (extracorporeal circuit) containing bilirubin oxidase covalently immobilized in agarose beads. These researchers obtained good results for the removal of the bilirubin in humans and in genetically jaundiced rats. Heparinase, an enzyme that degrades heparin into small polysaccharides, has also been immobilized into an extracorporeal device (artificial kidney bioreactor) to eliminate the anticoagulant properties of heparin (used to prevent clotting in the device) before the blood returns to the patient.

Diagnostic Assays and Biosensors

Isolated or combined enzymes are being used in medicine as useful tools for clinical analysis. About 50 different enzymes are used in different aspects of clinical diagnoses, and for most of these, much higher levels of purity are required than for most industrial enzymes. Two of the major enzymes used are peroxidase from horseradish and alkaline phosphatase from beef intestinal mucosa, both being required for immunoassays. The enzymes may be used in test strips, ELISA, biosensors, and autoanalysers. The serum uric acid concentration is an important index for clinical diagnosis of gout, leukemia, toxemia of pregnancy, and severe renal impairment.

A number of enzymes are assayed in serum and urine for diagnostic purposes; the more frequently used ones are discussed below. Alkaline phosphatases (ALP) are a group of enzymes found primarily in the liver (isoenzyme ALP-1) and bone (isoenzyme ALP-2). There are also small amounts produced by cells lining the intestines (isoenzyme ALP-3), the placenta, and the kidney (in the proximal convoluted tubules). What is measured in the blood is the total amount of alkaline phosphatase released from these tissues into the blood. As the name implies, this enzyme works best at an alkaline pH (pH 10), and thus the enzyme itself is inactive in the blood. Alkaline phosphatase acts by splitting off phosphorus (an acidic mineral), creating an alkaline pH.

The primary importance of measuring alkaline phosphatase is to check the possibility of bone or liver diseases. Another application of immobilized enzymes is the development of improved sensing devices. Because of their high specificity for given substances, enzymes and monoclonal antibodies are particularly suitable for use as sensors. The membrane-covered electrode described by Clark in 1959 is the dominating sensor for the measurement of dissolved oxygen. Numerous modifications of the original concept have been developed. For instance, biosensors using enzymes have been used to detect the presence of various organic compounds, and recent developments have proven to be both rapid and highly selective. They have been used in important applications such as in clinical laboratories, fermentation processes, and pollution monitoring. Most of them have used a free or immobilized enzyme and an ion-sensitive electrode that measures indirectly (*e.g.*, by temperature or colour changes produced by an enzymatic reaction) the presence of a product whose formation is catalyzed by the enzyme.

The biosensors usually have immobilized biological molecules attached to the surface of a transducer that allows an electronic or optical signal to be converted into an appropriate signal. This type of biosensor could be used to measure glucose, sucrose, lactose, *L*-lactate, galactose, *L*-glutamate, *L*-glutamine, choline, ethanol, methanol, hydrogen peroxide, starch, uric acid, etc., by using specific enzymes. Glucose oxidase is normally used to assay glucose concentration. Glucose sensors are the biosensors that have attracted much interest in both research and applications fields. One particularly important medical application of improved biosensors could be in the treatment of diabetic patients for whom proper levels of insulin and glucose must be maintained. For instance, small implantable devices for sampling blood to determine the levels of glucose and regulate the delivery of insulin could be developed using this enzyme.

A great variety of immobilization methods (*e.g.*, encapsulation, entrapment) and transducers have been developed to construct the glucose sensors with better performance and practicability since the work of Clark and Lyons in 1962. Several materials have been used, such as polyethylene terephthalate (PET), polyacrylamide, *N*- isopolyacrylamide, sol-gel, poly(2-hydroxyethyl methacrylate), alginate, artificial resins, glass, etc. Immobilized alkaline phosphatase on a nanoporous nickel-titanium film for sensor applications. A number of other enzymes have been described with great potential for medical applications, including carboxypeptidases, collagenase, fibrinolysin, pepsin, streptokinase, subtilisin, thrombin, tissue plasminogen activator, amylase, galactosidase, glucoamylase, lactase (-galactosidase), pectinase, pancreatin, phospholipases, cholesterol esterase and other DNases, RNases, phosphatases, esterases, sulfatases, isomerases, glucose isomerase, superoxide dismutase, cholesterol esterease, creatine kinase, and penicillin acylase.

Advantages and Disadvantages of Immobilized Enzymes

The use of immobilized enzymes normally offers several advantages over free enzymes, such as increased stability, localization, and retention of the molecules at the material surface, which enables easier handling, repeated use, and decreased cost. Other important advantages of using therapeutic immobilized enzymes are the prolonged blood circulation lifetime without the loss of specific activity and the lower immunogenicity. This advantage is particularly important for delivering enzymes or other biomolecules and may constitute an alternative and suitable method for the enzyme replacement therapy. However, some limitations have been attributed to the use of immobilized enzymes in biomedical applications, such as mass transfer resistances (substrate in and product out), adverse biological responses of enzyme support surfaces (*in vivo* or *ex vivo*), fouling by other biomolecules, greater potential for product inhibition, and sterilization difficulties.

Although the preparation of sterile immobilized enzyme systems may be complex, sterilization may be achieved by filtrating all the reagents and protein solutions through 0.2m filters and working under aseptic conditions. A very important issue regarding the use of enzymes or other products derived from biological or biotechnological processes in medical applications is to ensure that these therapeutic products do not contain any pyrogenic material, toxins, or infectious agents able to cause harmful effects. For that, it is necessary to perform a complete examination of the products to test their safety in terms of local tolerance, toxicity, carcinogenicity, and immunogenicity, among other pharmacological safety tests. Taking into account the diversity in the range of products and the uncertainty about the regulatory status of some of them, it is necessary to design safety evaluation programmes to provide useful information to the responsible of clinical trials and to ensure patient safety.

IMMOBILIZED ENZYMES IN BIOPROCESS

Immobilization of biocatalysts helps in their economic reuse and in the development of continuous bioprocesses. Biocatalysts can be immobilized either using the isolated enzymes or the whole cells. Immobilization often stabilizes structure of the enzymes, thereby allowing their applications even under harsh environmental conditions of pH, temperature and organic solvents, and thus enable their uses at high temperatures in nonaqueous enzymology, and in the fabrication of biosensor probes. In the future, development of techniques for the immobilization of multienzymes along with cofactor regeneration and retention system can be gainfully exploited in developing biochemical processes involving complex chemical conversions. The present review outlines some of the above aspects, and delineates the present status and future potentials of immobilized enzymes and nonviable cells in the emerging biotech industries.

Biotechnology is currently considered as a useful alternative to conventional process technology in industrial and analytical fields. This is mainly because, unlike the chemical catalysts, the biological systems have the advantages of accomplishing complex chemical conversions under mild environmental conditions with high specificity and efficiency. Biological systems help in ingredient substitution, processing aid substitution, more efficient processing, less undesirable products, increased plant capacity, increased product yields, and improved or unique products. The variety of chemical transformations catalysed by enzymes has made these catalysts a prime target of exploitation by the emerging biotech industries.

Despite these advantages, the use of enzymes in industrial applications has been limited by several factors, mainly the high cost of the enzymes, their instability, and availability in small amounts. Also the enzymes are soluble in aqueous media and it is difficult and expensive to recover them from reactor effluents at the end of the catalytic process. This restricts the use of soluble enzymes to batch operations, followed by disposal of the spent enzyme-containing solvent. Over the last few decades, intense research in the area of enzyme technology has provided many approaches that facilitate their practical applications. Among them, the newer technological developments in the field of immobilized biocatalysts can offer the possibility of a wider and more economical exploitation of biocatalysts in industry, waste treatment, medicine, and in the development of bioprocess monitoring devices like the biosensor.

Immobilization means associating the biocatalysts with an insoluble matrix, so that it can be retained in a proper reactor geometry for its economic reuse under stabilized conditions. Immobilization thus allows, by essence, to decouple the enzyme location from the flow of the liquid carrying the reagents and products. Immobilization helps in the development of continuous processes allowing more economic organization of the operations, automation, decrease of labour, and investment/capacity ratio. Immobilized biocatalysts offer several other advantages, notable among them is the availability of the product in greater purity. Purity of the product is very crucial in food processing and pharmaceutical industry since contamination could cause serious toxicological, sensory, or immunological problems. The other major advantages include greater control over enzymatic reaction as well as high volumetric productivity with lower residence time, which are of great significance in the food industry, specially in the treatment of perishable commodities as well as in other applications involving labile substrates, intermediates or products·

IMMOBILIZED NONVIABLE CELLS AS AN ECONOMICAL SOURCE OF ENZYMES

Biocatalysts can be immobilized using either the isolated enzymes or the whole cells or cellular organelles. Immobilization of whole cells has been shown to be a better alternative to immobilization of isolated enzymes· Doing so avoids the lengthy and expensive operations of enzyme purification, preserves the enzyme in its natural environment thus protecting it from inactivation either during immobilization or its subsequent use in continuous system. It may also provide a multipurpose catalyst, specially when the process requires the participation of number of enzymes in sequence. The major limitations which may need to be addressed while using such cells are the diffusion of substrate and products through the cell wall, and unwanted side reactions due to the presence of other enzymes. The cells can be immobilized either in a viable or a nonviable form. Immobilized nonviable cell preparations, which are normally obtained by permeabilizing the intact cells, for the expression of intracellular activity are useful for simple processes that require single-enzyme with no requirement for cofactor regeneration, like hydrolysis of sucrose or lactose· On the other hand immobilized viable cells, which serve as 'controlled catalytic biomass', have opened new avenues for continuous fermentation on heterogeneous catalysis basis by serving as self-proliferating biocatalysts· The present review will focus mainly on immobilized isolated enzyme systems, and the use of immobilized nonviable cells as a source of enzymes. For the aspects on the immobilized viable cell systems, the readers can refer to another article in this issue. Some of the aspects on the applications of immobilized viable cells in biochemical process development and monitoring have also been reviewed recently·

Most of the enzymes used at industrial scale are normally the extracellular enzymes produced by the microbes. This has been mainly due to their ease of isolation as crude enzymes from the fermentation broth. Moreover, the extracellular enzymes are more stable to external environmental perturbations compared to the intracellular enzymes. However, over 90% of the enzymes produced by a cell are intracellular. The economic exploitation of these, having a variety of biochemical potentials, has been limited in view of the high cost involved in their isolation. Also, compared to extracellular enzymes, the intracellular enzymes are more labile. Delicate and expensive separation methods are required to release the enzymes undamaged from the cell, and to isolate them. This increases the labour and the cost of the enzyme. These problems could now be obviated by the use of permeabilized cells as a source of enzyme.

Permeabilization of the cells removes the barrier for the free diffusion of the substrate/product across the cell membrane, and also empties the cell of most of the small molecular weight cofactors, etc., thus minimizing the unwanted side reactions. Side reactions, which can occur due to the presence of other enzymes in a cell, can also be minimized by inactivating such enzymes prior to or after immobilization· Such permeabilized cells, which are often referred to as nonviable or nongrowing cells, can be exploited in an immobilized form as a very economical source of intracellular enzyme for simple bioconversions like hydrolysis, isomerization and oxidation reactions that do not need a cofactor-regeneration system· The decision to immobilize cells either in a viable or nonviable form is very important and depends on their ultimate application. Thus the permeabilized *K. fragilis* cells, which are nonviable, convert lactose in milk to glucose and galactose, whereas the viable (nonpermeabilized) cells convert the lactose to ethanol and are useful in the complete desugaration of milk· A variety of physical, chemical, and enzymatic techniques have been developed for the permeabilization of cells· Some techniques like entrapment in polyacrylamide itself have been shown to result in permeabilization of the cells· A few typical examples of immobilized enzyme preparation obtained using permeabilized cells.

TECHNIQUES AND SUPPORTS FOR IMMOBILIZATION

A large number of techniques and supports are now available for the immobilization of enzymes or cells on a variety of natural and synthetic supports. The choice of the support as well as the technique depends on the nature of the enzyme, nature of the substrate and its ultimate application. Therefore, it will not be possible to suggest any universal means of immobilization. It can only be said that the search must continue for matrices which provide facile, secure immobilization with good interaction with substrates, and which conform in shape, size, density and so on to the use for which they are intended. Care has to be taken to select the support materials as well as the reagents used for immobilization, which have GRAS status, particularly when their ultimate applications are in the food processing and pharmaceutical industries.

Macromolecular, colloidal, viscous, sticky, dense or particulate food constituents or waste streams also limit the choice of reactor and support geometries. Commercial success has been achieved when support materials have been chosen for their flow properties, low cost, nontoxicity, maximum biocatalysts loading while retaining desirable flow characteristics, operational durability, ease of availability, and ease of immobilization· The variety of techniques and supports investigated for the immobilization have been reviewed in a number of articles and books· The objective of the preceding part of the paper is to review briefly, in the light of current developments, some approaches that have been used for the immobilization of biocatalysts with reference to the studies carried out in the author's laboratory as well as few others.

Techniques for immobilization have been broadly classified into four categories, namely entrapment, covalent binding, cross-linking and adsorption. A combination of one or more of these techniques has also been investigated. It must be emphasized that in terms of economy of a process, both the activity and the operational stability of the biocatalysts are important. They determine its productivity, which is the activity integrated over the operational time.

Entrapment

Entrapment has been extensively used for the immobilization of cells, but not for enzymes. The major limitation of this technique for the immobilization of enzymes is the possible slow leakage during continuous use in view of the small molecular size compared to the cells. Biocatalysts have been entrapped in natural polymers like agar, agarose and gelatine through thermoreversal polymerization, but in alginate and carrageenan by ionotropic gelation· A number of synthetic polymers have also been investigated. Notable among them are the photo-crosslinkable resins, polyurethane prepolymers, and acrylic polymers like polyacrylamide·

Among these, the most widespread matrix made from monomeric precursors is the polyacrylamide gel. Polyacrylamide may not be a useful support for use in food industry in view of its toxicity, but can have potentials in the treatment of waste and in the fabrication of analytical devices containing biocatalysts. One of the major limitations of entrapment technique is the diffusional limitation as well as the steric hindrance, especially when the macromolecular substrates like starch and proteins are used. Diffusional problems can be minimized by entrapment in fine fibres of cellulose acetate or other synthetic materials or by using an open pore matrix· Recently, the development of so-called hydrogels and thermoreactive water-soluble polymers, like the albumin-poly (ethylene glycol) hydrogel, have attracted attention in the field of biotechnology· In the area of health care, they offer new avenues for enzyme immobilization. Such gels with a water content of about 96% provide a microenvironment for the immobilized enzyme close to that of the soluble enzyme with minimal diffusional restrictions·

COVALENT BINDING

Covalent binding is an extensively used technique for the immobilization of enzymes, though it is not a good technique for the immobilization of cells.

The functional groups extensively investigated are the amino, carboxyl, and the phenolic group of tyrosine· Enzymes are covalently linked to the support through the functional groups in the enzymes, which are not essential for the catalytic activity. It is often advisable to carry out the immobilization in the presence of its substrate or a competitive inhibitor so as to protect the active site. The covalent binding should also be optimized so as not to alter its conformational flexibility. Some of these problems however, can be obviated by covalent bonding through the carbohydrate moiety when a glycoprotein is concerned. A number of industrially useful enzymes are glycoproteins wherein the carbohydrate moiety may not be essential for its activity. In general, functional aldehyde group can be introduced in a glycoprotein by oxidizing the carbohydrate moiety by periodate oxidation without significantly affecting the enzyme activity. The enzyme could then be covalently linked to a support containing an alky amine group through Schiffs base reaction. Enzymes like glucose oxidase, peroxidase, invertase, etc. have been immobilized using this technique·

Covalent binding has been extensively investigated using inorganic supports. Enzymes covalently bound to inorganic supports have been used in the industry· Enzymes have also been bound to synthetic membranes, thus integrating biconversion and downstream processing· Large-scale processes using such an approach have been demonstrated for the preparation of invert sugar using invertase·

Cross-linking

Biocatalysts can also be immobilized through chemical cross-linking using homo- as well as heterobifunctional cross-linking agents· Among these, glutaraldehyde which interacts with the amino groups through a base reaction has been extensively used in view of its GRAS status, low cost, high efficiency, and stability· The enzymes or the cells have been normally cross-linked in the presence of an inert protein like gelatine, albumin, and collagen· Studies from our laboratory have shown the use of raw hen egg white as an economic, easily available novel proteinic support rich in lysozyme for the immobilization of enzyme or nonviable cells either in a powder, bead or highly porous foam form, using glutaraldehyde as the cross-linker.

The unique feature of this support is the large concentration of lysozyme naturally present in hen egg white which gets co-immobilized, thus imparting the bacteriolytic property to the support· Adsorption followed by cross-linking has also been used for the immobilization of enzymes. The technique of cross-linking in the presence of an inert protein can be applied to either enzymes or cells. The technique can also be used for the immobilization of enzymes by cross-linking the cell homogenates· Osmotic stabilization of cellular organelles or halophilic cells prior to immobilization using cross-linkers has also shown promise.

Adsorption

This is perhaps the simplest of all the techniques and one which does not grossly alter the activity of the bound enzyme. In case of enzymes immobilized through ionic interactions, adsorption and desorption of the enzyme depends on the basicity of the ion exchanger. Moreover, a dynamic equilibrium is normally observed between the adsorbed enzyme and the support which is often affected by pH as well as the ionic strength of the surrounding medium.

This property of reversibility of binding has often been used for the economic recovery of the support. This has been successfully adapted in industry for the resolution of racemic mixtures of amino acids, using amino acid acylase· A variety of commercially available ion exchangers have been investigated for this purpose· One of the techniques, which has gained importance more recently, is the use of polyethylenimine for imparting polycationic characteristics to many of the neutral supports based on cellulose or inorganic materials· Enzymes with low pI, like invertase' urease' glucose oxidase' catalase' and other enzymes have been bound through adsorption followed by cross-linking on polyethylenimine-coated supports.

Immobilization of enzymes through hydrophobic interaction has also shown promise· One of the important features of this technique, which is of great significance, is that, unlike ionic binding, hydrophobic interactions are usually stabilized by high ionic concentrations, thus enabling the use of high concentrations of substrates as desired in an industrial process without the fear of desorption. Other types of strong interactive binding techniques have also been reported for the reversible immobilization of enzymes.

A typical example is the immobilization of soybean b -amylase on phenyl-boronate agarose, which can be reversed for the recovery of the support using sorbitol· Varieties of biospecific interactions have also been investigated for the reversible immobilization of enzymes by adsorption. Enzymes like acetyl choline esterase, ascorbic acid oxidase, invertase, peroxidase, glucose oxidase, etc. have been immobilized by biospecific-reversible immobilization on lectin-bound supports and invertase, using polyclonal antiinvertase antibodies·

Traditional enzyme immobilization procedures involve isolation of the enzyme, followed by use of several steps for the immobilization. Costs of enzyme purification, the immobilization procedure, bioreactor operational stability, and bioreactor regeneration are the major factors that determine the cost of a bioreactor process. Development of techniques for the simultaneous isolation and immobilization of enzymes from crude extracts has obviated these problems. Some typical examples include the immobilization of a streptavidin-b -galactosidase fusion protein expressed in *Escherichia coli* and bioselectively adsorbed from a crude cell lysate to biotin which is covalently immobilized on controlled pore glass' and the simultaneous purification and immobilization of D-amino acid oxidase from *Trigonopsis variabilis* cell lysate adsorbed on phenyl sepharose· The technique can have future potentials especially in the downstream processing and immobilization of enzyme/ proteins obtained by recombinant DNA technology.

Techniques for the adhesion of whole cells on polymeric surfaces are also currently gaining considerable importance. The major advantage for the cells immobilized through adhesion is reduction or elimination of the mass transfer problems associated with the commonly used gel entrapment method. The technique of immobilization usually being followed is the microbial colonization by recycling of the cell suspension along with nutrients, such that a biofilm is gradually formed. This often results in the immobilization of cells in a viable form for use in heterogeneous fermentations.

Useful techniques have been developed also for the immobilization of nonviable cells to be used as an enzyme source for simple chemical conversions. Notable among them include treating the cells or the support with trivalent metal ions like Al or Fe^{3+} or charged colloidal particles, and use of polycationic polymers like chitosan. Novel techniques have been developed to adhere cells strongly on a variety of polymeric surfaces including glass, cotton cloth, cotton threads, and other synthetic and inorganic surfaces using polyethylenimine. This technique has also been recently used for the simultaneous filtration and immobilization of cells from a flowing suspension, thus integrating downstream processing with bioprocessing. This technique may have future potentials for the immobilization of cells for food applications. Polyethylenimine is nontoxic and the United States of America Food and Drug Administration has permitted its use as a direct food additive under the Food Drug and Cosmetic Act.

7

Immune System

The immune system must maintain a delicate balance, with potent defensive responses capable of destroying large numbers of foreign cells and viruses while refraining from undue destruction of the host's body. When the immune system cannot mount a sufficient defense of the host, there is an immune deficiency; this is seen in HIV infection and SCID. If, on the other hand, the immune system acts too vigorously and begins to attack the host, we have autoimmunity. This is a defiance of the integral immune system property of self/nonself recognition. That is, the immune system begins attacking or forming antibodies against the host's own body tissues. Examples of autoimmune diseases include Graves' disease, Hashimoto's thyroiditis, *myaesthenia gravis* and type I diabetes mellitus. The human immune system recognizes non-self entities and mounts an effector response to neutralize the organism. A faster and stronger memory response may occur upon later exposure. The memory response has been used throughout history to confer immunity upon several populations, even previous to our understanding of the physiological basis for such a response. Thucydides wrote in his History of the Peloponnesian War that persons who had been exposed to plague previously could care for the sick without danger. In the 19th century, variolation was commonplace; this was the removal of smallpox (variola virus) skin pustules which were subsequently put into small cuts in the skin of healthy people. This was itself a crude form of vaccination, with the crusty dry pustules acting as an incubator of attenuated virus.

Edward Jenner would later use the cowpox virus to vaccinate (from *vacca*, Latin for "cow") patients against smallpox, and Louis Pasteur attenuated rabies and injected it into a small boy, naming this substance a vaccine in honor of Jenner's earlier studies in the science of immunology.

As immunology progressed, many people began to question how these vaccines worked. Why should exposure to plague in Thucycides' time confer protection only against plague and not all disease? Why should cowpox, a similar disease to smallpox but clearly a less severe virus, give milk maids sufficient immunity to resist full smallpox infection? In short, what has caused this memory response to be relatively (yet not absolutely) specific as well as selective?

THE BASICS

First, some vocabulary:

- Serum—liquid, noncellular component of blood after coagulation has occurred (and thus devoid of clotting factors).
- Immunoglobulin—a serum fraction (aka gamma globulin) that has antitoxin, precipitin, and agglutin factors (abbreviation: Ig)
- Antigen—"Antibody Generator"; a foreign organism or molecule that generates a humoral immune response, causing the release of antibodies (abbreviation: Ag)
- Epitope—the molecular sidechain of an antigen that each antibody attaches to; there can be many epitopes on a single antigen
- Antibody—refined, Y-shaped proteins that make up the immunoglobulin fraction of serum; antibodies are specific to certain foreign bodies; antibodies can be membrane-bound or free in the serum (abbreviation: Ab)

Note: Often, antibody and immunoglobulin are used interchangably.

There have been several competing theories of the immune response mechanism. The instructional theory of antigen interaction postulated that the antigen itself caused antibodies to fold around the antigen in a certain way; this theory was later disproven. The selection theory states that the body creates many different sidechains on antibodies, and the antigen "selects" the correct antibody; in other words, the body creates every possible permutation of chemical sidechain-specific antibodies, and when an antigen enters the body, it is matched up with antibodies that correspond to its epitopes. The current theory of immune response is known as the clonal selection theory, which states that an individual lymphocyte (specifically, a B cell) expresses receptors specific to the distinct antigen, determined before the antibody ever encounters the antigen. Binding of Ag to a cell activates the cell, causing a proliferation of clone daughter cells.

OVERVIEW OF THE IMMUNE SYSTEM

We are constantly being exposed to infectious agents and yet, in most cases, we are able to resist these infections. It is our immune system that enables us to resist infections. The immune system is composed of two major subdivisions, the innate or non-specific immune system and the adaptive or specific immune system. The innate immune system is our first line of defence against invading organisms while the adaptive immune system acts as a second line of defence and also affords protection against re-exposure to the same pathogen. Each of the major subdivisions of the immune system has both cellular and humoral components by which they carry out their protective function. In addition, the innate immune system also has anatomical features that function as barriers to infection. Although these two arms of the immune system have distinct functions, there is interplay between these systems (i.e., components of the innate immune system influence the adaptive immune system and vice versa).

Although the innate and adaptive immune systems both function to protect against invading organisms, they differ in a number of ways. The adaptive immune system requires some time to react to an invading organism, whereas the innate immune system includes defenses that, for the most part, are constitutively present and ready to be mobilized upon infection. Second, the adaptive immune system is antigen specific and reacts only with the organism that induced the response. In contrast, the innate system is not antigen specific and reacts equally well to a variety of organisms. Finally, the adaptive immune system demonstrates immunological memory. It "remembers" that it has encountered an invading organism and reacts more rapidly on subsequent exposure to the same organism. In contrast, the innate immune system does not demonstrate immunological memory.

All cells of the immune system have their origin in the bone marrow and they include myeloid (neutrophils, basophils, eosinpophils, macrophages and dendritic cells) and lymphoid (B lymphocyte, T lymphocyte and Natural Killer) cells, which differentiate along distinct pathways. The myeloid progenitor (stem) cell in the bone marrow gives rise to erythrocytes, platelets, neutrophils, monocytes/macrophages and dendritic cells whereas the lymphoid progenitor (stem) cell gives rise to the NK, T cells and B cells. For T cell development the precursor T cells must migrate to the thymus where they undergo differentiation into two distinct types of T cells, the CD4+ T helper cell and the CD8+ pre-cytotoxic T cell. Two types of T helper cells are produced in the thymus the TH1 cells, which help the CD8+ pre-cytotoxic cells to differentiate into cytotoxic T cells, and TH2 cells, which help B cells, differentiate into plasma cells, which secrete antibodies.

The main function of the immune system is self/non-self discrimination. This ability to distinguish between self and non-self is necessary to protect the organism from invading pathogens and to eliminate modified or altered cells (e.g. malignant cells). Since pathogens may replicate intracellularly (viruses and some bacteria and parasites) or extracellularly (most bacteria, fungi and parasites), different components of the immune system have evolved to protect against these different types of pathogens. It is important to remember that infection with an organism does not necessarily mean diseases, since the immune system in most cases will be able to eliminate the infection before disease occurs. Disease occurs only when the bolus of infection is high, when the virulence of the invading organism is great or when immunity is compromised. Although the immune system, for the most part, has beneficial effects, there can be detrimental effects as well. During inflammation, which is the response to an invading organism, there may be local discomfort and collateral damage to healthy tissue as a result of the toxic products produced by the immune response. In addition, in some cases the immune response can be directed toward self tissues resulting in autoimmune disease.

HISTOLOGICAL EXAMINATION OF THE IMMUNE SYSTEM

Even before the concept of immunity (from *immunis*, Latin for "exempt") was developed, numerous early physicians characterized organs that would later prove to be part of the immune system. The key primary lymphoid organs of the immune system are like thymus and bone marrow, and secondary lymphatic tissues such as spleen, tonsils, lymph vessels, lymph nodes, adenoids, and skin and liver. When health conditions warrant, immune system organs including the thymus, spleen, portions of bone marrow, lymph nodes and secondary lymphatic tissues can be surgically excised for examination while patients are still alive. Many components of the immune system are actually cellular in nature and not associated with any specific organ but rather are embedded or circulating in various tissues located throughout the body.

Classical Immunology

Classical immunology ties in with the fields of epidemiology and medicine. It studies the relationship between the body systems, pathogens, and immunity. The earliest written mention of immunity can be traced back to the plague of Athens in 430 BCE. Thucydides noted that people who had recovered from a previous bout of the disease could nurse the sick without contracting the illness a second time. Many other ancient societies have references to this phenomenon, but it was not until the 19th and 20th centuries before the concept developed into scientific theory. The study of the molecular and cellular components that comprise the immune system, including their function and interaction, is the central science of immunology. The immune system has been divided into a more primitive innate immune system, and acquired or adaptive immune system of vertebrates, the latter of which is further divided into humoral and cellular components. The humoral (antibody) response is defined as the interaction between antibodies and antigens.

Antibodies are specific proteins released from a certain class of immune cells (B lymphocytes). Antigens are defined as anything that elicits generation of antibodies, hence they are Antibody Generators. Immunology itself rests on an understanding of the properties of these two biological entities. However, equally important is the cellular response, which can not only kill infected cells in its own right, but is also crucial in controlling the antibody response. Put simply, both systems are highly interdependent. In the 21st century, immunology has broadened its horizons with much research being performed in the more specialized niches of immunology. This includes the immunological function of cells, organs and systems not normally associated with the immune system, as well as the function of the immune system outside classical models of immunity (Yemeserach 2010).

Clinical Immunology

Clinical immunology is the study of diseases caused by disorders of the immune system (failure, aberrant action, and malignant growth of the cellular elements of the system). It also involves diseases of other systems, where immune reactions play a part in the pathology and clinical features. The diseases caused by disorders of the immune system fall into two broad categories: immunodeficiency, in which parts of the immune system fail to provide an adequate response (examples include chronic granulomatous disease), and auto-immunity, in which the immune system attacks its own host's body (examples include systemic lupus erythematosus, rheumatoid arthritis, Hashimoto's disease and myasthenia gravis).

Other immune system disorders include different hypersensitivities, in which the system responds inappropriately to harmless compounds (asthma and other allergies) or responds too intensely. The most well-known disease that affects the immune system itself is AIDS, caused by HIV. AIDS is an immunodeficiency characterized by the lack of CD4+ ("helper") T cells and macrophages, which are destroyed by HIV. Clinical immunologists also study ways to prevent transplant rejection, in which the immune system attempts to destroy allografts.

Developmental Immunology

The body's capability to react to antigen depends on a person's age, antigen type, maternal factors and the area where the antigen is presented. Neonates are said to be in a state of physiological immunodeficiency, because both their innate and adaptive immunological responses are greatly suppressed. Once born, a child's immune system responds favourably to protein antigens while not as well to glycoproteins and polysaccharies. In fact, many of the infections acquired by neonates are caused by low virulence organisms like Staphylococcus and Pseudomonas. In neonates, opsonic activity and the ability to activate the complement cascade is very limited.

For example, the mean level of C3 in a newborn is approximately 65% of that found in the adult. Phagocytic activity is also greatly impaired in newborns. This is due to lower opsonic activity, as well as diminished up-regulation of integrin and selectin receptors, which limit the ability of neutrophils to interact with adhesion molecules in the endothelium. Their monocytes are slow and have a reduced ATP production, which also limits the newborns phagocitic activity. Although, the number of total lymphocytes is significantly higher than in adults, the cellular and humoral immunity is also impaired. Antigen presenting cells in newborns have a reduced capability to activate T cells. Also, T cells of a newborn proliferate poorly and produce very small amounts of cytokines like IL-2, IL-4, IL-5, IL-12, and IFN-g which limits their capacity to activate the humoral response as well as the phagocitic activity of macrophage. B cells develop early in gestation but are not fully active.

Maternal factors also play a role in the body's immune response. At birth most of the immunoglobulin is present is maternal IgG. Because IgM, IgD, IgE and IgA don't cross the placenta, they are almost undetectable at birth. Although some IgA is provided in breast milk. These passively acquired antibodies can protect the newborn up to 18 months, but their response is usually short-live and of low affinity. These antibodies can also produce a negative response.

If a child is exposed to the antibody for a particular antigen before being exposed to the antigen itself then the child will produce a dampened response. Passively acquired maternal antibodies can suppress the antibody response to active immunization. Similarly the response of T-cells to vaccination differs in children compared to adults, and vaccines that induce Th1 responses in adults do not readily elicit these same responses in neonates. By 6-9 months after birth, a child's immune system begins to respond more strongly to glycoproteins. Not until 12-24 months of age is there a marked improvement in the body's response to polysaccharides. This can be the reason for the specific time frames found in vaccination schedules.

During adolescence the human body undergoes several physical, physiological and immunological changes. These changes are started and mediated by different hormones. Depending on the sex either testosterone or 17-β-oestradiol, act on male and female bodies accordingly, start acting at ages of 12 and 10 years. There is evidence that these steroids act directly not only on the primary and secondary sexual characteristics, but also have an effect on the development and regulation of the immune system. There is an increased risk in developing autoimmunity for pubescent and post pubescent females and males.

There is also some evidence that cell surface receptors on B cells and macrophages may detect sex hormones in the system. The female sex hormone 17-β-oestradiol has been shown to regulate the level of immunological response. Similarly, some male androgens, like testosterone, seem to suppress the stress response to infection; but other androgens like DHEA have the opposite effect, as it increases the immune response instead of down playing it. As in females, the male sex hormones seem to have more control of the immune system during puberty and the time right after than in fully developed adults. Other than hormonal changes physical changes like the involution of the Thymus during puberty will also affect the immunological response of the subject or patient.

INNATE IMMUNITY

Innate immunity is basic and nonspecific. It includes:

- Phagocytic cells (macrophages, neutrophils; more generally, antigen-presenting cells (APCs))
- Barriers (e.g. skin)
- Antimicrobial compounds
- Inflammation

PHAGOCYTIC CELLS

Certain cells "eat up" foreign invaders; this is termed phagocytosis. Many of these cells are known as Antigen-Presenting Cells (APCs) because they break apart the ingested pathogen and display certain epitopes of the antigen on their surface. In this way, they localize the presentation of antigen, forming a vital link between the innate response and the adaptive response. Lymphocytes (such as B cells and T-helper cells) will have antigen presented to them, initiating the adaptive response. Monocyte-derived cells are common APCs, and they include tissue macrophages and monocytes within the blood; neutrophils are also APCs.

BARRIERS

The skin contains epidermis and dermis. The epidermis contains tightly packed epithelial (cytokeratin positive) cells with keratin waterproofing. The dermis contains connective tissue blood vessels hair follicles, sweat glands, and sebaceous glands. Sebaceous glands secrete sebum; this contains fatty acids and lactic acid, lowering skin pH to 3-5.

Mucous membranes contain normal flora, mucus, and cilia. Mucous membranes are found in the nose, eyes, mouth, urogenital, and anal regions of the body. Flora refers to bacteria that inhabit the human body in a relative steady-state; the gastro-intestinal tract contains a large number of these bacteria, and different areas of the world contain different flora. It is for this reason that people contract Traveler's Diarrhea; in short, flora from one region of the world are more dangerous because the body has not acclimated to their presence. Mucus contains certain mucin proteins, inorganic salts, and water; it is secreted from goblet cells. Cilia acts to sway back and forth during two phases, known as the power stroke and the recovery stroke; this allows mucus to be swept out of the body, either proximodistally from the lungs up the respiratory tract, or down the GI tract through the intestines, culminating in defecation. The enteric nervous system causes contractions of the gut, moving foodstuffs, waste, and bacteria/toxins down the digestive tract. This is just one example of disparate systems of the body working together with the immune system proper. Other examples are skeletal muscles, which squeeze blood along the veins and lymph along the lymph vessels, or the nervous system, which supports a rise in body temperature in response to infection.

Some bacteria can attach to mucous membranes via fimbriae or pili, which attach to special glyoproteins/glycolipids on the epithelial cells of mucous membranes.

ANTIMICROBIAL COMPOUNDS

Several antimicrobial compounds mediate the innate response.

- Lysozyme—hydrolytic enzyme in tears and mucous; cleaves peptidoglycan in bacterial cell walls
- Interferon—produced by virus-infected cells; binds to nearby cells (it is a paracrine factor), inducing a generalized antiviral state
- Complement—inactive circulating serum proteins that act on pathogen cell membranes
- Collectins—surfactant materials; kills bacteria by disrupting their lipid membranes or agglutinating them
- Toll-Like Receptors—(TLRs) membrane-bound receptors that react via pattern recognition to certain classes of molecules; TLR4, for example, recognizes lipopolysaccharide (LPS) on gram-negative bacteria

INFLAMMATION

Classical biology put several characteristics together as an inflammatory response. These included redness (erythema, *rubor*), heat (*calor*), swelling (edema, *tumor*), pain (*dolor*), and loss of function (*functio laesa*). The physiological processes that bring about these symptoms are central to the innate immune response.

Erythema results from constriction of post-capillary venules in tissue beds and vasodilation in pre-capillary arterioles/metarterioles. This results in an increase of hydrostatic pressure in the capillary bed, overcoming the osmotic pressure of the interstitial fluid and causing exudate (high-protein fluid with acute-phase proteins like C-reactive protein and macrophages) to flow into the interstitial tissue. This flowing of exudate also allows the factors of the clotting cascade to enter areas of tissue damage, forming clots and, eventually, scars. Certain inflammatory factors also cause phagocytic cells to enter the damaged tissue.

The now-permeable capillaries are traversed by these phagocytes in response to chemotaxis, the release of factors that lure the cells to the site of injury. The cells first approach the side of a capillary (margination), move through the spaces between capillary endothelial cells (diapedesis), and then enter the tissue itself. Histamine helps to mediate this response, and certain factors (such as bradykinin and possibly prostaglandins) stimulate skin pain receptors. Thus, blood flows into the tissue, causing redness, warmer steady-state temperature, swelling, and pain; loss of function is a secondary effect of these four states.

ADAPTIVE IMMUNITY

Adaptive immunity occurs in response to antigen exposure. It is specific, and it shows memory. As long as an antigen is made of the normal chemical elements we experience in biological systems (e.g. Carbon, Nitrogen, Sulfur, Hydrogen, Oxygen), we can form an adaptive response to it. This is how we fight off new diseases, and it has even been shown that we can create antigens in the laboratory that have never before existed on Earth, only to have animals mount competent immune reactions to them. As stated earlier, the adaptive immune system's specificity is tempered with an ability to differentiate between self and non-self antigen; simply put, the body doesn't attack its own cells, unless they have been invaded by virus and ask to be sacrificed for the sake of the host. When the immune system does attack the body, this is a disease state: autoimmunity.

The primary response takes 5-6 days, but the memory (secondary) response will be swifter and deadlier. It includes:

LYMPHOCYTES

Blood cells are made in the bone marrow of adults. Leukocytes are white blood cells (WBCs). This includes monocytes (which become myriad cell types in the body, most importantly macrophages), granulocytes (neutrophils, basophils, eosinophils), and lymphocytes. Of all the leukocytes, the lymphocyte class are the most preeminent in the adaptive immune response. On a peripheral blood smear, lymphocytes are approximately the size of erythrocytes (red blood cells, RBCs), although they can be larger if activated and have a characteristic "clock-face" nucleus if they are B cells. Lymphocytes are so named because fewer than 1% are present in the circulating blood; the rest lie in the lymph nodes, spleen and other lymphoid organs. T lymphocytes leave the bone marrow, travelling to the thymus gland, where they mature and gain their specificity for the diverse antigens the body might come into contact with; additionally, any T cells that react against the body's own epitopes are selected against (killed) in the thymus, in an effort to stop any possible autoimmunity. A similar process occurs in the bone marrow in the case of B lymphocytes. B cells are conveniently named ("B for bone marrow"), but this is just a coincidence; it turns out that they are named B cells after the Bursa of Fabricius, a small pouch in the cloaca (lower large intestine, *cloaca* Latin for "sewer") of birds.

B cells leave the bone marrow with their specific membrane bound Ig (antibody) already specified. The Ig itself is made up of two medial heavy chains (both identical) with two lateral light chains (also identical) attached to the "top" of the heavy chains, which form a Y shape. Before they encounter antigen, B cells are known as "naive." Once they encounter antigen, the naive B cells will undergo clonal expansion; an activated B cell will form some daughter memory B cells and some plasma cells. The memory cells will lie in wait for a second encounter with the antigen, while the plasma cells will begin a *massive* secretion of antibody (Ig). B cells can bind to antigen when it is free and unprocessed in the body, much like APCs can.

Before moving on to T lymphocytes, it should be noted that certain cell surface molecules distinguish each person's unique immune system profile. The Major Histocompatibility Complex(MHC, or HLA) is a type of protein expressed on the surface of host cells that interacts with T-cell receptors (TCRs) of T cells. Virtually all the body's cells, including APCs, express class I MHC (MHC-I) on their surface.

Only APCs express class II MHC (MHC-II). Thus, most body cells express MHC-I, while APCs express both MHC-I *and* MHC-II. When antigen enters a body cell and is broken down, the products of this breakdown are sent to the surface of the cell coupled with MHC-I. This forms the MHC-I/Ag complex, and usually occurs when a virus or bacterium enter a cell and are broken down by intracellular defenses. This can also occur in APCs, but APCs additionally process the antigen that they phagocytose, presenting it as an MHC-II/Ag complex on the surface of their cells.

T cells can be subdivided into two broad types: T helper cells and T cytotoxic cells. T helper cells express a T-Cell Receptor (TCR) that will interact with APC surfaces. Specifically, they interact with the MHC-II/Ag complex on the surface of APCs, and the TCR is stabilized in its binding by a CD4 receptor. CD stands for "cluster of differentiation," and is simply a class of cell receptor which occurs predominantly on T cells. T helper cells contain CD4 receptors and T cytoxic cells contain CD8 receptors. Upon binding, T helper cells release cytokines, which act as chemotactic agents to call for more T helper cells, T cytotoxic cells, APCs, and B cells.

T cytotoxic cells encounter body cells that have been invaded and are presenting MHC-I/Ag on their surface. Tc cells extend their TCR to the MHC-I/Ag complex and stabilize this interaction with their CD8 receptor arms.

Upon binding, the CD8+ cells differentiate, much like naive B cells, into memory T cells and cytotoxic T lymphocytes (CTLs), effector cells that cause the MHC-I/Ag-presenting cell (the "altered self cell") to die (apoptose). CTLs trigger apoptosis by secreting a perforin that allows the entry of a serine protease (Granzyme B) which activates intracellular executioner caspases. It should be noted that cancer cells can also become altered self cells, and CTLs are very important in the destruction of cancerous cells. T cells can interact with antigen *only* after it has been processed, either by a normal body cell (MHC-I) or by an APC (MHC-II).

Difference between the two types of T cells:

- T helper cells react to exogenous antigen, phagocytosed by APCs, presented on MHC-II, via binding with TCR and CD4.
- T cytotoxic cells react to endogenous antigen (such as viral or cancer proteins), broken down by lysosomes in many types of body cells, presented on MHC-I, via binding with TCR and CD8.

Note that it is both the antigen and the MHC that is presented to T cells; each person has a unique MHC. This is why we must type for bone marrow transplants—we don't want people producing tons of new immune cells in bone marrow that think every MHC in the body is actually just antigen; when this does occur, it is called graft-vs.-host disease. Not only would the new bone marrow make cells that attacked every body cell's MHC, but when the body did present Ag on MHC, the lymphocytes from the transplant would not be able to recognize this MHC-I/Ag complex (although the MHC-II complex cells would be made in the new bone marrow, and thus APCs could, in some cases, still present). Unfortunately, these APCs would be very busy presenting the host's own body cells; clearly MHC typing (also known as HLA typing) for bone marrow transplants is a necessary result of the elegant self-nonself recognition of the human immune system. It is also worth noting that we are only as good as our presenting molecules. Some people have MHC genes that are not as good at presenting antigen, and thus some people have more vigorous immune respones than others.

PRIMARY IMMUNODEFICIENCY DISEASES

The majority of primary immunodeficiency's are rare diseases, that are instructive for the contribution of individual parts of the human immune system to overall defence. Depending on the genes concerned, the following functions may be impaired:

- T- and B cell function: severe combined immunodeficiency (SCID)
- global or partial B cell response
- phagocytosis
- complement functions

In addition, the immune system is affected in a number of more complex syndromes. Here, only a select few primary immunodeficiencies are mentioned that should illustrate certain aspects of immune system functioning.

SCID

Severe combined immunodeficiency (SCID) is caused by several genetic defects, e.g., by a deficiency of RAG proteins. If neither immunoglobulin genes nor T cell receptor genes are rearranged, the result is a total loss of adaptive immunity. Only a few patients with defective RAG proteins have been described; more frequent causes for SCID are defects that primarily concern T cells. These patients' inability to produce antibodies underscores the importance of T cell help for immunoglobulin production.

Autosomal recessive deficiencies of two enzymes in purine metabolism, ADA and PNP, cause problems in T cell development. Adenosine deaminase (ADA) deaminates (deoxy-) adenosine to (deoxy-) inosine. Purine nucleoside phosphorylase (PNP) cleaves (deoxy-) inosine and (deoxy-) guanosine, resulting in ribose-1-phosphate plus base. Both defects lead to an accumulation of dAMP/dATP. High concentrations of dAMP/dATP cause substrate-inhibition of the enzyme ribonucleotide reductase, necessary for reducing ribonucleotides to their deoxy-forms. The resulting lack of dCTP, dGTP, and dTTP impairs synthesis of DNA. Why this is deleterious for T cells, but tolerable for other cells is not entirely clear. Maybe this effect is most pronounced in the thymus, as apoptotic death of more than 95% of thymocytes results in locally increased concentrations of dAMP/dATP due to breakdown of DNA. ADA deficiency is successfully treated by infusion of PEG-ADA. Obviously, this only works extracellularly, but intracellular concentrations are rapidly equilibrated by nucleoside transporters. ADA deficiency was the first disease where gene therapy was attempted. A functioning version of the gene was first introduced into T cells, later into hematopoietic stem cells. Expression of the enzyme was successful in part of the cells, but mostly that part was too small to cure immunodeficiency.

So far, gene therapy was most successful in attempts to cure another form of the disease, X-linked SCID. The disease got some public awareness from the "bubble boy", a little boy who was kept alive by isolation in a pathogen-free plastic bubble; he died in 1984 in an attempt to cure his disease by a bone marrow transplant. In X-linked SCID, T cells fail to develop due to a missing common γ-chain of interleukin receptors for IL 2, IL 4, IL 7, IL 9 and IL 15. For gene therapy, a sound copy of this X-chromosomal gene was introduced into stem cells of affected boys by a retroviral vector. The boys successfully developed T cells and responded favorably to typical immunizations against diphtheria, tetanus and polio. Initial euphoria quickly subsided, however, when two of the boys developed T cell leukemia. Further investigations indicated that the retroviral vector in both cases had inserted near the proto-oncogene LMO2, resulting in its activation. Overexpression of the encoded protein had caused leukemia. The trials were halted.

Immunoglobulin deficiency

An X-chromosomally inherited disease also exists at the B cell level: X-linked or Bruton's agammaglobulinemia. Missing in affected boys is Bruton's tyrosine kinase (BTK), a kinase necessary for B cell maturation. As a quality control step in pre-B cells, the product of a rearranged heavy chain is brought to the surface together with a temporary, "surrogate" light chain. This *ersatz*-immunoglobulin then signals via BTK, indicating successful rearrangement of a heavy chain, so that the cell now initiates light chain rearrangement. If no BTK signal is detected, it is assumed that heavy chain rearrangement failed (wrong reading frame on both alleles, or similar problems), making the cell useless; the cell undergoes apoptosis. A defect in BTK thus results in apoptosis of all pre-B cells and a general lack of immunoglobulins. The affected boys suffer from recurrent infections with pyogenic bacteria such as *Streptococcus pneumoniae*.

A third X-linked disease is hyper-IgM syndrome. Here, normal numbers of B and T cells combine with high serum levels of T cell-independent IgM but a lack of other immunoglobulin isotypes. The defective molecule is CD40-ligand, required for TH2 cells to help B cells, as well as for TH1 cells to activate macrophages. Without T cell help, class switch mostly doesn't occur. Children suffer from recurrent infections with extracellular bacteria, as well as with the parasite *Pneumocystis jirovecii* that is otherwise easily cleared by activated macrophages.

A range of additional causes for class switch problems exist. One of them is hyper IgM syndrome type 2, which is caused by a deficiency of the enzyme AID. Both class switch recombination and somatic hypermutation are initiated by this enzyme, which deaminates cytosine to form uracil. Uracil is removed from DNA by the enzyme UNG (uracil-N-glycosylase) followed by additional steps resulting in double strand breaks in the so called switch regions of the immunoglobulin heavy chain genes (described in section 2.5). This is necessary to switch a B cell's immunoglobulin production from IgM to IgG or another isotype. Accordingly, with a deficiency in AID, class switch is severely impaired, resulting in a lack of IgG and IgA associated with normal or high IgM.

The most frequent form of immunoglobulin deficiency, at 1:800, is selective IgA deficiency. As symptoms of affected children tend to be mild (all kids are sick from time to time...), the disease is underdiagnosed. More severely affected children suffer from recurrent mucosal infections such as otitis media, paranasal sinusitis and bronchitis, as well as pneumonia and intestinal infections. Allergy and autoimmune phenomena are increased. After repeated transfusions, a tendency to produce anti-IgA antibodies may become critical. The genetic basis of IgA deficiency is insufficiently understood, but some progress has recently been made.

Selective IgA deficiency might be a marginal form of another disease entity, common variable immunodeficiency (CVID), as both forms sometimes are found in different members of the same families. CVID is less frequent (approximately 1:25,000) and genetically heterogeneous. CVID is characterized by recurrent infections of respiratory or gastrointestinal tracts starting not before age three, but in most patients only during the second or third decade of life. Characteristically, responses to vaccinations are weak or lacking altogether. Lab work shows hypogammaglobulinemia with low IgG and IgA, but frequently normal IgM levels. A few genetic defects resulting in CVID have been elucidated, each accounting for only a small fraction of total patients. They concern transmembrane proteins that (in addition to CD40-ligand-CD40 contact) are required for germinal center processes including class switch and somatic hypermutation, which in turn are necessary for generation of memory B cells. CVID-causing defects were found in genes encoding ICOS, TACI and CD19. Intriguingly, in some families with TACI deficiency, homozygous members had CVID, while selective IgA deficiency was found in heterozygous members. TACI is a member of the TNF receptor superfamily (TNFRSF13B) and functions as a trimer. Heterozygous TACI deficiency would thus result in a strongly diminished number of functional trimers.

Defects in phagocytosis and complement

Cooperation between antibodies, complement and phagocytes is essential to eliminate pyogenic bacteria that, for their polysaccharide capsule, are not readily recognized by neutrophils. The significance of this cooperation is underscored by the fact that deficiencies of either of these components result in severe infections with this type of pathogens.

Several aspects of the complex processes of chemotaxis and phagocytosis can be affected by genetic defects. A deficiency of surface molecules such as integrins or the carbohydrate ligand of selectins prevents leukocytes from adhering to the endothelial vessel wall in inflamed tissue. Lack of an enzyme necessary to produce reactive oxygen species means pathogens are phagocytosed, but not killed. Examples are chronic granulomatous disease (deficiency of NADPH oxidase) or myeloperoxidase (MPO) deficiency. Chediak-Higashi syndrome is caused by the deficiency of a vesicle transport protein, so that phagosomes fail to fuse with lysosomes.

Symptoms of complement deficiencies depend on their location in the complement cascade. Defects in alternative or lectin pathways predispose to infections with pyogenic bacteria. Deficiencies in components C1, C2 or C4 result in impaired clearance of immune complexes, leading to type III diseases (explained in section 5.1). Defects in the membrane attack complex (C5-C9) predispose to severe infections with *Neisseria meningitidis*.

HIV / AIDS

By the spread of human immunodeficiency virus (HIV), AIDS has become a common disease. Two virus types exist: HIV-1 can be found around the globe, HIV-2 mainly in West Africa. Both seem to have jumped the species barrier form non-human primates to humans in Africa during the 20th century.

The problem of this disease is that HIV infection targets and kills CD4 T cells, over time depleting these cells that are main and center to the functioning of the immune system. Also dendritic cells, macrophages and several cell types of the central nervous system are infected. The CD4 molecule itself is the most important receptor; in addition, infection requires a chemokine receptor, either CCR5 or CXCR4. A defective CCR5 haplotype exists, CCR5-Δ32, that cannot be used by the virus to enter the cell, making homozygotes resistant to the most frequent strains. About 1% of Caucasian populations are homozygous, while the haplotype is far less common in African and Asian populations.

HIV is transmitted by body fluids, primarily during sexual intercourse. Other routes of infection are by contaminated needles or blood transfusions and from an infected mother to her child before or during childbirth or by breast feeding. The virus particle docks via its gp120 envelope protein to CD4 and the chemokine receptor, fuses its membrane with that of the cell and enters its capsid containing its RNA genome, reverse transcriptase and other enzymes into the cell. Using reverse transcriptase, the virus transcribes its single stranded RNA genome into cDNA and inserts this copy into a host cell chromosome with the help of integrase. The integrated cDNA is called provirus. In this state, the virus can remain latent and unassailable for years. The viral genome consists of retrovirus-typical long terminal repeats (LTRs) flanking the usual gag-, pol-, env- and six additional smaller genes. Virus activation is induced by transcription factor NFκB, in parallel and by the same mechanism as T cell, dendritic cell and macrophage activation. Several transcripts are made and spliced, giving rise to the different viral components. Some of its proteins are synthesized as polyprotein precursors that are subsequently cleaved to their final form by viral protease. After all components have been produced in sufficient quantities, they are packaged by self-assembly and leave the cell by budding.

Three mechanisms contribute to destroying infected CD4 T cells:

- cytotoxic CD8 T cells doing their normal job of killing virus factories
- direct cytopathic effects of the virus
- an increased propensity of activated T cells to enter apoptosis

In the absence of treatment, the disease typically develops in three phases:

Acute HIV infection during the first few weeks, characterized by fever, lymphadenopathy, pharyngitis and aphthae (not rhinitis). Obviously, these symptoms are very unspecific. Differential diagnosis has to take into account many potential causes, e.g. Epstein-Barr-Virus (EBV) infection (mononucleosis or kissing disease). At this time, a diagnosis can only be established by PCR, as antibodies appear only after several weeks or months: 6 months after infection, 99% are antibody-positive. Asymptomatic latent infection. The duration of this phase depends on several factors, e.g., age, nutrition and other infections. In young people with a good standard of living, this phase can last ten and more years. Diagnosis is mostly established by positive Western blot following less specific ELISA screening tests. At the end of this phase, infections become more frequent.

AIDS phase. The onset of the complete clinical picture of AIDS depends on the number of remaining CD4 T cells. Normally, this is 500-1000 CD4 T cells per cubic millimeter (or μl) of blood. When this number falls below a critical threshold of 200, adaptive immunity is so weak that the patient starts to suffer from opportunistic infections with, e.g., *Candida albicans* (esophagitis), *Pneumocystis jirovecii* (pneumonia), cytomegalovirus, herpes zoster (shingles) etc. In the event that a previous infection with tuberculosis was not completely eliminated, the TH1-dependent granuloma walls now start to crumble, giving rise to an acute flare-up. Another possibility is Kaposi's sarcoma, which is caused in endothelial cells by another virus from the herpes group, human herpes virus type 8 (HHV8). In the end, the patient succumbs to one of these infections.

There is neither a vaccine nor a cure for HIV, yet efficient therapy has nevertheless been developed in the form of HAART (highly active antiretroviral therapy). The line of attack is to inhibit enzymes necessary for virus replication, such as reverse transcriptase (RT) and protease. Nucleoside analogue RT inhibitors lead to incorporation of "wrong" nucleotides into the growing cDNA, blocking RT progress, while non-nucleoside inhibitors block RT by other means. A range of protease inhibitors is used; a virus strain may develop resistance by point mutations in its protease gene. Detection of these mutations by sequence analysis is a means to rationally adjust therapy. As a third principle, virus entry inhibitors may be used. The drug cocktail is effective but frequently causes unwanted side effects including mitochondrial dysfunction resulting in myopathy or pancreatitis or lipodystrophy (esthetically displeasing changes in the distribution of subcutaneous fat).

Under optimal conditions, HAART is able to suppress virus load to extremely low levels, sharply reducing morbidity and mortality. Yet, it is not able to eliminate the virus, and therefore has to be taken for the rest of the patient's life. This is not a question of popping two pills a day; therapy protocol is complex, requiring the patient to take medications at exact time points distributed over the entire day. If the protocol is not exactly adhered to, the virus gets breathing space to replicate. This promotes resistance: in contrast to DNA-dependent DNA polymerases with proofreading function, reverse transcriptase lacks proofreading, leading to frequent misincorporations. If this is paired with intermittent selective pressure by therapy, resistant strains will emerge rapidly. Enzymes modified by point mutations cease being inhibited by the drugs, making therapy ineffective.

ADAPTIVE IMMUNITY AND IMMUNIZATION

Adaptive immunity, acquired immunity, obtained in some manner other than heredity. provides specific, nonhereditary defence and protection after exposure to a specific pathogen. Naturally acquired adaptive immunity is obtained by having a specific disease. The immune system respond to antigens, activates T cells, produces antibodies to protect against the same disease. Colostrum is the first fluid secreted by the mammary glands after childbirth. It contains a lot of antibodies. Only protect for a short time then disappear. Naturally acquire passive immunity refers to the antibodies across the placenta and in breast milk.

Artificially acquired adaptive immunity is obtained by the injections of vaccine that produces immunity. All animals have an innate immune system. Only vertebrates have an adaptive immune system.

Active and Passive Immunity

In active immunity, an individual's own immune system makes antibodies. In passive immunity, ready-made antibodies are introduced into the body. Naturally acquired passive immunity is by receiving antibodies from planceta or in colostrums. Artificially acquired passive immunity is by receiving injection of gamma globulin or immune serum.

CHARACTERISTICS OF THE IMMUNE SYSTEM

Antigens

An antigen is a foreign substance that can elicit a specific immune response. Most antigens are proteins, but some are polysaccharides, nucleoproteins, or glycoproteins. Each antigen has several epitopes (antigenic determinants), or antigenic determinants.

Hapten: a small molecules can acts as an antigen only if it binds to a larger protein molecule.

Generating antibody diversity

The diversity of B cell receptors is sufficient to recognize epitopes on any pathogen we are likely to encounter in our life time. B-cell diversity is generated by recombining epitopes on any pathogen we are likely to encounter in our life time.

B-cell diversity is generated by recombining bits of DNA in our genome to generate 100 millions different antibody genes.

Cells and Tissues of the Immune System

Lymphocytes develop from lymphoid stem cells in the bone marrow. Lymphocytes differenciate into B cells in the bone marrow or into T cells in the thymus.

Subsequent differentiatoion of T cells produces four different kinds of cells.

1. Cytotoxic (killer) T cells
2. Delay-hypersensitivity T cells
3. Helper T cells
4. Regulatory T cells.

Natural killer cells (NK cells), non B, non T cells, which nonspecifically kill cancer cells and cells infected with virus without having to utilize the specific immune responses. By releasing cytotoxic molecules to create holes in the target's cell membrane, leading to lysis.Other molecules enter the target cell and fragment DNA, causing apoptosis(programmed cell death). NK cells are also affected by interferons.

Dual nature of the Immune Syetem

The dual roles of the immune system consist of humoral immunity, which is carried out mainly by antibodies produced by B cells and plasma cells, and cell mediated immunity, which is carried out mainly by certain T cells.

General Properties of Immune Responses

When our first line of defence is breached and our innate immune system fails to control an infection, our adaptive immune system is activated.

Recognition

Major Histocompatibility Complex

MHC Class 1: receptors on all cells, identify as self, present antigen for systemic. Recognized by CD8 Cytotoxic T cells. MHC Class 2: leukocytes only, intercelluar communication. Recognized by CD4 Helper T cells.

MHC Class 3: all protein produces as listed above.

Self: Class1 receptors on Neutrophils/Macrophage must match

Nonself: Class 2 receptors do not matchà phagocytosis

Recognition of self versus nonself

Embryos contain many different lymphocytes, each genetically programmed to recognize a particular antigen and make antibodies to destroy it. A lymphocyte divides repeatedly to produce a clone. According to the clonal selection theory, B cells recognize specific epitopes on antigens according to the particular antibody present on the B cell plasma membrane. When a B cell detects an antigen with which it can react, it binds to the antigen, engulfs it, process it, displays a peptide fragment as MHC class II to T_H2 cells, and divides many times. Clone of genetically identical B cells, which differentiate into many plasma cells and some memory cells, is produced. Tolerance is B or T cells encounters its programmed antigen as part of a normal host substance (self).

Specificity

Immune system fully matures at age 2 to 3.

Specificityrefers to the ability of immune responses to respond to and distinguish among different antigens and epitopes. Cross-reaction, reactions of a particular antibody with very similar antigens.

Heterogeneity (Diversity)

Diversity refers to the ability of immune responses to produce many different antibodies and cell substances on the basis of the different antigens they encounter.

Memory

Ability of T and B cells can recognize substances it has previously encountered- memory cells. Anamnestic (secondary) response is prompt due to recall of memory cells.The power and rapid secondary immune response initiated by memory B cellsacts within few days, sometimes suppressing all symptoms.

The action of vaccines depends upon their stimulating the formation of memory B cells.

SPECIFIC IMMUNE RESPONSE

Cell Mediated – involves different cells and their secretory products
Antibody mediated – production, secretion and action of antibodies

HUMORAL IMMUNITY

B-cell activation

Native B cells are fully differentiated B cells that have not yet encountered a complementary antigen; they bear membrane–bound IgM and IgD on their outer surface. Clonal selection begins when native B cells bind to their complementary antibodies: Then they multiply; produce soluble IgM and IgD, and undergo a high rate of mutation in their DNA-encoding region, resulting in affinity maturation.

Later, class switching occurs: Various activated B cells begin to make all classes of antibodies. Some members of the clone become short-lived effectors (plasma) cells, which make large quantities of antibody; others become long-lived memory cells, which fight future infections by the same pathogen.

Lymphocytes, including interleukins-1,-2,-4,-5, and -6, produced by NK cells help B-cell activation stimulated by most antigens. Some T-independent antigens stimulate B-cell activation without help. B cells are selected to respond to specific antigens in accordance with the particular antibody present on the B cell membrane prior to encountering an antigen.

When a B cell detects an antigen with which it can react, it binds to the antigen (sensitized or activates) and divides many times to produce a clone of many plasma cells and some memory cells. Many B cells require helper T cells to proliferate and differentiate into both plasma and memory B cells. Plasma cells synthesize and release large numbers of antibodies. Memory cells remain in lymphoid tissues ready to respond to subsequent exposure to the same antigen.

Properties of Antibodies (Immunoglobulins)

Antibodies are glycoproteins because carbohydrate groups are attached to them.Structually, antibodies consist of two heavy and two light polypeptide chains, joined by disulfide bonds to form a Y shape. The upper ends of the Y, consists of variable region in both the light and heavy chains, differ from antibody to antibody. These variable regions form the two antigen-binding sites (part of the Fab fragment), which are responsible for the specificity of the antibody. The remaining part of the molecule consists of constant regions (Fc) that are similar in all antibodies of a particular class. The constant region binds to and activates phagocytes and complement.

Class of Immunoglobins

Five classes of immunoglobulins have been identified. Each class has a particular kind of constant region which gives that class its distinguishing properties. IgD, IgE, and IgG are secreted as monomers, IgM as pentamers, and IgA as monomers or dimmers.

IgG

The main class of antibodies found in blood, accounts 20 percent of all plasma proteins.

IgG is produced in large quantities during a secondary response. The antigen-binding sites of IgG attach to antigens on microorganisms and their tissue-binding sites attach to receptors on phagocytic cells. IgG is the only immunoglobulin that can cross the placenta from mother to fetus and provide antibody protection. IgG – monomer, predominant type, specific immune response, transmammary, transplacental.

IgA

Occurs in small amounts in blood and in larger amounts in body secretions such as tears, milk, saliva, mucus and linings of GI tracts. Secretory IgA, which consists of two monomer units held together by a J chain, has an attached secretory component, which protects the IgA from proteolytic enzymes and facilitates its transport. Mucosal surfaces such as in the respiratory, GI tracts are major sites for invasion by pathogens. The main function of IgA is to bind antigens on microorganisms before they invade. It also activates complement, which helps to kill the microorganisms. IgA – monomer (plasma), dimer (mucus/saliva/tears/milk/intestinal secretions), inhibit microbe attachment to epithelia

IgM

IgM is found as a monomer on the surface of B cells and is secreted as a pentamer by plasma cells. IgM consists of five units connected by their tails to a J chain and has 10 peripheral antigen-binding sites. As IgM binds to antigens, it also activates complement and causes microorganisms to clump together. High levels of IgM indicate recent infection or expose to antigen. IgM – monomers (antigen receptor) penerates (plasma), nonspecific immune response, blood typing

IgE

IgE has a special affinity for receptors of basophils in blood or mast cells in tissue. It binds to these cells by tissue binding sites, leaving antigen-binding sites free to bind antigens which can develop allergy. When IgE binds antigens, the basophils and mast cells secrete histamine which produce allergy symptoms. IgE are elevated in patients with allergies and worm parasites.

IgE – monomer, tonsils, skin, mucus membranes. Stimulates basophils to release histamine

IgD

IgD, monomer, is found mainly on B-cell membranes and function is unknown.Only trace amount.

Primary and Secondary Responses

Primary responses are the immune system's first encounter with foreign antigens. First produce IgM, then IgG. Memory cells remain in lymphoid tissue, ready to respond to subsequent exposure to the same antigen. The primary response of B cells can occur by two mechanisms.

- T-independent antigens - B cells can be activated by binding antigen, proliferating and forming plasma cells.
- T-dependent antigens – B cells becomes an antigen presenting cell, and activate the T helper cell (T_H). The activated T_H cells then secretes lymphokines that further activate the B cell causing it to differentiate and proliferate producing B memory cells and T_H memory cells.

Secondary responses bring fast and efficient destruction of antigens recognized by B and T memory cells. IgM is produced in a smaller quantities and IgG is produced in much larger quantities than in primary response.

Kinds of Antigen-Antibody Reactions

Humoral immunity depends upon the production of antibodies by B cells. Antibodies protect against invading microbes by neutralization, opsonization, and activating complement. Humoral immunity is most effective against bacteria, antigen-antibody reaction result in agglutination (clumping).

1. Neutralization of pathogens and toxins by IgA or IgG.
2. Opsonization of bacteria by IgG, which is activated by antigen-bound IgM or IgG, produces complement C_3b, a powerful opsonin, they coat microbes so that they can be phagocytized.
3. Cell lysed by complement after or directly by IgMs or IgG immune complexes

MONOCLONAL ANTIBODIES

Monoclonal antibodies are antibodies produced in the laboratory from a clone of cultured cells that make one specific antibody to one specific epitope. Specific monoclonal antibodies can be used in some diagnostic tests, and methods to use them in treating infectious diseases and cancer are being developed.

Lymphocytes

Lymphocytes are small, smooth, round leukocytes that lack granules.

There are three kinds of leukocytes: B cells, T cells, and natural killer (NK) cells. B cells and T cells are triggered into defensive action when they encounter antigens, which are foreign molecules we lack. B cells respond to antigen by producing antibodies. Cytotoxic T (CD_8) cells respond to antigen by killing our own pathogen-infected cells. Helper T (CD_4) cells respond by stimulating phagocytes and B cells.

Lymphoid Tissues

Primary lymphoid tissues are the bone marrow where blood cells, including lymphocytes, are formed and where B cells differentiate; the thymus is where T cells differentiate. Differentiated B and T cells travel to secondary lymphoid tissues, where they are stored and interact. Secondary lymphoid tissues include lymph nodes, spleen, tonsils, adenoids, appendix, and Peyer's patches. Spleen is the largest lymph organ. Lymph is formed in our tissues as fluid from blood leaks across capillary walls. A system of lymph vessels collects lymph and returns it to th thoracic duct that empties into the heart. Lymphocytes circulate through blood and lymph. Lymphatic vessels and lymph nodes usually become swollen and inflamed during infections that activate an immune response.

The Cell-Mediated Immune Reaction

Cell mediated Response:

Cytotoxic (CD_8) T cells, with the aid of helper (CD_4 or Th cells) T cells, are responsible for cell-mediated immunity. Activated T helper cells in turn activate other cells and processes clonal expansion of undifferentiated B cells.

B cell differentiation creates memory B and Plasma B

Plasma B synthesize and secrete antibodies which neutralize antigen

Memory T cells

Nature Killer cells

Cytotoxic T cells

Leukokine production of all leukocytes.

Involves T cells, requires presentation of the antigen on the surface of cell with major histocompatibility complex (MHC) proteins by antigen presenting cells.

Antigen Recognition

1. All nucleated cells have major MHCI proteins on their surfaces.
2. Dendritic cells and macrophages (antigen presenting cells) also have MHCII on their surfaces. They phagocytize pathogens, they digest and present pieces of the foreign peptides on their surface by MHCII proteins to T cells that have the proper antigen receptor.
3. T helper (T_H) cells are activated by antigen present by MHCII from antigen-presenting cells.

4. Cytotoxic (killer) T (T_C) cells are activated by antigen-presented by MHCI – cells infected by virus, intracellular bacterial pathogens, transformed cancer cells, or foreign tissues (organ transplant).
5. Macrophages that have processed an antigen secrete lymphokine interleukin-1 (IL-1), which activates T helper cells, in turn, T help cells secrete lymphokines such as interleukin-2 (IL-2) and gamma interferon.
6. IL-1 from macrophages and IL-2 from T helper cells activate B cells to make antibody and other T cells, such as cytotoxic (killer) T (T_C) cells and memory cells.
7. T_C cells can be recognized by a CD8 glycoprotein on their membrane. T helper cells can be recognized by a CD4 glycoprotein on their membrane.
8. IL-1, Il-2 and gamma interferon together cause undifferentiated cells to become natural killer (NK) cells.

Processed antigens on MHC class II molecules bind with T cell receptors. Next, IL-1 secreted from macrophages and IL-2 secreted from T helper cells activate the T helper cells, which can then differentiate into T_H1 cells and T_H2 cells. Certain pathogenic bacteria can grow in macrophages after they have been phagocytized. T_H1 cells can release γ-interferon, a cytokine that causes such infected macrophages to become resensitized to other T cells – cell-mediated immunity.

T_H2 can activated B cell to produce antibody – humoral immunity.

How Killer Cells Kill

T_C and NK cells destroy target cells by releasing the lethal protein perforin. Peforins bores holes in the target cell membranes, so that essential molecules leak out and the cells die.

The Role of activated Macrophages

Certain pathogenic bacteria can grow in macrophages after phagocytosis, such as those that cause tuberculosis, leprosy. The lymphokine macrophage activating factor helps stimulate antimicrobial processes by increasing production of hydrogen peroxideand other enzymes to kill the organismsWhen macrophages fails to kill pathogens, the pathogens are walled off in granulomas.

Superantigens

Such as staphylococcal toxins that cause food poisoning, can bind to MHCII molecules and T cells but does not involve specificity. The superantigens binds to T cells with different specificities, it is polyclonal. This activates T cells at up to 100 times the normal rate and then secrete immense amounts of IL-2. The excess IL-2 gets to the blood stream, and cause nausea, vomiting, fever, malaise and symptoms of shock.

INNATE IMMUNE SYSTEM

Microorganisms or toxins that successfully enter an organism encounter the cells and mechanisms of the innate immune system. The innate response is usually triggered when microbes are identified by pattern recognition receptors, which recognize components that are conserved among broad groups of microorganisms, or when damaged, injured or stressed cells send out alarm signals, many of which (but not all) are recognized by the same receptors as those that recognize pathogens. Innate immune defenses are non-specific, meaning these systems respond to pathogens in a generic way. This system does not confer long-lasting immunity against a pathogen. The innate immune system is the dominant system of host defence in most organisms.

ADAPTIVE IMMUNE SYSTEM

The adaptive immune system evolved in early vertebrates and allows for a stronger immune response as well as immunological memory, where each pathogen is "remembered" by a signature antigen. The adaptive immune response is antigen-specific and requires the recognition of specific "non-self" antigens during a process called antigen presentation. Antigen specificity allows for the generation of responses that are tailored to specific pathogens or pathogen-infected cells. The ability to mount these tailored responses is maintained in the body by "memory cells". Should a pathogen infect the body more than once, these specific memory cells are used to quickly eliminate it.

Lymphocytes

The cells of the adaptive immune system are special types of leukocytes, called lymphocytes. B cells and T cells are the major types of lymphocytes and are derived from hematopoietic stem cells in the bone marrow. B cells are involved in the humoral immune response, whereas T cells are involved in cell-mediated immune response. Both B cells and T cells carry receptor molecules that recognize specific targets. T cells recognize a "non-self" target, such as a pathogen, only after antigens (small fragments of the pathogen) have been processed and presented in combination with a "self" receptor called a major histocompatibility complex (MHC) molecule. There are two major subtypes of T cells: the killer T cell and the helper T cell. Killer T cells only recognize antigens coupled to Class I MHC molecules, while helper T cells only recognize antigens coupled to Class II MHC molecules. These two mechanisms of antigen presentation reflect the different roles of the two types of T cell. A third, minor subtype are the $\gamma\delta$ T cells that recognize intact antigens that are not bound to MHC receptors. In contrast, the B cell antigen-specific receptor is an antibody molecule on the B cell surface, and recognizes whole pathogens without any need for antigen processing. Each lineage of B cell expresses a different antibody, so the complete set of B cell antigen receptors represent all the antibodies that the body can manufacture.

Killer T cells

Killer T cells are a sub-group of T cells that kill cells that are infected with viruses (and other pathogens), or are otherwise damaged or dysfunctional. As with B cells, each type of T cell recognizes a different antigen. Killer T cells are activated when their T cell receptor (TCR) binds to this specific antigen in a complex with the MHC Class I receptor of another cell. Recognition of this MHC:antigen complex is aided by a co-receptor on the T cell, called CD8. The T cell then travels throughout the body in search of cells where the MHC I receptors bear this antigen. When an activated T cell contacts such cells, it releases cytotoxins, such as perforin, which form pores in the target cell's plasma membrane, allowing ions, water and toxins to enter. The entry of another toxin called granulysin (a protease) induces the target cell to undergo apoptosis. T cell killing of host cells is particularly important in preventing the replication of viruses. T cell activation is tightly controlled and generally requires a very strong MHC/antigen activation signal, or additional activation signals provided by "helper" T cells.

Helper T cells

Helper T cells regulate both the innate and adaptive immune responses and help determine which immune responses the body makes to a particular pathogen. These cells have no cytotoxic activity and do not kill infected cells or clear pathogens directly. They instead control the immune response by directing other cells to perform these tasks.

Helper T cells express T cell receptors (TCR) that recognize antigen bound to Class II MHC molecules. The MHC:antigen complex is also recognized by the helper cell's CD4 co-receptor, which recruits molecules inside the T cell (e.g., Lck) that are responsible for the T cell's activation. Helper T cells have a weaker association with the MHC:antigen complex than observed for killer T cells, meaning many receptors (around 200–300) on the helper T cell must be bound by an MHC:antigen in order to activate the helper cell, while killer T cells can be activated by engagement of a single MHC:antigen molecule. Helper T cell activation also requires longer duration of engagement with an antigen-presenting cell. The activation of a resting helper T cell causes it to release cytokines that influence the activity of many cell types. Cytokine signals produced by helper T cells enhance the microbicidal function of macrophages and the activity of killer T cells. In addition, helper T cell activation causes an upregulation of molecules expressed on the T cell's surface, such as CD40 ligand (also called CD154), which provide extra stimulatory signals typically required to activate antibody-producing B cells.

B lymphocytes and antibodies

A B cell identifies pathogens when antibodies on its surface bind to a specific foreign antigen. This antigen/antibody complex is taken up by the B cell and processed by proteolysis into peptides. The B cell then displays these antigenic peptides on its surface MHC class II molecules. This combination of MHC and antigen attracts a matching helper T cell, which releases lymphokines and activates the B cell. As the activated B cell then begins to divide, its offspring (plasma cells) secrete millions of copies of the antibody that recognizes this antigen. These antibodies circulate in blood plasma and lymph, bind to pathogens expressing the antigen and mark them for destruction by complement activation or for uptake and destruction by phagocytes. Antibodies can also neutralize challenges directly, by binding to bacterial toxins or by interfering with the receptors that viruses and bacteria use to infect cells.

HUMORAL AND CHEMICAL BARRIERS

Complement System

The complement system is a biochemical cascade that attacks the surfaces of foreign cells. It contains over 20 different proteins and is named for its ability to "complement" the killing of pathogens by antibodies. Complement is the major humoral component of the innate immune response. Many species have complement systems, including non-mammals like plants, fish, and some invertebrates.

In humans, this response is activated by complement binding to antibodies that have attached to these microbes or the binding of complement proteins to carbohydrates on the surfaces of microbes. This recognition signal triggers a rapid killing response. The speed of the response is a result of signal amplification that occurs following sequential proteolytic activation of complement molecules, which are also proteases. After complement proteins initially bind to the microbe, they activate their protease activity, which in turn activates other complement proteases, and so on. This produces a catalytic cascade that amplifies the initial signal by controlled positive feedback. The cascade results in the production of peptides that attract immune cells, increase vascular permeability, and opsonize (coat) the surface of a pathogen, marking it for destruction. This deposition of complement can also kill cells directly by disrupting their plasma membrane.

Inflammation

Inflammation is one of the first responses of the immune system to infection. The symptoms of inflammation are redness, swelling, heat, and pain, which are caused by increased blood flow into tissue. Inflammation is produced by eicosanoids and cytokines, which are released by injured or infected cells. Eicosanoids include prostaglandins that produce fever and the dilation of blood vessels associated with inflammation, and leukotrienes that attract certain white blood cells (leukocytes). Common cytokines include interleukins that are responsible for communication between white blood cells; chemokines that promote chemotaxis; and interferons that have anti-viral effects, such as shutting down protein synthesis in the host cell. Growth factors and cytotoxic factors may also be released. These cytokines and other chemicals recruit immune cells to the site of infection and promote healing of any damaged tissue following the removal of pathogens.

Cellular Barriers

Leukocytes (white blood cells) act like independent, single-celled organisms and are the second arm of the innate immune system. The innate leukocytes include the phagocytes (macrophages, neutrophils, and dendritic cells), mast cells, eosinophils, basophils, and natural killer cells. These cells identify and eliminate pathogens, either by attacking larger pathogens through contact or by engulfing and then killing microorganisms. Innate cells are also important mediators in the activation of the adaptive immune system.

Phagocytosis is an important feature of cellular innate immunity performed by cells called 'phagocytes' that engulf, or eat, pathogens or particles. Phagocytes generally patrol the body searching for pathogens, but can be called to specific locations by cytokines. Once a pathogen has been engulfed by a phagocyte, it becomes trapped in an intracellular vesicle called a phagosome, which subsequently fuses with another vesicle called alysosome to form a phagolysosome. The pathogen is killed by the activity of digestive enzymes or following a respiratory burst that releases free radicals into the phagolysosome. Phagocytosis evolved as a means of acquiring nutrients, but this role was extended in phagocytes to include engulfment of pathogens as a defence mechanism. Phagocytosis probably represents the oldest form of host defence, as phagocytes have been identified in both vertebrate and invertebrate animals.

Neutrophils and macrophages are phagocytes that travel throughout the body in pursuit of invading pathogens. Neutrophils are normally found in the blood stream and are the most abundant type of phagocyte, normally representing 50% to 60% of the total circulating leukocytes. During the acute phase of inflammation, particularly as a result of bacterial infection, neutrophils migrate toward the site of inflammation in a process called chemotaxis, and are usually the first cells to arrive at the scene of infection. Macrophages are versatile cells that reside within tissues and produce a wide array of chemicals including enzymes, complement proteins, and regulatory factors such as interleukin 1. Macrophages also act as scavengers, ridding the body of worn-out cells and other debris, and as antigen-presenting cells that activate the adaptive immune system.

Dendritic cells (DC) are phagocytes in tissues that are in contact with the external environment; therefore, they are located mainly in the skin, nose, lungs, stomach, and intestines. They are named for their resemblance to neuronal dendrites, as both have many spine-like projections, but dendritic cells are in no way connected to the nervous system. Dendritic cells serve as a link between the bodily tissues and the innate and adaptive immune systems, as they present antigen to T cells, one of the key cell types of the adaptive immune system. Mast cells reside in connective tissues and mucous membranes, and regulate the inflammatory response. They are most often associated with allergy and anaphylaxis. Basophils and eosinophils are related to neutrophils. They secrete chemical mediators that are involved in defending against parasites and play a role in allergic reactions, such as asthma. Natural killer (NK cells) cells are leukocytes that attack and destroy tumor cells, or cells that have been infected by viruses.

8

Nutrition

Nutrition (also called nourishment or aliment) is the provision, to cells and organisms, of the materials necessary (in the form of food) to support life. Many common health problems can be prevented or alleviated with a healthy diet. The diet of an organism is what it eats, which is largely determined by the perceived palatability of foods. Dietitians are health professionals who specialize in human nutrition, meal planning, economics, and preparation. They are trained to provide safe, evidence-based dietary advice and management to individuals (in health and disease), as well as to institutions. A poor diet can have an injurious impact on health, causing deficiency diseases such as scurvy, beriberi, and kwashiorkor; health-threatening conditions like obesity and metabolic syndrome; and such common chronic systemic diseases as cardiovascular disease, diabetes, and osteoporosis.

Nutrition science investigates the metabolic and physiological responses of the body to diet. With advances in the fields of molecular biology, biochemistry, and genetics, the study of nutrition is increasingly concerned with metabolism and metabolic pathways: the sequences of biochemical steps through which substances in living things change from one form to another. Nitrogen is needed by animals to build proteins. Carnivore and herbivore diets vary in their source of nitrogen, which is a limiting nutrient for both. Herbivores consume plants to get nitrogen and carnivores consume other animals to obtain nitrogen. Nitrogen is a common element in the atmosphere but exists in a state that is not usable by most living organisms, certain fungi and bacteria are able to convert atmospheric nitrogen into a form plants can adsorb and utilize.

The human body contains chemical compounds, such as water, carbohydrates (sugar, starch, and fibre), amino acids (in proteins), fatty acids (in lipids), and nucleic acids (DNA and RNA). These compounds in turn consist of elements such as carbon, hydrogen, oxygen, nitrogen, phosphorus, calcium, iron, zinc, magnesium, manganese, and so on. All of these chemical compounds and elements occur in various forms and combinations (e.g., hormones, vitamins, phospholipids, hydroxyapatite), both in the human body and in the plant and animal organisms that humans eat.

The human body consists of elements and compounds ingested, digested, absorbed, and circulated through the bloodstream to feed the cells of the body. Except in the unborn foetus, which receive processed nutrients from the mother, the digestive system is the first system involved in breaking down food prior to further digestion.

Digestive juices, excreted into the lumen of the gastrointestinal tract, break chemical bonds in ingested molecules, and modulate their conformations and energy states. Though some molecules are absorbed into the bloodstream unchanged, digestive processes release them from the matrix of foods. Unabsorbed matter, along with some waste products of metabolism, is eliminated from the body in the feces. Studies of nutritional status must take into account the state of the body before and after experiments, as well as the chemical composition of the whole diet and of all material excreted and eliminated from the body (in urine and foeces).

Comparing the food to the waste can help determine the specific compounds and elements absorbed and metabolized in the body. The effects of nutrients may only be discernible over an extended period, during which all food and waste must be analysed. The number of variables involved in such experiments is high, making nutritional studies time-consuming and expensive, which explains why the science of human nutrition is still slowly evolving. In general, eating a wide variety of fresh, whole (unprocessed), foods has proven favourable for one's health compared to monotonous diets based on processed foods. In particular, the consumption of whole-plant foods slows digestion and allows better absorption, and a more favourable balance of essential nutrients per Calorie, resulting in better management of cell growth, maintenance, and mitosis (cell division), as well as better regulation of appetite and blood sugar. Regularly scheduled meals (every few hours) have also proven more wholesome than infrequent or haphazard ones, although a recent study has also linked more frequent meals with a higher risk of colon cancer in men.

There are six major classes of nutrients: carbohydrates, fats, minerals, protein, vitamins, and water. These nutrient classes can be categorized as either macronutrients (needed in relatively large amounts) or micronutrients (needed in smaller quantities). The macronutrients include carbohydrates, fats, protein, and water. The micronutrients are minerals and vitamins. The macronutrients (excluding water) provide structural material (amino acids from which proteins are built, and lipids from which cell membranes and some signaling molecules are built), energy.

Some of the structural material can be used to generate energy internally, and in either case it is measured in Joules or kilocalories (often called "Calories" and written with a capital C to distinguish them from little 'c' calories). Carbohydrates and proteins provide 17 kJ approximately (4 kcal) of energy per gram, while fats provide 37 kJ (9 kcal) per gm., though the net energy from either depends on such factors as absorption and digestive effort, which vary substantially from instance to instance.

Vitamins, minerals, fibre, and water do not provide energy, but are required for other reasons. A third class of dietary material, fibre (*i.e.*, non-digestible material such as cellulose), is also required, for both mechanical and biochemical reasons, although the exact reasons remain unclear.

Molecules of carbohydrates and fats consist of carbon, hydrogen, and oxygen atoms. Carbohydrates range from simple monosaccharides (glucose, fructose, galactose) to complex polysaccharides (starch). Fats are triglycerides, made of assorted fatty acid monomers bound to glycerol backbone. Some fatty acids, but not all, are essential in the diet: they cannot be synthesized in the body. Protein molecules contain nitrogen atoms in addition to carbon, oxygen, and hydrogen. The fundamental components of protein are nitrogen-containing amino acids, some of which are essential in the sense that humans cannot make them internally. Some of the amino acids are convertible (with the expenditure of energy) to glucose and can be used for energy production just as ordinary glucose in a process known as gluconeogenesis.

By breaking down existing protein, some glucose can be produced internally; the remaining amino acids are discarded, primarily as urea in urine. This occurs normally only during prolonged starvation. Other micronutrients include antioxidants and phytochemicals, which are said to influence (or protect) some body systems. Their necessity is not as well established as in the case of, for instance, vitamins.

Most foods contain a mix of some or all of the nutrient classes, together with other substances, such as toxins of various sorts. Some nutrients can be stored internally (e.g., the fat soluble vitamins), while others are required more or less continuously. Poor health can be caused by a lack of required nutrients or, in extreme cases, too much of a required nutrient. For example, both salt and water (both absolutely required) will cause illness or even death in excessive amounts.

ANTI AGING NUTRITION

Although ageing is unavoidable, physical decay is not. More than a few of the external manifestations of growing old can be decelerated - and life can even be protracted - by sustaining a wise and meaningful approach to diet. As individuals grow old, their bodies' process nutrients less competently, leading to the requirement for individuals to enhance their nutrient ingestion. For instance, Vitamin D is a nutrient necessary for the deterrence of osteoporosis. The human body produces vitamin D when the skin is subjected to sunlight, but when individuals are in their 70s their bodies manufacture merely 40 per cent of it. An ample quantity of vitamin D for people in their

20s is 200 IU; for individuals who are grown-up, 400 IU to 600 IU is required for performing the identical amount of work.

Requirement related to age:

- The requirement for B vitamins rises with age also.
- Three B vitamins—folic acid, vitamin B6 and vitamin B12—are crucial for maintaining levels of a compound described as homocysteine at a low level in the blood; if permitted to increase, homocysteine plays a part in heart-disease threat and probably memory loss, as indicated by studies in the relevant sphere.
- When individuals age it is imperative for them to augment their B6 amount from 2 mg to 5 mg; augment B12 in due course from 2 mcg to 10 mcg.
- Women ought to receive 400 mcg of folic acid every day; pregnant women ought to ingest 800 mcg everyday in order to thwart neural-tube flaws in the foetus.

Women, specifically, should be conscious that their calcium ingestion should augment as they age to avert osteoporosis. Individuals who eat diets rich in fresh fruits and vegetables have reduced ailment rates, additional energy and less danger for weight gain that can result in health troubles in comparison to those who omit these foods.

With the omission of avocados, olives and coconuts, fresh fruits and vegetables are devoid of fat, cholesterol or sodium. Fresh fruits and vegetables are furthermore rich in fibre. Eight servings of fruits and vegetables consumed everyday offers roughly 27 gm. of fibre, which is well inside the everyday requirement of 25 gm. to 35 gm.. Foods rich in fibre reduce an individual's threat for developing age-linked ailments for example heart disease, cancer, diabetes and hypertension. Foods rich in fibre are moreover reduced in calories, nonetheless satisfying; consequently they assist in filling up individuals without filling him or her.

Fresh fruits and vegetables are rich in nutrients, making available sufficient quantities of calcium, iron, magnesium, vitamin C, beta carotene and folic acid, and they are short in calories. Fruits and vegetables also contain longevity-boosting compounds referred to as antioxidants which comprise vitamins C and E in addition to beta carotene. Antioxidants resist free radicals, oxygen fragments that strike and harm cell membranes, life-supporting proteins and even genetic code of cells that trigger aging and ailments. Diets high in antioxidants check ailment and untimely aging. Antioxidants, moreover, invigourate the immune system and defend the nervous system and brain from the oxidative harm connected with age-linked memory loss.

DIET DURING MENOPAUSE

Menopause is the expression employed to signify the cease of the period of probable sexual reproduction, as demonstrated by the termination of menstrual periods. This is completely usual and happens between 45 to 55 years. Menopause does not reduce women's physical capability, sexual energy and capacity to enjoy life. Menopause fetches specific freedoms to female life. Such as no longer they have to be anxious about the monthly bleeding and about birth control.

Still then in this period a woman suffers a great deal of emotional strain. Ovaries discontinue making oestrogen and this result in oestrogen deficiency. This hormonal disparity may trigger short-term symptoms and long term health threat.

Adhering to a healthy diet:

- Sticking to a healthy diet can be beneficial for vitality in general and this in the process assists in providing a woman with better resistance against any troubles during menopause.
- Hot flushes can be diminished by acquiring vitamin E rich foods similar to wheat germ, nuts, eggs and olive oil.
- Vitamin A, D, calcium, phosphorous and magnesium can assist in checking osteoporosis.
- Fish, drumstick leaves, ragi and dairy products are outstanding suppliers of calcium. During menopause women should shun consuming uncooked bran, which hinders calcium absorption and should reduce the consumption of tea, coffee which advances the excretion of calcium.

Many women gain weight which augments the blood cholesterol level. To regulate body weight and blood cholesterol one should choose low fat dairy foods and reduce the consumption of saturated fats such as butter and cheese. Studies indicate that substances from plants identified as phytoestrogens can assist in diminishing the acuteness of hot flushes and additional indications of menopause.

Phytoestrogens, which imitate human oestrogen, are obtained chiefly from soybeans and alfalfa sprouts. Soya bean flour can be combined with wheat flour for making bread. Additional soya foods are soymilk, tofu, as well as soy sauce. Carrot in addition to beet root juice is considered to be extremely helpful in menopausal maladies. Oats, corn, barley, brown rice, whole wheat are as well outstanding sources of phytoestrogens.

Frequent exercise is necessary in this period. This stage is to be considered as a usual period in a women's life. Nonetheless the family members ought to understand that this stage in the life of a woman requires a great deal of thought, attention and sympathy to make the daily life stress free. A nourishing diet can not only diminish the indications of menopause, but also bring about an improved lifestyle and augmented energy levels, enabling one to do the types of activities that will support an improved body and an improved lifestyle in general. Foods rich in calcium and supplements are an excellent suggestion if one intends to lead a vigourous way of life during menopause. Calcium is mainly helpful when used with additional vitamins and minerals such as magnesium and vitamin D, which supports superior skeletal health.

NUTRITION FOR HEALTHY BONES

Healthy bones provide the body with power and firmness to defend the internal organs from damage. Significant as the skeleton is, it accounts for merely around 12 per cent of a human being's entire body weight.

Bones are living tissues and are continuously experiencing remodeling since old bone tissue gets worn-out and fresh bone is produced. A subtle equilibrium has to be kept between these two procedures. If that equilibrium is disrupted, problems can take place, including brittle bones and osteoporosis.

Ensuring healthy bones needs:

- Frequent physical exercise and a nourishing, stable diet are equally vital.
- Evading threat factors similar to smoking and strain also add to healthy bones.
- Every individual can do a bit to shape the health of their bones.
- By indulging in weight-bearing physical work out one can promote bone mineralization.
- By eating correctly one can make certain that the essential nutrients are obtainable for bone metabolism.
- These major nutrients are Calcium and Vitamin D.

Calcium is essential for strong teeth and bones and furthermore, plays a vital part in a variety of bodily systems, for example, the health and performance of nerves and muscle tissue. Superior sources of calcium comprise dairy foods and leafy green vegetables, even though calcium from milk and milk products is more effortlessly assimilated and present in larger quantities. Individuals at various life phases require dissimilar quantities of calcium. Adolescents, young children in addition to elderly women all have bigger than normal necessities. Calcium is one of the vital nutrients required for healthy growth of bone.

Ample calcium ingestion is crucial for the accom-plishment of peak bone mass in the late teen. This is the maximum stage of bone strength which takes place at cease of growth, and consequently results in tough healthy bones, which will prolong the consequences of ageing on the skeleton. It is vital to have a proportionate diet with sufficient quantities of dairy products, which are the principal sources of foods high in calcium.

Vitamin D is the broad name for a cluster of steroid-like substances with anti-rachitic action. Vitamin D is obtained only from animals and there are simply some foods which have vitamin D like oily fish, fish oils, butter and eggs. Different from other vitamins, individuals can in reality produce vitamin D in their bodies as an outcome of contact to sunlight, on condition that ingestion of ascorbic acid is sufficient. The two major members of this vitamin cluster are ergocalciferol or vitamin D2 and cholecalciferol or vitamin D3. Vitamin D is necessary for the assimilation of calcium and phosphorus from the small intestine, their re-assimilation in the kidneys, and the mineralization procedure of the bones. It therefore promotes healthy bone development. It, furthermore performs a vital role in the appropriate working of muscles, nerves, blood clotting, cell growth and energy utilization.

BREAST CANCER DIET

When an individual is coping with breast cancer, appropriate nutrition becomes especially vital. Nonetheless, a proportionate diet and breast cancer treatment may not be effective simultaneously. The side effects from breast cancer treatments, together with the emotional stress of being afflicted by the disease can make it tough for patients to eat. By pursuing recommen-dations to deal with the side effects when eating; selecting beneficial foods to eat; and drawing assistance and support from others, breast cancer patients can preserve a well-balanced diet during cure and recuperation. While taking into account the diet in breast cancer treatment, one should initiate a hunt for cancer-combating nutrients. Foods with such nutrients should assist in satisfying the palate of the breast cancer patient. One may start by looking for foods that are rich in selenium.

Selenium effective in fighting cancer:

- Selenium is equally an indispensable natural mineral and a helpful cancer-fighting antioxidant.
- Foods rich in selenium require being an ingredient of a diet for breast cancer treatment.
- Selenium aids to fortify the immune system, and helps in easing the action of the thyroid. It also assists in stimulating specific enzymes.
- One can acquire a required ingestion of selenium by consuming tuna, eggs, wheat grain, chicken, liver, garlic as well as Brazil nuts.

The benefits of selenium make it suitable for addition in the diet for breast cancer cure. Studies indicate a connection between the capacity of selenium to initiate the action of specific enzymes and the mineral's ability to defend against cancer threats. An every day ingestion of no less than 70 microgm. of selenium should be the aim of every woman who is concerned with regard to her diet for breast cancer treatment. If an individual consumes a quantity far higher than 100 microgm. every day, then this surplus quantity of selenium can trigger nausea, awful breath, skin complaints, giddiness, feebleness and cold indications.

Ingestion of above 60 microgm. of selenium every day is unsuitable for pregnant women. An excessive ingestion of selenium seems connected with birth problems. However, certain women can include extra selenium to a diet in breast cancer treatment. One more nutrient that should be an ingredient of every diet in breast cancer treatment is indole-3-carbinol, a chemical that is as well described as 13C. This phytochemical, a crucial component of the chemical composition of broccoli and additional cruciferous vegetables has been revealed to be a natural means of averting specific cancers.

The useful 13C utilizes three dissimilar means for supplying its beneficial effects. Initially, 13C disrupts the cancer cell phase; a broken up cycle checks cell division. Second, 13C checks the development of blood vessels in the tumor; with no blood vessels, the tumor cells can not obtain required nutrients. Third, 13C work in a way that is capable of preparing the phase for the death of the cancer cells. The maximum worth of 13C stems from the anticipated product formed when an individual has consumed a diet high in 13C.

NUTRIENT ADDITIONS TO FOOD PRODUCTS

The following is not an exhaustive review of all food products which are fortified. Food products have been selected for discussion on the basis of their nutritional impact or their applicability to fortification programmes in developing countries.

RICE AND OTHER WHOLE CEREAL GRAINS

Among the cereals, rice presents unique problems in fortification. These are due to the fact that it is most commonly consumed as a whole grain and also in many countries, extensive washing of the grain prior to cooking is the normal practice. The earliest methods of rice enrichment involved the production of parboiled and converted rice. By this means nutrients from the bran layer were transferred to the starchy endosperm.

The parboiling process involved the soaking of the rough rice, the application of heat followed by drying and milling. It was demonstrated that in this way 50-90 per cent of the thiamine was retained. The process for converted rice developed by Huzenlaub was similar to that for parboiling, but also employed pressure differences to facilitate transfer of nutrients. Acid parboiling, described by Kondo was similar to the parboiling except that it was carried out in the presence of acetic acid. After parboiling and converting, the next methods of enrichment involved the actual addition of nutrients to the milled products. Techniques used for this have been classified into two main groups 'powder type' and 'grain type' enrichment.

In powder type enrichment, a powdered pre-blended mixture of vitamins and minerals has been added at a rate of 1, 0.5. or 0.25 oz. per 100 lbs of rice (a w/w ratio of 1:1600, 1:3200 or 1:6400). For white parboiled rice, the normal practice has been to add the premix soon after milling as the heat and moisture at the grain surface at this point facilitates adherence of the powder. A major disadvantage of this method of nutrient addition has been that 20-100 per cent of the nutrients are lost on washing. In the USA, rice enriched in this way must bear a label stating 'to retain vitamins do not rinse before or drain after cooking'.

In the second major type of enrichment, a powdered nutrient mixture has been applied to the milled rice grains followed by coating with a water insoluble substance. A fortified rice premix produced in this way has then been added to milled rice at a rate of 0.5 per cent, to yield an enriched product conforming to the required standard of identity. In the United States of America, this standard requires between 2.0–4.0 mg of thiamine, 1.2–2.4 mg of riboflavin, 16–32 mg of niacin or niacinamide and 13–26 mg of iron per 100 lbs. of rice. It may contain 250 -1000 USP units of vitamin D and 500 -1000 mg of calcium.

The requirement regarding riboflavin has been stayed pending final action. The objections regarding the use of this vitamin in the premix have been due to its role in the colour deterioration of the enriched grain. 'Grain type' or 'coated grain' enrichment have also been carried out by spraying the premix solution onto the rice which is contained in a rotary cylinder, followed by hot air drying of the rice, application of a water insoluble sealant, addition of the iron compound and finally a second application of the water insoluble compound.

Water insoluble coatings which have been reported include an ethanol or isopropanol solution of zein, palmitic or stearic acid and abeitic acid. Other coating materials have contained ethyl cellulose. They dissolve at the elevated temperatures employed during cooking. Using this 'coated grain' procedure, Cort et al and Rubin et al. were able to successfully enrich rice with niacin, thiamine, pyridoxine, vitamin A, vitamin E, folic acid, iron, and zinc using expanded versions of the procedures already described. The water insoluble vitamins and minerals were added in different layers with intermittent coatings of shellac. Stability to rinsing treatments were reportedly high, with vitamin losses in the range of 0.2-1.1 per cent. Ricegrowers Co-operative Ltd. (RCL) in Australia revised the HLR rice fortification procedure for the enrichment of rice with thiamine, niacin and iron. The HLR process involved dissolution of vitamins in dilute sulphuric acid, application of vitamin solution to rice, application of a water insoluble substance, application of ferric pyrophosphate and talc followed by reapplication of a water insoluble layer and iron mixture before screening and packaging.

Due to increased cost of raw materials, low factory output, use of hazardous materials, reported problems with browning of enriched grains, RCL modified the procedure to include the application of the nutrient mixture to the rice followed by drying, screening and packing. The nutrient mixture was composed of ferric pyrophosphate suspended in an acid solution of the vitamins. They reported that the browning found in the HLR premix on storage was due to the formation of ferric sulphates and that this problem was eliminated with the use of an alternative acid. Acid hydrolysis at the surface of the rice grain provided a sufficiently sticky surface to ensure adhesion of the iron compound. Cost of the process was reduced and the factory output was doubled. There was no mention, however, of the stability of premix to nutrient loss through rinsing.

In the Philippines, attempts at fortifying rice by application of vitamin A followed by a water insoluble layer were abandoned since washing losses of between 10–20 per cent were recorded. Rice fortification with iron has also been tried in the Philippines using the coating method. The fortificant used in this programme was anhydrous ferrous sulphate. Some discolouration was found after 20 weeks of storage at room temperature and washing losses were 9 per cent. Ferric orthophosphate (white iron) is a recommended form of iron for use in the fortification of rice. This iron compound is almost water insoluble and has been preferred for mixing with milled rice due to its white colour. When it is oxidised or contains excessive moisture it may become tan, yellow, purple and or black.

Hurrel reported that the bioavailability of ferric orthophosphate varied widely from batch to batch and was highly negatively correlated with particle size as was also found with elemental iron. This fortificant compound is more expensive than anhydrous ferrous sulphate. According to Hurrel the cost of ferric orthophosphate was about six times that of anhydrous ferrous sulphate, for the same level of total iron. In Japan, a multinutrient enriched rice has been on the market since 1981. The first step in this procedure was acid parboiling in the presence of thiamine, riboflavin, niacin, pantothenic acid and pyridoxine. The second step involved coating the grain with separate layers of vitamin E, calcium and iron in separate layers and finally a protective coating material. Modified atmosphere packaging, utilising aluminium laminate and carbon dioxide, ensured the stability of vitamin E during storage. The yellow or brown colour of the premix was not a problem in this case as it was used in the fortification of 'brown' rice.

Apart from these well established procedures there have been other innovations regarding alternative methods for rice fortification. Joseph et al. described an enrichment procedure using a premix containing thiamine, riboflavin, niacin and pyridoxine. This procedure involved soaking the milled rice in an acid medium containing the water soluble vitamins followed by the cross linking of starch granules in the enriched grains. The cross linking procedure itself was demonstrated to have caused significant vitamin loss, but the added vitamins were highly cook and wash stable. It is possible that this method could have some utility in the future.

Use of fortified simulated grains has featured prominently in attempts at rice fortification in the developing countries. Murphy et al. described the production of such a fortified rice product for use in the Philippines. The synthetic rice grains were produced by extrusion of rice flour in a pasta machine. The best formulation contained vitamin A stabilised by a mixture of tocopherol, ascorbate and lipids with a low level of unsaturation. Retinyl palmitate stabilised in an acacia matrix, type 250 SD was the fortificant used. Retention of vitamin A after washing was reported to be 100 per cent. Vitamin retention after cooking, however, ranged from 60–94 per cent. The formulations which demonstrated the better storage stability, particularly at high humidity, suffered greater cooking losses. Drawbacks with this technology have been reported to exist with respect to blending with the natural product and in the consistency of the simulated grains after cooking. Field trials in Brazil showed no problems with acceptability of rice enriched in this way.

The enrichment of whole wheat grains with vitamin A has been attempted. The fortificant used was a premix comprised of concentrated vitamin A attached to wheat grains, for mixing at a level of 0.25 per cent with wheat grains. The feasibility of this procedure was not determined. The fortification of whole grain cereals with soluble iron compounds is difficult because they promote oxidation of the lipid component of the grain, thus reducing the shelf life.

CEREALS AND CEREAL BASED PRODUCTS

This is one of the most important areas for consideration of food fortification technologies. Foods from this category form the major component of diets around the world. This is especially true in the case of developing countries where dietary diversity is limited.

On average, cereals provide 52 per cent of caloric intake globally. For Africans and Asians they represent 60-75 per cent of the caloric intake; for Latin Americans, 50 per cent; and in the United States of America, 26 per cent. In developing countries 95 per cent of the population consume cereals as a dietary staple which also provide about 47 per cent of the per capita protein intake.

Milling of cereal grains prior to their consumption is a common practice. During the milling process a substantial proportion of the nutrients are lost from the refined product. The fortification/enrichment of cereal grains can therefore be rationalised in more than one way. One valid reason is to restore to refined products, nutrients which have been removed during the milling process. Another reason is to improve the nutrient intake levels of target populations which are at risk of micronutrient deficiency.

FLOURS, CORNMEAL AND BREAD AND PASTA

In the fortification of flour the required nutrient mixture is mixed with an appropriate diluent to produce a premix, which is then accurately metered into the flour. The addition of vitamins B_1 and B_2, niacin, iron and calcium to wheat flour is a common practice in many developed countries. It is technologically feasible to add other vitamins and minerals as well. Cort et al successfully used two premixes to fortify wheat flour; the first comprised vitamin A, pyridoxine, folic acid, tocopherol acetate, thiamine, riboflavin, niacin and iron while the second contained calcium, magnesium and zinc.

The vitamin/iron premix demonstrated excellent stability on storage. The flour enriched with both premixes also demonstrated excellent stability on storage at room temperature. Under conditions of accelerated storage at elevated temperature (45 °C), however, there was substantial loss of vitamin A beyond 4 weeks of storage. Parrish et al. also reported good stability of enriched wheat flour stored at room temperature, but about 50 per cent losses in flour stored at 40 °C for 6 months.

The form of vitamin A most commonly used in the fortification of flour was dry stabilised vitamin A palmitate (type 250-sd) powder form. The water soluble vitamins (thiamin, riboflavin, niacin, pyridoxine, folate and calcium pantothenate) are used in pure crystalline form. The mononitrate salt of thiamine is preferred for this use.

Iron is normally used in its reduced elemental form. In the work of Parrish et al. ferrous sulphate was also used as the source of iron and there was no significant difference demonstrated between the flour enriched with ferrous sulphate and that enriched with elemental iron. It has been recommended that ferrous fumarate be used in the fortification of wheat flour within the P.L. 480 programme, due to its greater bioavailability. When mixing dry fortificants with dry foods, careful selection of the physical characteristics of the fortificant compound is important so as to ensure adequate mixing and to minimise segregation on storage.

Rubin et al. investigated the stability of bread made from flour enriched by the 6 vitamins and 4 minerals described by Cort et al. They found that the inclusion of calcium and magnesium adversely affected vitamin A retention during the baking process. There was also an off-flavour detected in bread which included magnesium after 5 days of storage. Crumb colour and grain were also negatively affected by the nutrient multimixes tested by Rubin et al., with the greater affect occurring with the nutrient mix which included magnesium. Fortification of bread with zinc salts exerted no adverse effect on loaf volume, flavour or any other index of bread quality.

Iodisation of bread has been carried out in the Netherlands and Tasmania by the addition of 2-4 ppm KIO_3 to the bread improver which was already in general use. KIO_3 has been used in bread production in the past, not as a fortificant but as an oxidising agent to improve dough quality.

The fortification of corn meal with 6+4 vitamin/iron mixture was shown to be technically feasible (Rubin et al., 1977; Parrish et al., 1980). Losses of vitamin A on storage at room temperature for up to six months were greater than with wheat flour but still remained below 20 per cent.

In many countries pasta or noodles are commonly eaten and these can therefore be important vehicles for fortification. The manufacture of such products involves the production of a dough which is then extruded and dried. Enrichment can be through the use of enriched flour or alternatively, wet addition of a dispersion of the required vitamins can be carried out at the dough-making stage.

Vitamin losses during production depend largely on the drying conditions employed. Dexter et al. enriched durum wheat flour with a vitamin mixture containing riboflavin, thiamine mononitrate and niacin for the production of spaghetti. They reported that high temperature drying treatments resulted in significant losses of riboflavin, whereas the other vitamins were stable to the processing conditions. In all cases cooking losses, estimated between 40–50 per cent, exceeded those experienced during processing.

BIO NUTRIENTS

ENERGY

The main flow of energy that helps shape conditions on Earth's surface comes from space; and when the energy's work here is done, to space it returns. With respect to energy, then, Earth is an open system. With respect to its chemical endowment, however, Earth is a closed system. That is, the amounts of carbon, hydrogen, oxygen, iron, gold, and other elements in the planet-atmosphere system do not change with time; the chemical arrangement and physical distribution of these elements can and do vary, but essentially nothing enters and nothing leaves the system. The elements in this closed system that are essential to life are called *nutrients*. They can be divided into three categories: the 4 main chemical building blocks of living matter (carbon, oxygen, hydrogen, nitrogen); 7 *macronutrients*, of which smaller but still significant quantities are required for life; and 13 *micronutrients*, or trace elements, of which tiny quantities perform essential functions. The suitability of any terrestrial or aquatic environment for the support of life depends on the availability of nutrients in appropriate forms and quantities. The processes that govern this availability (or lack of it) are known collectively as *nutrient cycles*, because of the way the individual basic stocks of physical material move cyclically through the living and nonliving parts of the physical world.

These cycles function in support of life not merely by making nutrients continuously available—in other words, by maintaining the fertility of the environment -but also by limiting the accumulation of material in quantities, forms, and places in which it would damage organisms. It is important to recognize in this connection that the same elements and compounds that serve as nutrients for some organisms in some concentrations are often toxic for other organisms, or even for the same organisms at higher concentrations. For example, molecular oxygen (O_2) is toxic to anaerobic organisms and, at high enough concentrations, even to mammals. Ammonia (NH_3) is an important source of nutrient nitrogen for many plants but is toxic to people.

Hydrogen sulfide (H_2S) is a nutrient for certain types of bacteria but is extremely toxic to mammals.

DYNAMICS OF NUTRIENT CYCLING

The steps that make up all nutrient cycles involve two basic processes: physical transport and chemical transformation.

They are:

- The hydrologic cycle, including evaporation and precipitation as well as the flow of streams, groundwater, and ice;
- Winds;
- Ocean currents;
- Geologic movement, especially the upward and downward motions at the boundaries of tectonic plates and geologic uplifting on continents.

A fifth agent of physical transport that is important in a few circumstances is the movement of organisms. For example, fish-eating birds that deposit their excrement on land provide a significant pathway by which nitrogen and phosphorus are transferred from the sea to the land. Fish such as salmon, that feed mostly in the ocean but migrate up freshwater rivers to spawn and die, perform a similar function, as do the ocean-caught fish consumed by continent-dwelling human beings. How the various mechanisms of transport link the principal "compartments" of the physical world. Obviously, the ease (and, hence, the rate) with which a given substance can move through these pathways depends on its physical state (solid, liquid, gas), on other physical properties (solubility, volatility), and on the possibility of chemical transformations that can change those properties. Such chemical transformations take place in living organisms and in the nonliving environment.

They are considered in more detail below. Each of the compartments indicated often is divided into subcompa-rtments for the purpose of tracing nutrient flows. Thus, for example, it is customary to consider three subcompartments of material in both the soil and the oceans: living organic matter, dead organic matter, and the inorganic medium itself.

These subcompartments can be further subdivided—for example, organic matter, into plants, animals, and microorganisms; and seawater, into the well-mixed surface layer and the stratified deeper layers. Finally, within each subdivision of each subcompartment, the magnitudes of the stocks, or pools, of the important elements and compounds are of interest, as are the magnitudes of the flows into and out of the pools (where flows can mean physical movement or chemical transformation or a combination of the two). For any pool that is in steady state—that is, for which inflows balance outflows so the size of the pool does not change with time—an average residence time for a compound can be calculated as the ratio of the size of the pool to the sum of the inflows or outflows. Examination of the residence times for important pools provides quick insight into the dynamics of nutrient cycling, and, indirectly, into the probable vulnerability of specific components of nutrient cycles to disruption by human intervention. For example, a long residence time means the annual flows are small compared to the pool, which means in turn that unless human disruptions cause a very large alteration in inflow or outflow they will affect the size of the pool only very slowly. A short residence time, by contrast, means that significant changes in inflow or outflow can affect the size of the pool very quickly.

CHEMISTRY OF NUTRIENT CYCLES

Two kinds of chemistry play crucial roles in the principal global nutrient cycles: the chemistry of the sedimentary cycle, associated with the very slow transformation of very large pools of material, and the chemistry of life, associated with the more rapid transformation of much smaller pools. Both kinds of chemistry depend very heavily on the very special roles of water as a source of hydrogen (H^+) and hydroxyl (OH^-) ions for many chemical reactions, as a medium for holding other chemical compounds in solution where they can react, and as a vehicle for physical transport of compounds from place to place.

Sedimentary Cycle

Nutrients are mobilized in the sedimentary cycle when water comes in contact with rock. Rainwater is slightly acidic, having absorbed some atmospheric carbon dioxide to form weak carbonic acid:

$$H_2O + CO_2 \rightarrow H_2CO_3$$

$$H_2CO_3 \rightarrow H^+ + HCO_3^-$$

The pH of rainfall in areas free of industrial pollution is around 5.7. (Recall from basic chemistry that pH is the negative of the log to base 10 of the concentration of hydrogen ions in moles per litre. The pH of distilled water is 7.0, corresponding to 10-7 moles of hydrogen ion per litre.) This slight acidity is enough to facilitate greatly the dissolution of many minerals present in exposed rock—a process called weathering. Among the most important weathering reactions is that of the common mineral feldspar (a constituent of granite):

$$\underset{\text{feldspar}}{2KAlSi_3O_8} + \underset{\text{carbonic acid}}{2(H^+ + HCO_3-)} + \underset{\text{water}}{H_2O} \rightarrow$$

$$\underset{\text{kalinite}}{Al_2Si_2O_5(OH)_4} + \underset{\text{dissolved silica}}{4SiO_2} + 2K^+ + 2K^+ + \underset{\text{dissolved potassium andbicarbonate ion}}{2K^+ + 2HCO_3^-}$$

Kaolinite is a clay; it is left behind in this process as a contribution to the formation of soil, whereas the dissolved silica, potassium ion, and bicarbonate ion are carried away in the run-off. The mobilization and removal of positive ions such as (in this case) potassium is called leaching. Note that the reaction consumes some water and that it reduces the acidity of the solution (that is, it removes hydrogen ions) as it proceeds. These properties are characteristic of weathering reactions. Essentially the same reaction also occurs in the weathering of the aluminosilicates of sodium and calcium ($NaAlSi_3O_8$, $CaAl_2Si_2O_8$). Other silicate minerals also weather in the same general manner to produce kaolinite or other clays.

The weathering of certain other minerals leads to their complete dissolution; no solid trace is left behind to be incorporated into soil. The most important of these minerals are the limestones, or carbonate rocks, which undergo the following reactions:

$$CaCO_3H^+ + HCO_3^- \rightarrow Ca^{++} + 2HCO_3^-$$

$$CaMg(CO_3)_2 + 2(H^+ + HCO_3^-) \rightarrow Mg^{++} + Ca^{++} + 4HCO_3^-$$

Still other weathering reactions consume atmospheric oxygen in the oxidation of iron minerals:

$$4FeSiO_3 + O_2 + 2H_2O \rightarrow 4FeO(OH) + 4SiO_2$$

$$4Fe_3O_4 + O_2 \rightarrow 6Fe_2O_3$$

$$4FeS_2 + 15O_2 + 8H_2O \rightarrow 2Fe_2O_3 + 8H_2SO_4$$

Most of the products of the weathering reactions mentioned here eventually reach the sea, either as dissolved ions or as suspended solids. Between the time they are freed from rock and the time they reach the sea, of course, some of these elements have made one or more passages through the living part of the biosphere as nutrients. In the sea, too, the nutrient substances are captured from the water by living organisms, eventually to be returned to the water or to the sediments on the ocean floor.

The reverse of many of the weathering reactions also can occur in seawater in the absence of intervention by living organisms, forming solid precipitates that join the organic debris on the ocean bottom. Apparently these outputs of the major nutrient elements from the oceans roughly balance the inputs, because considerable physical evidence indicates that the salinity of seawater has been approximately constant for at least the past 200 million years.

The sedimentary cycle is finally closed by the geologic processes that is, by the processes of transformation and transportation that turn seafloor sediments into continental rocks, and by volcanic action.

Living Matter

Ecologist Edward S. Deevey, Jr., has pointed out that living matter can be regarded as a chemical compound with the empirical formula

$$H_{2960}O_{1480}C_{1480}N_{16}P_{1.8}S.$$

This enormous (and completely intentional) oversimpli-fication serves both to call attention to the chemical side of life the side concerned -and, more specifically,the relative proportions in which the six most abundant elemental constituents of living material exist.

Actually, there is substantial variation in the ratios of the chemical elements of life in different kinds of organisms, the above formula being a weighted average in which woody plants predominate. The numbers are, in any case, not known with great accuracy. The basic reactions that link the water cycle, the carbon dioxide cycle, and the oxygen cycle through living matter are photosynthesis:

$$6CO_2 + 12H_2O + \text{solar energy} \rightarrow + 6O_2 + C_6H_{12}O_6 + 6H_2O$$

and its opposite, respiration, or oxidative metabolism:

$$6O_2 + C_6H_{12}O_6 \rightarrow 6CO_2 + 6H_2O + \text{heat}$$

The compound ($C_6H_{12}O_6$, or $(CH_{2O})_6$, is glucose, the simplest carbohydrate produced by photosynthesis. The photosynthesis reaction is written with what appear to be six extra H_2O molecules on each side because twelve molecules of H_2O are actually broken up for each molecule of $C_6H_{12}O_6$ synthesized; radioactive-tracer experiments show that all twelve oxygen atoms from the 12 H_2O end up in the $6O_2$, and the oxygen in the six new water molecules comes from the CO_2. The amount of energy transformed from light into stored chemical energy by the photosynthesis reaction is 112 kilocalories (469 kilojoules) per mole of carbon fixed. The net result of the respiration reaction is to transform the same 112 kilocalories per mole of carbon from chemical energy into heat. Photosynthesis is carried out not only by higher green plants (such as oak trees, grasses, and ferns) but also by various kinds of algae (green, brown, red, blue-green) and by certain bacteria. In the bacteria, substances other than water serve as the sources of hydrogen for the reduction of carbon. For example, the green sulfur bacteria use hydrogen sulfide instead of water, according to the equation

$$12H_2S + 6CO_2 \rightarrow 6CO_2 + \text{light} \rightarrow C_6H_{12}O + 6H_2O + 12S$$

producing elemental sulfur rather than oxygen. (In fact, these bacteria are obligate anaerobes—they would be poisoned by oxygen.) The reverse reaction (respiration) consumes sulfur and carbohydrate and liberates heat, carbon dioxide, and hydrogen sulfide. Still other photosynthesis reactions use hydrogen, obtained from H_2O or H_2S or other hydrogen donors, to reduce substances other than carbon dioxide. For example, most higher plants can reduce nitrate, instead of CO_2, in photosynthesis:

$$9H_2O + 2NO_3^- + \text{light} \rightarrow 2NH_3 + 6H_2O + 4\tfrac{1}{2}O_2$$

and nitrogen-fixing photosynthetic organisms can reduce atmospheric nitrogen:

$$6O_2 + C_6H_{12}O_6 \rightarrow 6CO_2 + 6H_2O + \text{heat}$$

These variations notwithstanding, the great bulk of photosynthesis on Earth consists of the reduction of carbon dioxide, with water as the hydrogen source. The photosynthesis and respiration reactions, as we have written them here, actually represent only the end points of vastly more complicated sequences of chemical events. For example, although burning glucose in air is described essentially completely by the respiration reaction given

$$6O_2 + C_6H_{12}O_6 \rightarrow 6\ CO_2 + 6H_2O + \text{heat},$$

to achieve the same result in a living cell (at a temperature far below that of combustion) requires many intermediate steps. The intermediate steps in both respiration and photosynthesis rely heavily on three kinds of organic compounds: the energy carrier ATP (adenosine triphosphate); the electron carrier NADP (nicotinamide adenine dinucleotide phosphate); and various catalysts called enzymes. (Catalysts speed up chemical reactions without themselves being consumed.)

A closer look at these compounds provides some clues to the importance of the elements nitrogen, phosphorus, and sulfur in the chemistry of life. All biochemical reactions of importance (not just photosynthesis and respiration) require enzyme catalysts if they are to proceed at significant rates. All enzymes are *proteins*. Proteins other than enzymes also serve as hormones; as vehicles to transport oxygen; as important structural components of skin, tendons, muscle, cartilage, and bone; and in numerous other roles. All proteins are made up of building blocks called *amino acids,* and all amino acids contain the amino group—NH_2. A single protein contains from fifty to some tens of thousands of amino acids, and hence at least that many nitrogen atoms. (Some amino acids contain more than one nitrogen.) The shape and stiffness of many proteins—properties essential to their functions—are governed and maintained by bonds in which sulfur plays a crucial role, so it, too, is essential to the chemistry of life.

ATP and NADP are examples of *nucleotides,* compounds consisting of a five-carbon sugar to which one or more phosphate groups ($-PO_4$) and a *nitrogen* base are attached. A nitrogen base is a ring compound containing both nitrogen and carbon in the ring structure. Nucleotides not only perform critical energy-transfer and electron-transfer functions in the chemistry of living cells, but they are also the building blocks of *nucleic acids*. Each nucleic acid molecule contains hundreds to thousands of nucleotides, the arrangement of which is a chemical means of storing information.

Nucleic acids control the synthesis of proteins and a variety of other processes in living cells. The nucleic acid DNA (deoxyribonucleic acid) is the *genetic material*—the bearer of the coded information that is passed from parent cells and organisms to their offspring to "tell" them what to become. The nucleic acid RNA (ribonucleic acid) plays essential roles in carrying out the instructions coded in DNA molecules. In addition to its role as a constituent of DNA and RNA, phosphorus, in the form of phosphate, is one of the principal anions (ions with a negative charge) that maintains electric neutrality in the fluids of living organisms. Along with calcium, it is also an important component in shells and bony structures. Sulfur, besides playing its role in proteins, in the form of sulfate (SO_4) is another principal anion in body fluids and cells. The third principal anion is the chloride ion (Cl). The principal cations (positively charged ions) are sodium, potassium, calcium, and magnesium. Magnesium is an essential constituent of many enzymes and of the green pigment chlorophyll. Iron is a key component of many enzymes and of the oxygen-transporting protein hemoglobin.

Oversimplifying considerably, one can say that the quantity of living matter in a given environment is *limited by the stock of that requisite of life that is in shortest supply*. This rather obvious proposition carries the name *Liebig's law of the minimum*. Of course, the limiting factor varies from one type of organism to another and from one environment to another. Generalizations are difficult, but it is fair to say that under many circumstances the quantity of plant material in a given environment is in fact controlled by a *limiting nutrient* (although the density of a given kind of plant is often controlled by a herbivore). In desert regions the limiting nutrient is generally water. Under many conditions where water is not limiting, as in freshwater lakes, the limiting nutrient turns out to be phosphorus; in the open ocean it may be iron. In some richly productive environments, nitrogen and phosphorus are jointly limiting—producing any more plant material would require supplying more of both nutrients. Obviously, the limiting nutrient in a given biological system represents a leverage point where the effects of natural or human-induced perturbations manifest themselves rapidly.

MICRONUTRIENT NUTRITION

A brief discussion of micronutrient functions and nutrient deficiency symptoms in plants and soil conditions affecting micronutrient availability serves to help understand their importance in crop production and to recognize symptoms of possible deficiencies.

Boron

A primary function of boron is related to cell wall formation, so boron-deficient plants may be stunted. Sugar transport in plants, flower retention and pollen formation and germination also are affected by boron. Seed and grain production are reduced with low boron supply. Boron-deficiency symptoms first appear at the growing points. This results in a stunted appearance (rosetting), barren ears due to poor pollination, hollow stems and fruit (hollow heart) and brittle, discolored leaves and loss of fruiting bodies. Boron deficiencies are mainly found in acid, sandy soils in regions of high rainfall, and those with low soil organic matter. Borate ions are mobile in soil and can be leached from the root zone. Boron deficiencies are more pronounced during drought periods when root activity is restricted.

Copper

Copper is necessary for carbohydrate and nitrogen metabolism, so inadequate copper results in stunting of plants. Copper also is required for lignin synthesis which is needed for cell wall strength and prevention of wilting. Deficiency symptoms of copper are dieback of stems and twigs, yellowing of leaves, stunted growth and pale green leaves that wither easily. Copper deficiencies are mainly reported on organic soils (peats and mucks), and on sandy soils which are low in organic matter. Copper uptake decreases as soil pH increases. Increased phosphorus and iron availability in soils decreases copper uptake by plants.

Iron

Iron is involved in the production of chlorophyll, and iron chlorosis is easily recognized on ironsensitive crops growing on calcareous soils. Iron also is a component of many enzymes associated with energy transfer, nitrogen reduction and fixation, and lignin formation. Iron is associated with sulfur in plants to form compounds that catalyze other reactions. Iron deficiencies are mainly manifested by yellow leaves due to low levels of chlorophyll. Leaf yellowing first appears on the younger upper leaves in interveinal tissues. Severe iron deficiencies cause leaves to turn completely yellow or almost white, and then brown as leaves die. Iron deficiencies are found mainly on calcareous (high pH) soils, although some acid, sandy soils low in organic matter also may be iron-deficient. Cool, wet weather enhances iron deficiencies, especially on soils with marginal levels of available iron. Poorly aerated or compacted soils also reduce iron uptake by plants. Uptake of iron decreases with increased soil pH, and is adversely affected by high levels of available phosphorus, manganese and zinc in soils.

Manganese

Manganese is necessary in photosynthesis, nitrogen metabolism and to form other compounds required for plant metabolism. Interveinal chlorosis is a characteristic manganese-deficiency symptom. In very severe manganese deficiencies, brown necrotic spots appear on leaves, resulting in premature leaf drop. Delayed maturity is another deficiency symptom in some species. Whitish-gray spots on leaves of some cereal crops and shortened internodes in cotton are other manganese-deficiency symptoms. Manganese deficiencies mainly occur on organic soils, high-pH soils, sandy soils low in organic matter, and on over-limed soils. Soil manganese may be less available in dry, well-aerated soils, but can become more available under wet soil conditions when manganese is reduced to the plant-available form. Conversely, manganese toxicity can result in some acidic, high-manganese soils. Uptake of manganese decreases with increased soil pH and is adversely affected by high levels of available iron in soils.

Molybdenum

Molybdenum is involved in enzyme systems relating to nitrogen fixation by bacteria growing symbiotically with legumes. Nitrogen metabolism, protein synthesis and sulfur metabolism are also affected by molybdenum. Molybdenum has a significant effect on pollen formation, so fruit and grain formation are affected in molybdenum-deficient plants. Because molybdenum requirements are so low, most plant species do not exhibit molybdenum-deficiency symptoms.

These deficiency symptoms in legumes are mainly exhibited as nitrogen-deficiency symptoms because of the primary role of molybdenum in nitrogen fixation. Unlike the other micronutrients, molybdenum-deficiency symptoms are not confined mainly to the youngest leaves because molybdenum is mobile in plants. The characteristic molybdenum-deficiency symptom in some vegetable crops is irregular leaf blade formation known as whiptail, but interveinal mottling and marginal chlorosis of older leaves also have been observed. Molybdenum deficiencies are found mainly on acid, sandy soils in humid regions. Molybdenum uptake by plants increases with increased soil pH, which is opposite that of the other micronutrients. Molybdenum deficiencies in legumes may be corrected by liming acid soils rather than by molybdenum applications. However, seed treatment with molybdenum sources may be more economical than liming in some areas.

Zinc

Zinc is an essential component of various enzyme systems for energy production, protein synthesis, and growth regulation. Zinc-deficient plants also exhibit delayed maturity. Zinc is not mobile in plants so zincdeficiency symptoms occur mainly in new growth. Poor mobility in plants suggests the need for a constant supply of available zinc for optimum growth. The most visible zinc-deficiency symptoms are short internodes (rosetting) and a decrease in leaf size. Chlorotic bands along the midribs of corn, mottled leaves of dry bean and chlorosis of rice are characteristic zinc-deficiency symptoms.

Loss of lower bolls of cotton and narrow, yellow leaves in the new growth of citrus also have been diagnosed as zinc deficiencies. Delayed maturity also is a symptom of zinc-deficient plants. Zinc deficiencies are mainly found on sandy soils low in organic matter and on organic soils. Zinc deficiencies occur more often during cold, wet spring weather and are related to reduced root growth and activity as well as lower microbial activity decreases zinc release from soil organic matter. Zinc uptake by plants decreases with increased soil pH. Uptake of zinc also is adversely affected by high levels of available phosphorus and iron in soils.

Chloride

Because chloride is a mobile anion in plants, most of its functions relate to salt effects (stomatal opening) and electrical charge balance in physiological functions in plants. Chloride also indirectly affects plant growth by stomatal regulation of water loss. Wilting and restricted, highly branched root systems are the main chloride-deficiency symptoms, which are found mainly in cereal crops. Most soils contain sufficient levels of chloride for adequate plant nutrition. However, reported chloride deficiencies have been reported on sandy soils in high rainfall areas or those derived from low-chloride parent materials. There are few areas of chloride-deficient soils in the U. S., so this micronutrient generally is not considered in fertilizer programmes. In addition, chloride is applied to soils with KCl, the dominant potassium fertilizer. The role of chloride in decreasing the incidence of various diseases in small grains is perhaps more important than its nutritional role from a practical viewpoint. Plants differ in their requirements for certain micronutrients. The following table shows the estimate of the relative response of selected crops to micronutrients. The ratings of low medium and high are used to indicate the relative degree of responsiveness.

THE ELEMENTS OF COMPLETE PLANT NUTRITION

Complete nutrition results in superior plant growth. Why choose anything less for your plants? There are 20 elements necessary for optimum plant growth. Air and water supply carbon, hydrogen and oxygen. Macronutrients are required by plants in large amounts. Micronutrients are required in smaller amounts. Eliminate any of these elements,and plants will display abnormal growth, deficiencies or may not reproduce. The following is a brief guide to the role played by each of these essential nutrient elements.

Macronutrients

- (*N*)-*Nitrogen*: Component of proteins, hormones, chlorophyll, vitamins and enzymes. Promotes stem and leaf growth. The ammoniacal and nitrate forms are used directly by plants for stem and leaf growth. The urea form of nitrogen must be broken down by soil borne microorganisms or urease before it can be utilized by the plant. Urea can cause leaf tip and root burn. Deficiency (Def.): reduced yields, yellowing of leaves, stunted growth. Excess nitrogen can delay fruiting and flowering.
- (*P*)-*Phosphorus*: Seed germination, photosynthesis, protein formation, overall growth and metabolism, flower and fruit formation. Def.: purple stems and leaves, retarded growth and maturity, poor flowering and fruiting. Large amounts without zinc cause zinc deficiency. Low pH (<4) ties up phosphates in organic soils.
- (*K*)-*Potassium*: Formation of sugars, carbohydrates, proteins, cell division. Adjusts water balance; improves stem rigidity and cold hardiness; enhances flavour, colour and oil content of fruits; important for leafy crops. Def.: spotted, curled or burned look to leaves; lower yields.
- (*Ca*)-*Calcium*: Activates enzymes; structural part of cell walls; influences water movement, cell growth and division Required for uptake of nitrogen and other minerals. Leached from soil by watering.

Immobile-requires a constant supply for growth. Def.: stunting of new growth in stems, flowers, roots; black spots on leaves and fruit; yellow leaf margins.

- (*Mg*)-*Magnesium*: Critical component of chlorophyll; needed for functioning of enzymes for carbohydrates, sugars and fats; fruit and nut formation; germination of seeds.
- Def.: yellowing between veins of older leaves; chlorosis; leaf droop. Leached by watering. Foliar spray to correct deficiencies.
- (*S*)-*Sulfur*: Component of amino acids, proteins, vitamins, enzymes. Essential for chlorophyll. Imparts flavour to many vegetables. Def.: light green leaves. Water supply may contain sulfur. Leached by watering.

Micronutrients

- (*B*)-*Boron*: Affects at least 16 functions: flowering, pollen germination, fruiting, cell division, water relationships, movement of hormones, cell wall formation, membrane integrity, calcium uptake, movement of sugars. Immobile; easily leached. Def.: terminal bud die back causes rosette of thick, curled, brittle leaves or brown, discolored, cracked fruits, tubers and roots.
- (*Cl*)-*Chlorine*: Involved in osmosis (movement of water or solutes in cells), ionic balance necessary to take up mineral elements and photosynthesis. Def.: wilting, stubby roots, yellowing, bronzing. Scents in some plants may be decreased. Leached by watering.
- (*Co*)-*Cobalt*: Required by nitrogen fixing bacteria; formation of B12 vitamin; formation of DNA. Will extend life of cut flowers such as roses. Def.: may result in nitrogen deficiency.
- (*Cu*)-*Copper*: Necessary for nitrogen metabolism; component of enzymes - may be part of enzyme systems that use carbohydrates and proteins. Bound tightly in organic matter. May be deficient in highly organic soils. Not readily lost from soil but may be unavailable. Def.: die back of shoot tips; terminal leaves develop brown spots. Excess is toxic.
- (*Fe*)-*Iron*: Enzyme functions; catalyst for synthesis of chlorophyll; essential for new growth. Def.: pale leaves, yellowing of leaves and veins. Leached by water and held in lower parts of soil. High pH soils may have iron present but unavailable to plants.
- (*Mn*)-*Manganese*: Enzyme activity for photosynthesis, respiration and nitrogen metabolism. Def.: young leaves are pale with green veins similar to iron deficiency; advanced stages-leaves are white and drop; brown, black or gray spots may appear next to veins. Plants in neutral or alkaline soils often show def. Acid soils may increase uptake causing toxicity.

- *(Mo)-Molybdenum*: Structural part of enzymes that reduce nitrates to ammonia for amino acid development essential to protein formation; required by nitrogen fixing bacteria. Def.: pale leaves with rolled, cupped margins. Seeds may not form. Nitrogen deficiency may occur if plants are lacking Mo.
- *(Ni)-Nickel*: Recently recognized as essential. Required for the urease enzyme to break down urea into usable nitrogen and for iron uptake. Seeds require nickel to germinate.
- *(Na)-Sodium*: Improves nitrogen metabolism in many plants, involved in osmotic (water movement) and ionic balance in plants. Def.: yellowing of leaves and leaf tip burn; may inhibit flower formation.
- *(Si)-Silicon*: Component of cell walls; enhances resistance to sucking insects and fungi.Foliar sprays reduce populations of aphids on some plants. Enhances leaf presentation; improves heat, drought and cold tolerance; improves photosynthesis; extends bloom life. Def.: wilting, poor fruit and flower set, increased susceptibility to insects and disease. Disease resistance is enhanced by regular foliar feeding.
- *(Zn)-Zinc*: Functional part of enzymes including auxin (growth hormone) synthesis, carbohydrate metabolism, protein synthesis, stem growth. Def.: mottled leaves, irregular yellow areas. Zinc deficiency leads to iron deficiency. Occurs in eroded soils; least available at pH of 5.5-7.0. Lower pH can cause availability to the point of toxicity.

PLANT NUTRIENTS

The nutrients available in a given soil ultimately depends on the rock from which the soil was made. If the plants grown from this soil die and decay where they have grown then their nutrients are recycled. However, if the plants have been grown for agriculture then they are removed from the area in which they have grown and their nutrients cannot be recycled. So in soils that are used for cropping, essential nutrients continually have to be replaced.

REPLACEMENT OF PLANT NUTRIENTS

This can be done either using manure or compost (which are relatively poor sources of nutrients) or a synthetic fertilizer. Synthetic fertilizers are usually used to replace nitrogen (in the form of urea, ammonium sulfate, ammonium nitrate or diammonium hydrogen phosphate), phosphorous, sulfur and potassium. The nitrogen is obtained from the air, the phosphorous is mined as $Ca_3(PO_4)_2$, the sulfur is mined in elemental form and reacted with oxygen to form sulfate, and the potasium is obtained from seawater. Other minor nutrients are also sometimes replaced using specialised artificial fertilizers.

SOIL ANALYSIS

If soil has an inappropriate composition for a given plant species then that plant will not thrive (or maybe will not grow at all) in that soil. To prevent this being a problem, samples of soil are often analysed so that nutrients can be added before seeds are planted. Soil is commonly analysed for potassium, magnesium, calcium, nitrogen and phosphorous (the exact choice of elements to monitor depends on which plants are to be grown) and also for pH. Some of these tests are so simple they can be carried out at home and others require specialised equipment. Plants grow in the thin upper layer of the Earth's crust known as soil. Soil is formed over very long times from igneous or sedimentary rock, volcanic ash, sand or peat. With the exception of peat, these are all silicate minerals, so soil is largely made up of silicates and organic material known as humus. Plants use this soil for anchorage and as a source of nutrients and water.

WHY PLANTS NEED SOIL

Green plants are very effective photochemical factories: they use the energy of light to combine carbon dioxide from the air with water from the soil to make carbohydrates:

$$6CO_2(g) + 6H_2O(l) + \text{energy} \rightarrow C_6H_{12}O_6(aq) + 6O_2(g).$$

This reaction occurs in the chloroplasts of all green plants, using the green pigment chlorophyll to trap the energy from sunlight. The basic building blocks of glucose are then used to build up more complex carbohydrates such as cellulose. However, plants need more than just carbon dioxide, water and sunlight to keep alive. The elemental composition of a 'typical' plant. It can be seen that the composition of plants is not much different from that of animals, and includes significant amounts of nitrogen and phosphorous, as well as trace amounts of many other elements.

Although there is enormous variation between the abundance of the most common element, hydrogen, and the least common, molybdenum, both are very important. For example, plants can only use nitrogen in the form of the ammonium ion. Nitrates are converted to ammonium ions by the enzyme *nitrate reductase*, and as this enzyme cannot function without molybdenum, the presence of molybdenum is essential if the plant is to receive nitrogen in a form that it can use. These trace elements, generally called plant nutrients, normally come from the soil in which the plant is growing. The plants absorb their nutrients from the soil solution (the film of water which covers all the particles in moist soil), usually in very simple forms. A typical composition of the main constituents of the soil solution from which plants must obtain all their requirements for growth. It is possible to grow plants without soil, using *hydroponics*. The plant is grown in a solution which contains all the required plant nutrients, in the correct proportions. It is possible to determine just how much of any particular nutrient a plant needs by starting with a balanced nutrient solution, and then varying the amount of each nutrient present, one nutrient at a time. When insufficient of the nutrient is present the plant will not thrive. As the amount of nutrient is increased the growth improves, until all that the plant can utilize is present, when adding more of that nutrient produces no additional growth. If the amount of the nutrient is increased still further, the excess may eventually cause toxic effects, and growth is retarded. For some nutrients, *e.g.* potassium, plants can tolerate a large excess before ill effects occur; for others, *e.g.* boron, the optimum range is narrow.

REPLACEMENT OF PLANT NUTRIENTS

Removing crops, or animals, from an area steadily drains nutrients from the soil, and these must be replaced if the soil is to remain fertile. The types of plants which are grown in modern agriculture usually require higher levels of nutrients for optimum growth than are usually present in New Zealand soils. To increase soil fertility to the level required by the "high growth" plants used in modern agriculture, and to maintain this level with continual crop removal, essential plant nutrients have to be regularly added to the soil.

The main nutrients required are nitrogen, phosphorus, potassium, sulfur, calcium and magnesium. In particular areas certain trace elements, such as magnesium, sulfur, copper, zinc, boron and molybdenum may also be lacking, and may have to be added to produce optimum growth. It should always be kept in mind that additional plant nutrients can produce improved plant growth only if the growth is being limited from the potential determined by the plant genetics and climate by a deficiency of some nutrient, or nutrients, and if the fertilizer makes up the deficiency. There are two main options for adding these nutrients; the *organic way*, using only "natural" materials, and the use of *fertilizers*, which generally contain relatively simple inorganic compounds.

The term "inorganic fertilizer" is commonly used to mean "manufactured", or "not natural", though the chemical meaning is simply "not carbon-based". In chemical terminology "organic" means carbon based, whether originating naturally or by manufacture. The organic (carbon-based) compounds of plants are made in the plant from the carbohydrates formed from carbon dioxide by photosynthesis, and plants can not use the organic components of composts or manures. Plants absorb their inorganic nutrients as simple anions and cations, so complex organic materials have to be broken down by the process of *mineralization*, releasing the nutrients as simple ions, before they become available to plants.

COMPOSTS AND MANURES

In *organic* agriculture nutrients are replenished using only natural materials, though what is and what is not natural is often debated. The main "organic" nutrient sources are composts and manures.

Composts

Composts are formed from plant residues, and have much the same chemical composition as the plants from which they are made. Composts are *not* a rich source of plant nutrients, and need to be used in large amounts if they are the only source of nutrient replenishment. Their main use is to boost the amount of soil organic matter (humus), and composts should be regarded primarily as *soil conditioners*. There is also a philosophical problem that if compost is made from "imported" plant material, then the area from which the material was obtained has been depleted in nutrients. As example, continual use of lawn mowings to make compost depletes the lawn area of nutrients, which need to be replenished to keep the lawn healthy. Therefore, while all available plant material should be converted into compost to maintain humus levels and to recycle nutrients, this does not really solve the nutrient deficit problem. An additional problem arises if a region is deficient in some particular nutrient. The plants (and animals) grown in the region will also be deficient in this nutrient, and so also will be composts made from plants grown in the area.

Manures

Manures are usually the faeces of herbivores. In the digestion process the animal absorbs as much as it can of the nutrient elements for its own needs, so the residue finally excreted consists of plant material depleted in nutrients. The nutrients not incorporated into the animal are excreted in the animal's urine, which is a better source of plant nutrients than the faeces. As with compost, manures are useful mainly as soil conditioners. They are not a rich source of plant nutrients, and need to be used in large amounts if they are the only source of nutrient replenishment.

The old gardening advice "to dig in six inches of well rotted stable manure" is fine in principle, but not very practicable for most modern gardeners, and there remains the problem that the area where the animal was feeding has been depleted of the nutrients. Animal residues, such as "blood and bone" form another type of "natural" fertilizer. These can contain high levels of nutrients, which are generally released slowly into the soil solution as the complex compounds are broken down, which has advantages. Again these nutrients have been "mined" from the soil where the animals fed.

FERTILIZERS

The only answer to the "nutrient mining" problem is to bring fresh nutrients into the area from some richer source, and fortunately natural processes have concentrated the major nutrients in locations around the world. All "synthetic" nitrogen fertilizers obtain their nitrogen from the atmosphere. This is "fixed" using the *Haber process*, in which dihydrogen and dinitrogen at high temperature and pressure are combined over a catalyst, which speeds up the reaction forming ammonia:

$$N_2(g) + 3H_2(g) \rightarrow 2NH_3(g).$$

The hydrogen required for the Haber process is normally made using methane separated from natural gas. Methane is reacted with water and air at high temperatures over catalysts to form a mixture of dihydrogen, dinitrogen (from the air) and carbon dioxide. Carbon dioxide and water are removed (the carbon dioxide is used later to make urea), leaving dihydrogen and dinitrogen in the correct proportion to combine to form ammonia (unbalanced equation)

$$CH_4(g) + H_2O(g) + [O_2(g) + N_2(g)] \rightarrow [N_2(g) + 3H_2(g)] + CO_2(g).$$

The ammonia obtained from the Haber process is converted into a form more useful as fertilizer. It can be combined with carbon dioxide to form urea, which is chemically identical to the urea excreted in the urine of animal:

$$2NH_3(g) + CO_2(g) \rightarrow O{=}C(NH_2)_2(s) + H_2O(l)$$

or combined with sulfuric, nitric or phosphoric acid to form ammonium sulfate, ammonium nitrate or diammonium hydrogen phosphate $(NH_4)_2(HPO_4)$. The nitric acid is formed by oxidizing ammonia over a catalyst

$$NH_3(g) + 2O_2(g) \rightarrow HNO_3(g) + H_2O(l)$$

Phosphorus

Phosphorus is a relatively abundant element in the earth's crust, and occurs naturally in marine deposits of *phosphate rock,* as calcium phosphate, $Ca_3(PO_4)_2$, or the minerals *apatite,* $CaF_2.3Ca_3(PO_4)_2$, or *hydroxy apatite,* $Ca(OH)_2.3Ca_3(PO_4)_2$ (hydroxy apatite is the mineral that forms bones and teeth). The phosphate rock from the Pacific region is found on coral islands where flocks of sea birds roost. It is thought that the phosphate in their excreta, derived from their fishy diet, combines over time with coral to form the phosphate rock. Some phosphate rock is sufficiently soluble so that the phosphorus from the finely ground rock is available to plants. Other phosphate rock is too insoluble for direct use, and is commonly treated with sulfuric acid to make a mixture of calcium sulfate and the more soluble calcium hydrogen phosphate, called *superphosphate.*

$$Ca(OH)_2.3Ca_3(PO_4)_2 + H_2SO_4 \rightarrow CaSO_4 + Ca(HPO_4)_2$$

Phosphate rock can also be reacted with excess sulfuric acid to make phosphoric acid, H_3PO_4, which can be converted into ammonium hydrogen phosphate, $(NH_4)_2(HPO_4)$ for use in soluble fertilizers, or into sparingly soluble magnesium ammonium phosphate $Mg(NH_4)(PO_4)$ (magamp) for use as a slow-release fertilizer.

Sulfur

Sulfur is also an essential nutrient, and some areas of New Zealand are sulfate deficient. Use of superphosphate, or ammonium, potassium or calcium sulfate (gypsum) as fertilizer supplies the needed sulfate. The sulfuric acid needed to make superphosphate is obtained by burning sulfur to form sulfur dioxide

$$S(s) + O_2(g) \rightarrow SO_2(g)$$

which is oxidised over a catalyst to form sulfur trioxide

$$2SO_2(g) + O_2(g) \rightarrow SO_3(g)$$

which reacts with water to form sulfuric acid

$$SO_3(g) + H_2O(l) \rightarrow H_2SO_4(aq).$$

The sulfur is obtained from deposits of elementary sulfur found in volcanic areas of the world, or is extracted from natural gas, which in some places contains hydrogen sulphide. Sulfur dioxide is released into the atmosphere from volcanoes. This is ultimately converted into sulfuric acid in the atmosphere; this is washed out by rain, so providing a "natural" source of sulphur. The 1995 eruption of Ruapehu "topdressed" much of the central North Island with economically useful amounts of sulphate. Sulfur dioxide is also released into the atmosphere when coal or oil containing sulfur is burned in power stations, and the resulting sulfuric acid in the atmosphere causes "acid rain" in some industrialized areas.

Potassium

Potassium is the other main nutrient needed, and this is obtained from sea water, directly or indirectly. When sea water evaporates sodium chloride (common salt) crystallises first. If the liquid remaining (bittens) is evaporated potassium chloride crystallises. In some places this occurred over geological times when oceans were isolated by geological processes and then evaporated, leaving large deposits of salt (which crystallised first) and of potassium chloride (which crystallised later). The potassium chloride can be used directly as a fertilizer, and is satisfactory for use at low rates, but this has the disadvantage that chloride is also added, and many plants are not very tolerant of high levels of chloride.

It is therefore usual to convert the potassium chloride to potassium sulfate (which also occurs in natural deposits) for use as a fertilizer. The so-called "un-natural" inorganic fertilizers are thus derived from the very natural sources of the air (plus natural gas as a hydrogen source) for nitrogen, coral rock plus sea-bird droppings (plus sulfur from volcanoes or "sour" natural gas to make superphosphate) for phosphorus, and the ocean for potassium.

QUANTITY AND QUALITY OF PLANT PROTEIN

Efforts to improve the protein content and quality of staple foods have been underway for decades. The main focus is crops grown in developing countries, where nutrient shortfalls are widespread and dietary diversity limited. Foods such as potato and cassava, staple foods in several parts of South America and Africa, have less than one per cent protein. Efforts to improve protein quality strive to increase the amount of limiting essential amino acids provided by the protein in the food. The amino acids most often present in inadequate amounts are lysine, tryptophan, and methionine.

Improvements in protein quality benefit both human and animal nutrition and increase the feed efficiency of crops fed to food animals. For example, corn is widely fed to cattle but it is limiting in lysine and methionine. Corn with higher levels of these amino acids would significantly improve feed efficiency and lower input costs to farmers. Improved corn varieties consumed by humans would also have nutritional benefits.

There are various ways of improving protein quantity and quality. One is to increase the total amount of protein produced by selecting germplasm with an altered balance of seed proteins. This may be done by traditional cross breeding or genetic engineering. Another approach is to introduce genes from other sources for proteins that have a favourable balance of essential amino acids. An example is the introduction into potato of a gene for seed albumin protein from amaranth. A third approach seeks to increase the production of specific amino acids such as lysine.

This approach was used in the development of Quality Protein Maize. William Folk and his team at the University of Missouri, Columbia, MO, pioneered another approach to improve seed protein quality. Their strategy was to substitute more desirable and scarce amino acids for more abundant ones in certain seed proteins. They applied this concept to rice by increasing the production of lysine, an essential amino acid, at the expense of the non-essential amino acids, glutamine, asparagine and glutamic acid. *Cassava*: A staple food for some 500 million people in tropical and sub-tropical parts of the world, cassava, also known as yucca or manioc, thrives in marginal lands having little rain and nutrient-poor soils. It is widely consumed in Africa, and parts of Asia and South America. Cassava root has less than 1% protein and poor nutritional value.

However, the leaves are also consumed and these are a good source of beta-carotene, the precursor of vitamin A. In 2003, Zhang and colleagues reported using a synthetic gene to increase the protein content in cassava. The gene is for a storage protein rich in nutritionally essential amino acids. When the gene was expressed in cassava, transformed plants expressed the gene in roots and leaves, both of which are consumed in human diets.

The experiment demonstrated the feasibility of increasing the quantity and quality of protein in cassava. Cassava also contains cyanogenic glucosides that can produce chronic toxicity if not eliminated or reduced by grating, sun-drying, or fermenting. Efforts to develop cassava varieties low in these toxicants is a high research priority. *Corn*: Corn is the predominant staple food in much of Latin America and Africa. Although some varieties may contain appreciable quantities of protein, its quality is poor because of low lysine and tryptophan content. In 1964, it was discovered that corn bearing a gene known as opaque-2 contained increased concentrations of lysine and tryptophan and had significantly improved nutritional quality. However, opaque-2 corn proved to have low yields, increased susceptibility to diseases and pests, and inferior functional characteristics. At the International Maize and Wheat Improvement Centre in Mexico, work with the opaque-2 gene continued using both traditional breeding and molecular methods.

After at least 12 years' work, CIMMYT researchers succeeded in developing hardy corn varieties that contained twice the lysine and tryptophan content as traditional varieties, but were disease-resistant and high-yielding. Scientists Surinder K. Vasal and Evangelina Villegas of CIMMYT were awarded the World Food Prize in 2000 for their work developing 'Quality Protein Maize'. Quality Protein Maize varieties have been adapted to and released in over 40 countries in Latin America, Africa, and Asia. Recent researchers at CIMMYT reported the development of transgenic corn with multiple copies of the gene from amaranth that encodes for the seed storage protein amarantin. Total protein in the transgenic corn was increased by 32% and some essential amino acids were elevated 8% to 44%. In 2004, a team of researchers at the University of California, Riverside, reported that transgenic corn with increased production of the plant regulating hormone, cytokinin, had nearly twice the content of protein and oil as conventional corn.

This development resulted from an unusual change in the way the plant developed. Normally, corn ears develop flowers in pairs, one of which usually dies. Under the influence of the additional cytokinin, both flowers developed but yielded only a single kernel. These kernels contained more protein and oil than conventional corn. Pursuing a different strategy to improve protein quality, researchers at Monsanto, St. Louis, MO, used genetic engineering techniques to reduce the amount of zein storage proteins. These storage proteins constitute over half the protein in corn and are deficient in lysine and tryptophan. Increased production of other proteins in the corn led to higher levels of lysine, tryptophan, and methionine. The agronomic and nutritional properties of these lines are currently being evaluated. Researchers at the Max Planck Institute, Germany, have focused on methionine, another limiting amino acid. They elucidated several key steps in methionine metabolism in plants.

This work, currently in the preliminary stage, could pave the way for using genetic engineering techniques to improve the methionine content of plants. *Potato*: Potato is a dietary staple throughout parts of Asia, Africa, and South America. Typically, potatoes contain about 2% protein and 0.1% fat. It was reported in 2000 that, as in cassava, transfer of the gene for seed albumin protein from *Amaranthus hypochondriacus* to potato resulted in a "striking" increase in protein content of the transgenic potatoes. In 2004, researchers at the National Centre for Plant Genome Research, India, reported the development of a nutritionally improved potato line with 25% higher yields of tubers and 35%–45% greater protein content. Dubbed the "protato," the protein-rich potato had significant increases in lysine and methionine, which enhance the quality of the additional protein. In February 2004, this potato was reported "approaching release" to farming communities. It should be noted that while potatoes are known for their high starch content, it has been possible to genetically engineer potatoes that contain fat. In July 2004, Klaus and colleagues at the Max Planck Institute of Molecular Plant Physiology demonstrated increased fatty acid synthesis in potatoes. *Rice*: Almost half the world's population eats rice, at least once a day. Rice is the staple food among the world's poor, especially in Asia and parts of Africa and South America. It is the primary source of energy and nutrition for millions.

Thus, improving the nutritional quality of rice could potentially improve the nutritional status of nearly half the world's population, particularly its children. Commodity rice contains about 7% protein, but some varieties, notably black rice, contain as much as 8.5%. The most limiting amino acid in rice is lysine. Efforts to increase the nutritional value of rice target protein content and quality along with key nutrients often deficient in rice-eating populations, such as vitamin A and iron.

The International Rice Research Institute, Philippines, is a primary centre for rice research and development of improved varieties. In 1999, Dr. Momma and colleagues at Kyoto University, Japan, reported a genetically engineered rice having about 20% greater protein content compared with control rice. Transgenic plants containing a soybean gene for the protein glycinin contained 8.0% protein and an improved essential amino acid profile compared with 6.5% protein in the control rice. Dr. William Folk and his team genetically modified rice to increase its content of the amino acid lysine. They did so by modifying the process of protein synthesis, rather than by gene transfer or the expression of new proteins. They achieved an overall 6% increase in lysine content in the grain. Although lysine content remained below optimum levels, the scientists suggested that additional transformations and modifications could further boost lysine levels. Perhaps the most famous genetic transformations in rice are those in "Golden Rice" involving the vitamin A precursor, beta-carotene, and iron.

The lead scientist in the golden rice project, Dr. Ingo Potrykus, now retired from the Swiss Federal Institute of Technology, was also involved in applying biotechnology for the improvement of rice protein. Although details are sparse, Potrykus described the work of Dr. Jesse Jaynes, who synthesized a synthetic gene coding for an ideal high-quality storage protein with a balanced mixture of amino acids. The gene, named Asp-1, was transferred to rice with the appropriate genetic instructions for its production in the endosperm or starchy part of the rice grain.

The transgenic rice plants accumulated the Asp-1 protein in their endosperm in a range of concentrations and provided essential amino acids but data are not yet available on the concentrations achieved or their nutritional relevance. Precedent for the expression of a synthetic gene in rice grown in cell culture suggests that Jaynes' approach is viable.

9

Constituents of Blood

Blood is a tissue made up of cells in suspension and of plasma containing substances in solution. Forty-five percent of blood volume consists of the formed elements (red cells, white cells, and platelets). Fifty-five percent of blood volume is composed of plasma, which consists mainly of water, some proteins, and the various mineral salts necessary to life.

HOW IS BLOOD MANUFACTURED

Blood is assembled by the body from many different sources. With the exception of injections that may be necessary for treatment, all constituents of blood ultimately come from substances taken by mouth. The plasma is formed from water, proteins, and salts, all of which are absorbed from the digestive tract. The proteins are modified by the metabolism of the body into the type needed for the human species.The cells of the blood come from diverse sources. The various organs and tissues that collaborate in the manufacture of blood cells are known as the hemopoietic (blood-forming) system.

This includes the bone marrow, spleen, lymph glands, stomach, liver, and the reticuloendothelial system (a network of special tissue cells plus the lining of the blood vessels). The reticulo-endothelial cells are widely distributed throughout the body, but the most important sites are the lining of the bone marrow and spleen. They are phagocytic; that is, they can ingest micro-organisms and foreign particles of all sorts, thus removing them from the circulating blood.

DETERMINATION OF BLOOD GROUPS IN THE LABORATORY

Blood groups are determined in the laboratory by a simple technique utilizing the known facts about the A and B substances in the red cells, and the anti-A and anti-B substances in the plasma. On the left side of a clean glass slide, the technician places a drop of serum known to contain anti-A substance; on the right, a drop of serum known to contain anti-B substance. He then places a drop of blood (obtained from the person whose group he wishes to determine) into each of the specimens of anti-A and anti-B sera. After a minute or two, the technician can observe one of four possible events take place by which he can determine the blood group:

- If the anti-A serum alone causes the red cells to clump, the blood sample is labeled *group* A.
- If the anti-B serum alone causes the red cells to clump, the blood sample is labeled *group* B.
- If *both* anti-A and anti-B sera cause the red cells to clump, the blood sample is labeled *group* AB.
- If *neither* anti-A nor anti-B serum causes the red cells to clump, the blood sample is labeled *group* O.

The clumping, or absence of it, is clearly visible to the naked eye in a few moments. Thus, in a well-organized laboratory the technician can determine the groups of a large number of samples in a few hours' time. For safety, the results of blood group determination are checked in the laboratory in another procedure called proof of blood group. This utilizes samples of known group A and group B red cells, against which samples of the unknown serum are tested. Once again, four observable results may be noted: the group A cells alone may be clumped, the group B cells alone may be clumped, both group A and B cells may be clumped, or no clumping at all will occur. One can easily work out the blood group that would occur with each of these events. Still another use is made of these facts when the bottle of blood, properly identified as to group, reaches the hospital and is ready to be transfused into the injured or ill patient. Even though the blood is the same group as the patient's, another precaution must be taken to ensure that bloods of donor and recipient are compatible.

This is a laboratory test, very similar to those described above, known as crossmatching. A drop of the donor's red cells is mixed with a drop of the recipient's serum; this is known as the major side of the crossmatching test. The mixture of recipient's cells with donor's serum is known as the minor side of the test. If agglutination occurs on the major side, the blood is incompatible and must never be given because the donor's cells would be agglutinated and hemolyzed in the recipient's blood stream, causing dangerous transfusion reaction. If agglutination occurs on the minor side, as it will when group O blood is given to A, B, or AB recipients, it does not mean that the blood cannot be given. At the present time, however, it is the practice to give group-specific blood (group A to group A recipients, B to B recipients, and so on) wherever possible and to use group O donors as universal only in emergencies.

HOW ARE BLOOD GROUPS INHERITED

Blood groups are inherited as dominant characteristics according to the Mendelian law: for every inherited characteristic, including blood group, there is a pair of genes, one contributed by each parent. The blood group that a child may inherit is therefore dependent on the combinations resulting from genes contributed by both parents. Owing to intermarriage of people with different blood groups, a variety of combinations can result; but if the parents' blood groups are known, those possible for the children can be predicted.

MEDICOLEGAL IMPLICATIONS OF BLOOD GROUPS

Blood groups are inherited and never change. For this reason, they may have legal importance in identifying bloodstains or determining parentage. But the evidence so provided is useful only in a negative sense. For example, if a murder suspect had his clothing stained with group A blood, and the victim was of group O, this is evidence that the suspect was not stained by the victim's blood. However, even if the stains proved to be group O, it still would not be conclusive evidence, for the suspect may have had his clothing stained with the blood of any person who was group O. This logic also applies to paternity cases. The determination of paternity in disputed cases is very complicated. Such cases must be studied by experts in forensic medicine.

THE RH FACTOR

The Rh factor is an inherited antigenic substance in the red cells (similar in nature to the group A and B substances) that occurs in 85 percent of the white population of this country. Those who possess this factor are said to be "type Rh positive." Those who do not possess it are "type Rh negative." It occurs in about the same percentage in all blood groups; that is, of 100 group A bloods, approximately 85 will be Rh positive and 15 Rh negative; of 100 group B bloods, approximately 85 will be Rh positive and 15 Rh negative. These same percentages hold for all four blood groups.

The Rh factor was named in 1940 by Landsteiner and Wiener, who were doing work with the red cells of the rhesus monkey. They discovered that the antiserum that agglutinated the cells of the rhesus species also agglutinated 85 percent of the donors they examined; it failed to agglutinate the other 15 percent. The word rhesus was shortened to Rh in discussions of this new factor and thus entered the literature. In 1939 Levine and Stetson reported a severe transfusion reaction in a woman delivered of a dead baby, and discovered that her blood would agglutinate 80 percent of donors who were of her own blood group.

They assumed that the infant had inherited from the father a blood factor that was not present in the mother's blood. Small quantities of the fetal red blood cells had seeped through the placental membranes and caused antibody production in the mother. The brilliant scientific work by which these groups of research workers correlated their separate findings and laid the foundations for the important work on the Rh factor is worthy of additional reading.

SIGNIFICANCE OF THE Rh FACTOR

The Rh factor is of significance in blood transfusion and may be significant in childbirth. The Rh substance can cause a reaction if injected into the blood stream of a person who is Rh negative; that is, his body will have a tendency to break down the Rh-positive cells. This may clog the kidney with the residue of broken-down red cells and, if severe, can result in death. However, the first tranfusion may not cause this reaction because the recipient may become only sensitized; in other words, he develops anti-bodies—but not enough to produce severe symptoms. However, having been sensitized, he might suffer a severe reaction from subsequent transfusions.

In pregnancy the Rh-positive child of an Rh-positive father may be carried by an Rh-negative mother. In a small percentage of these cases, Rh-positive cells from the infant's blood stream cross the placental barrier membrane and seep into the mothers' circulation. Because these cells of the infant carry an antigenic substance, Rh, they may evoke in the mother's body an antibody response against the cells. This has little effect on the mother because the amount of the infants' cells in her blood stream is very small compared with the volume of her circulating blood. However, if the antibodies to the Rh-positive cells generated by the mother are carried back into the blood stream of the infant, they may do a great deal of damage. The concentration of the antibodies in the small body of the infant is so great that it may damage a large percentage of the red cells. Depending on the degree of severity with which this happens, the infant at birth may be merely jaundiced, a condition that may clear up in a few days; he may be quite ill, and to save his life may require an exchange transfusion (removal of baby's blood and simultaneous replacement with Rh-negative blood). In a very few cases, the baby is born dead.

Because the Rh factor has received a great deal of publicity during the past few years, many women fear the possibility of giving birth to an infant affected by hemolytic disease of the newborn (erythroblastosis). Actually, much of this apprehension is unfounded. It should be understood that Rh is a normal inherited characteristic, just as colour of eyes and hair. Again, marriages with an Rh-negative woman and an Rh-positive man may occur in approximately 13 percent of marriages (15% x 85% = 13%).

Most women who are Rh negative do not produce anti-Rh antibodies, and those who do usually have at least two babies that are unaffected. Even in marriages with an Rh-negative woman and an Rh-positive man, only 1 in about 26 has an infant affected, and many of these children only mildly so. With the present knowledge about blood, there is a much better outlook for detecting any Rh incompatability and treating the affected infant.

FUNCTIONS OF THE BLOOD IN THE BODY

Blood, a complex fluid serving many purposes in the human body, preserves the internal environment required for normal activity of the living cells. Following are some of its main functions:

Respiratory. Blood transports oxygen from the lungs to all the body tissues, and simultaneously carries carbon dioxide from these tissues to the lungs, where it is expired. (External respiration.)

Nutritive. Blood conveys food materials from the alimentary canal and food depots to all the tissues of the body. Excretory. Blood assists in the removal of waste products of metabolism by conveying them to the kidneys, where they are excreted in the urine. The maintenance of the water content of the tissues. Fluid lymph, containing oxygen and food materials, leaves the vascular channels and comes in contact with the tissues. Here it picks up carbon dioxide and other waste products and then flows back into the vascular channels. (Internal respiration.) The complex chemical processes of the body are enabled to take place because of the high dissolving and ionizing property of water in the blood and lymph.

Regulation of body temperature. The body owes its ability to regulate temperature largely to the water content of blood and of the tissue fluid. Heat from the deeper regions of the body is brought to the skin and lungs by the blood stream and dispelled. Protective and regulatory. Blood and lymph contain complex chemical substances such as antibodies, antitoxins, and lysins—all of which protect the body against injurious substances. The circulating blood also brings hormones from the ductless glands into direct contact with the tissues that respond to such stimuli.

HOW MUCH BLOOD IS IN THE BODY

The average healthy individual has a blood volume equal to about 8 percent of his body weight. Thus the mean blood volume for a person weighing 110 pounds would be 4,400 ml.; for a person weighing 150 pounds, it would be 6,000 ml.; for a person weighing 200 pounds, it would be 7,600 ml. As 1,000 ml. is equal to 2.1 pints, it will be easy for one to estimate the number of pints of blood in his body. The volume for men will be slightly higher than that for women of the same weight because men have an agerage of 7½ percent more red cells than do women.

The amount of blood in the human body has been the subject of careful study. The methods used to calculate blood volume are based on this fact: When a known amount of concentrated substance is introduced into the blood stream, the extent to which the concentrate is diluted is a measure of blood volume. Substances used must be harmless in small quantities and easily identifiable in blood samples. Depending upon the type used, they attach themselves either to the red cell or to a plasma protein. Red cell volume has been measured by tagging the red cells with radioactive molecules; plasma volume has been measured by using a harmless blue dye. Information gained in this way can be used to calculate the whole blood volume.

The blood volume may vary within certain normal limits, depending upon the needs of the body. At rest, the circulating volume is at a minimum. But the blood volume may be increased by several conditions, including hot weather, high altitudes, muscular exercise, emotional excitement, and pregnancy. On the other hand, an abnormal decrease in blood volume may occur during anemia, hemorrhage, loss of plasma due to extensive burns, or by various forms of dehydration.

RED CELLS

Red cells, or erythrocytes, are small, solid, disc-shaped particles whose main function is to transport oxygen from the lungs to the tissues and to transport carbon dioxide from the tissues to the lungs. The red cells contain hemoglobin, a chemical substance that has an amazing affinity for oxygen and carbon dioxide. Compared to other cells in the body, red cells are incomplete and cannot reproduce themselves because they lose their nuclei shortly before being released from the bone marrow into the blood stream. However, they are considered to be "alive" in the sense that they are able to perform their functions in the body for as long as 120 days. After that time they wear out and are then removed from circulation and destroyed by the spleen. The various components of protein and iron are salvaged for reuse by the body.

Red cells are created in vast numbers, mainly in the marrow of the short, flat bones of the body. It is estimated that they are worn out and removed from the circulation at the rate of 10 billion red cells an hour. The means by which the bone marrow must produce this same amount in the same period of time is part of the marvelous balance of supply and demand carried on by the normal physiology of the body. Red cells appear fragile in structure under the microscope but are actually very durable. They must be able to withstand a good deal of hammering about while being pumped through the blood vessels, and are able to squeeze through a capillary, which has a smaller diameter than the cell itself. In order to do this, the red cells must fold and bend in a flexible manner and be able to repeat the process many times.

While in the lungs the red cells are exposed to oxygen for only a brief time, but, owing to a chemical balance in the surrounding plasma, they become saturated with oxygen in 1 second. These oxygenated cells are then carried from the lungs to the heart and pumped out into the body. In the tissues, a delicate chemical change takes place in the plasma, and the oxygen is released for use. The carbon dioxide, created by the metabolism of the body cells, is picked up immediately by the red cell in exchange for oxygen and carried to the lungs. Here, another chemical change takes place in the plasma, a change enabling the red cell to release the carbon dioxide and almost instantly pick up more oxygen. This cycle is constantly repeated. Red cells are very small, measuring about 7.5 microns in diameter and 1.95 microns in thickness.

(A micron is about 1/25,000 of an inch.) It would require approximately 3,000 red cells lined up in their widest diameter to cover the space of an inch. Yet, because there are so many red cells in the human body, they would encircle the earth four times if lined up in a beadlike chain. In spite of the smallness of the individual red cell, the surface area of the combined cell mass is very large. In a man of average size, the surface area of the red cells represents a space as large as the surface of a football field.

Because one-fourth of the circulating blood is in the lungs at any one time, a combined cell surface area of 1,200 square yards is available to air. In 1 second, about 370 square yards of red cell tissue surface is exposed to the air, binding the oxygen to the hemoglobin. Only through these unique properties of the red cell, can manwith his relatively large size and weight-obtain enough oxygen to satisfy the needs of his complex system. From the heavy work of the muscles to the delicate functions of the brain, all cells and tissues depend on oxygen for life.

WHITE CELLS

The white cells, or leukocytes, are the protective cells of the blood stream. They are two or three times as large as the red cell and possess a nucleus. In a normal person, there are about 7,000 white cells per cubic millimeter * of blood (or about one white cell to 600 or 700 red cells).

There are several varieties of white cells, and each variety has a different function. When a drop of blood is placed on a glass slide, spread into a thin film, and allowed to dry, it may be stained with a dye so that each type of white cell has a distinctive appearance and colour. White cells may thus be classified into two major groups: those that have easily visible granules in the cytoplasm (the body of the cell exclusive of the nucleus) and those that do not. The most common type of white cell in the circulating blood has granules and is thus called a granulocyte. This variety comprises 60 to 70 percent of the white cells. The granulocvtes have the ability to migrate, like an amoeba, along the wall of a blood vessel and out of the capillaries into the body tissues, to fight bacteria. Also like an amoeba, they are able to engulf foreign substances. (This property is common to the granulocytes, monocytes, and reticulo-endothelial cells—all sometimes referred to as phagocytes.) When the body is threatened by infection, the granulocytes quickly gather at the point of invasion and set up a first line of defence. As an example, when you prick your finger and the bacteria enter and start to multiply, the body will mobilize many white cells at that point. They call sometimes be seen as a yellow accumulation called pus.

Lymphocytes, a second type of white cell, comprise 25 to 35 percent of the white cell population. These cells—formed by the spleen, lymph nodes, and other lymphoid tissue of the body—do not appear to be so aggressive in combating acute infection, but cluster in the tissues where chronic infection exists. There they form a barrier against the spread of infection to other parts of the body. Lymphocytes may undergo certain chemical changes that are protective in nature, and they may have some function in the development of immunity.

Monocytes, a third type of white cell found in the blood, are larger than the other white cells and relatively few in number. Like the granulocyte, they have the ability to migrate to the site of infection and ingest bacteria. Like a lymphocyte, they tend to wall off the infected area. During the healing process, they are capable of growing into another type of tissue repair cell, strengthening the weakened area.

The disease of leukemia is an abnormal, unrestrained overgrowth of white-cell-forming tissue. In other diseases, there may be an increase or decrease of white cells in the circulating blood. By obtaining a laboratory count of the number of white cells in a cubic millimeter of blood, a doctor has a valuable clue as to the state of health or disease of the body.

PLATELETS

Platelets are small, colorless, cell-like bodies that have an important function in the clotting of blood. They are a source of thromboplastin, a substance that is one of the links in the chain of clot formation. Platelets are liberated into the blood stream as fragments of a large parent cell in the bone marrow. They are smaller than the red cell and fewer in number (one platelet to 10 to 20 red cells). The normal range in platelet counts is quite wide, varying from 200,000 to 400,000 per cubic millimeter of blood.

PLASMA

Plasma is the fluid part of the blood in which the red cells, white cells, and platelets are suspended. It is composed of 92 percent water and 7 percent proteins; the remaining 1 percent consists of fat, carbohydrates, and many mineral salts necessary for life. It also contains hormones, vitamins, and enzymes. Many of the plasma proteins have specific uses in the treatment of disease. The most well known of these plasma derivatives are gamma globulin, serum albumin, and fibrinogen. Fresh frozen plasma is also prepared for the treatment of certain bleeding problems, especially hemophilia. Plasma is separated from whole blood by centrifugation; the red cells, being heavier, settle to the bottom of the blood-collecting container. The plasma can then be drawn off into separate containers and processed as fresh frozen plasma or sent to commercial firms for fractionation into specific derivatives.

BLOOD DERIVATIVES

Blood derivatives are those substances that can be separated from whole blood and used for specific purposes. Plasma itself is a component of whole blood. From the plasma there are a number of proteins that can be separated, and these fractions are usually called derivatives. Serum albumin, which in the body maintains plasma volume, is used for shock, low blood protein, and edema (the loss of water from the blood stream into the tissues). Antibodies against disease are concentrated in the gamma globulin.

Fibrinogen is one of the proteins essential for the formation of a blood clot. In addition to serum albumin, gamma globulin, and fibrinogen, a number of other plasma proteins are under study. These may prove valuable additions to the derivatives mentioned above. The red cells left in the bottle after the withdrawal of plasma can also be used to great advantage in certain diseases.

BLOOD CLOT

The clotting of blood is the result of a delicate and complex series of biochemical events, some of which are not fully understood. This mechanism must satisfy two rigorous requirements if life is to be sustained:

- Blood must never clot while it is performing its functions within the intact circulatory system;
- Blood must always clot when any damage occurs that causes a break in the blood vessels. The body must be able to vary this complex response to bleeding ranging from a slight cut to a massive hemorrhage.

When tissues are damaged, thromboplastin is released from the nearby cells and platelets. This initiates a chain of reactions in which calcium plays an important part. The end result is that the soluble protein (fibrinogen) is converted to a gelatinous solid (fibrin).

Fibrin forms in threadlike strands, making a webbed mesh in which the sticky platelets, red cells, and white cells are trapped, forming a jellylike clot. As the platelets distintegrate, they cause the fibrin web to shrink and have a constricting action on the walls of the torn blood vessel, causing it to clamp down and narrow the tear through which blood is escaping.

This knowledge of clot formation is applied in the collection of blood for transfusions. The calcium, which plays a key role, is inactivated when it is mixed with a citrate solution in the blood container during collection.

ANTIBODY

An antibody is a substance in the plasma that reacts in a highly specific way with a foreign substance (called an antigen) when it is introduced into the body. Although this reaction is usually protective, it may sometimes result in injury to the body. As an example of its protective function, if the body is invaded by a virulent strain of bacteria, two main lines of defence are set up. The white cells attempt to wall off and engulf the invaders; the bacteria, being a foreign substance, are antigens, and stimulate the reticulo-endothelial cells of the body to produce an antibody that will assist in repelling them. This antibody, contained in the globulin fraction of the plasma, may act on the bacteria in one of several ways. It may dissolve them; it may cause them to clump so that they cannot circulate freely through the blood stream; it may soften their protective membrane and inactivate them so that they will fall easy prey to the phagocytic white cells.

So it is that this type of antibody forms a protective agent against disease. Furthermore, as most of us are exposed to a large number of disease-causing microorganisms during our lifetime, we develop a concentration of various antibodies that may remain in the blood stream for a long time, even after we are well.

If we donate blood, the globulin fraction containing these protective substances can be separated from the rest of the plasma components and concentrated to about 25 times its normal strength. This concentrated substance, called gamma globulin or immune globulin, can then be injected into others to help protect them against certain diseases.

The antigen-antibody reaction also has an important application in giving blood transfusions. Red blood cells, if they are not the proper group and type, may act as antigens in the blood stream of the recipient. In this case, oddly enough, the antibodies in the A, B, AB, and O groups do not have to be developed by the introduction of red cells but exist in the blood stream in the natural state. In contrast to this, the antibodies to the Rh factor do not exist in the natural state and must be developed by a transfusion of incompatible Rh-type cells. In the case of the ABO groups or the Rh types, the incompatible antigenic red cells are attacked by the circulating antibodies and caused to clump, or agglutinate, in the blood stream. If this reaction is strong enough, the red cells may be broken up, or hemolysed, by the antagonistic antibody. If enough of the transfused, incompatible red cells are clumped or broken up in the recipient's body, they have a damaging effect on the tubules of the kidney, and serious illness or even death may result from the inability of the kidney to excrete waste products. The steps by which these hazards are avoided will be discussed in a later section on laboratory work.

BLOOD GROUPS

Blood groups are inherited patterns into which human blood may be divided for scientific purposes. There are four major blood groups: A, B, AB, and O.* Groups A, B, and O were discovered by Dr. Karl Landsteiner in 1900, and a fourth, AB, was described by another scientist in 1902. Although the percentage of people belonging to each group tends to vary in different countries (as do other inherited characteristics, such as colour of eyes and hair), the following distribution represents the general spread in this country: 45 percent—group O; 40 percent—group A; 10 to 12 percent —group B; 3 to 5 percent—group AB. About the turn of the century, there was great interest in learning how to use blood transfusions as part of the practice of medicine; but doctors were held back by a forbidding obstacle: while about half the transfusions given to treat illness had a remarkable therapeutic effect, the other half seemed to make the patient much worse and sometimes resulted in death.

In seeking the clue to this riddle, Landsteiner performed a series of tests in the course of which he placed the red cells of one person in suspension in the serum of another person. From this crucial experiment, he noted that in a number of the combinations of red cells and sera, the red cells remained undamaged and freely suspended; in other combinations, the red cells were clumped together in sticky clusters—with resultant damage to the cells.

From this evidence, he deduced that certain combinations of cells and sera were incompatible and that it was these unfortunate combinations that were causing illness and death in some patients following transfusion (transfusion reaction).Further investigation of the factors causing clumping, or agglutination, of the red cells when exposed to incompatible sera revealed that there are two main antigenic substances (agglutinogens) in the red cells and two main antibodies (agglutinins) in the plasma. The antigens in the red cells were named A and B, and the antibodies in the plasma were given names to correspond with the group of red cells that they would agglutinate. Therefore, the plasma antibody causing group A cells to clump was called anti-A; the plasma antibody causing group B cells to clump was called anti-B.

As noted in the discussion on antibodies, the anti-A and anti-B substances are contained in the globulin protein of the plasma. These basic facts enable us to understand the four simple combinations that are the bases for the major blood groups:

- *Group A blood:* Has A substance in the red cells, and anti-B substance in the plasma.
- *Group B blood:* Has B substance in the red cells, and anti-A substance in the plasma.
- *Group AB blood:* Has both A and B substance in the red cells, and neither anti-A nor anti-B substances in the plasma.
- *Group O blood:* Has neither A nor B substance in the red cells, and has both anti-A and anti-B substances in the plasma.

It will be noted that each blood group is named for the type of antigenic substance that is an inherited characteristic of the red cell.

10

Synthetic Drugs

The term "Synthetic," as used in the National Drug Control Strategy (NDCS) Annual Report and the National Synthetic Drugs Action Plan, refers to drugs whose origins are not primarily organic, but rather are produced via chemical synthesis. For clarification, heroin, cocaine, and marijuana are in the organic category, while LSD, MDMA, and methamphetamine are synthetics. This report deals primarily with amphetamine-type stimulants often termed "designer drugs" or "club drugs " and drugs such as oxycodone or benzodiazepine, which are produced commercially by drug manufacturers for valid medical purposes, and are diverted from legal channels or produced illegally in clandestine laboratories for illicit markets worldwide. Regardless of source, commercial or clandestine, amphetamine-type stimulants (ATS) are emerging as a class of drugs that are widely abused in the United States and pose a serious threat to our youth. Some of the most prevalent synthetic drugs of abuse are:

METHAMPHETAMINE

Methamphetamine is a central nervous system stimulant with serious health implications that include violent behaviour, extreme paranoia, and other psychotic episodes. The drug is clandestinely manufactured in crude laboratories primarily located in Mexico and the United States by street chemists referred to as "cooks." Usually injected or snorted, the drug can also be smoked or ingested orally. The manufacture of methamphetamine poses serious environmental hazards.

3-4 Methylenedioxymethamphetamine (MDMA)

Commonly referred to as Ecstasy or XTC, MDMA is a psychoactive substance with both stimulant and mild hallucinogenic properties. Most often found in tablet form, although occasionally distributed as a crystalline powder. Taken orally, the health risks include severe hyperthermia, dehydration, and long term learning impairment. MDMA is manufactured in laboratories primarily located in Western Europe at this time.

Paramethoxyamphetamine (PMA)

PMA is a tablet form amphetamine derivative similar to MDMA (Ecstasy) but more lethal even in smaller doses. More than 50 milligrams may be fatal. PMA producers sometimes use the same imprinted logos on PMA that are used on MDMA tablets in order to market the product as MDMA to users, resulting in fatalities from overdose or mixing of the drug with MDMA.

Gamma Hydroxybutyrate (GHB)

GHB is a central nervous system depressant usually sold as an odorless, colorless liquid in spring water bottles or as a powder mixed with beverages and soft drinks, reportedly used in date-rapes.

Gamma Butyrolactone (GBL)

Often found in industrial cleaners, GBL is the precursor chemical for the manufacture of GHB. In addition, it has been marketed as a nutritional supplement in health food stores and over the Internet in both powder and capsule form. GBL is synthesized by the body to produce GHB. Ingestion of GBL often causes a severe physical reaction, usually through the violent regurgitation of the fluid. These chemicals increase the effects of alcohol, and can cause respiratory distress, seizures, coma, and death.

Ketamine

A prescription general anesthetic with some physical effects similar to PCP and visual effects of LSD, ketamine is primarily marketed for veterinary use. Sold as both a liquid and a powder, use in humans can cause delirium, amnesia, depression, long-term memory and cognitive difficulties. Due to its disassociative effects, ketamine is reportedly used as a date-rape drug.

Rohypnol

Rohypnol, a sleep aid that has never been manufactured or marketed in the United States, it is still produced in Mexico and Europe and available by prescription in many countries. It is smuggled into the United States from Mexico. Capable of producing extreme lethargy and significantly reducing recall capability of the brain; it has been often mentioned in relation to numerous date rapes (though the data do not support the contention that it is widely used for this purpose). Abuse of rohypnol is generally episodic use among teenagers and young adults as an "alcohol extender" and disinhibitory agent, most often in combination with beer.

Lysergic Acid Diethylamide (LSD)

LSD is an extreme hallucinogen most commonly marketed on blotter paper or in gelcap form. It has been known since the 1960's for its hallucinogenic properties and the adverse psychotic side effects that frequently occur during and after use (flashbacks). It is a clandestinely manufactured crystalline powder reduced to a clear liquid for dosing purposes. Generally it is produced by rogue chemists, some with extensive academic backgrounds in chemical synthesis, in relatively professional laboratory surroundings. Chemistry background, laboratory control knowledge, and an educated understanding of the process is generally required to successfully and safely conduct the synthesis.

Phencyclidine (PCP)

Commonly known as "angel dust," PCP was originally developed as an animal tranquilizer. It is now clandestinely manufactured in environments known as "bucket labs" by street chemists. In both a powder and liquid form, it is almost always smoked by placing it on a marijuana cigarette or regular menthol cigarette, known as a "kool-dip." It is an extremely dangerous substance causing coma, convulsions, and psychotic delusions. Users are known to become extremely violent and aggressive, commonly causing injuries to themselves and others.

OxyContin

OxyContin is a synthetic, opiate, prescription pain medicine often used in the treatment of pain related to cancer and other debilitating conditions. OxyContin contains the drug oxycodone, which is a common drug used in pain relievers such as Percocet and Percodan. However unlike other forms of oxycodone, OxyContin is available in higher dose units, as a time-release formulation. Law enforcement sources have reported an increase in the diversion of OxyContin and other medications containing oxycodone. This increase in illegal use has been especially apparent on the East Coast. The increase in the abuse of OxyContin has led to an increased number of pharmacy robberies and health care fraud incidents.

Precursor Chemicals

Necessary precursors, despite efforts to restrict them, are widely available worldwide. Thirty-four chemicals are listed by federal statute, Section 812 of the Controlled Substances Act, as controlled substances on the Drug Enforcement Agency's (DEA) controlled substance schedules.

DRUGS

Anthropologists have studied its use and its role in social structure in a number of cultures. Jamaica, for example, has a high rate of regular users of cannabis, making it an excellent place to study the issues that unfold around the use and abuse of this drug, known to Jamaicans as ganja.In Jamaica, ganja use is integrally linked to all aspects of working-class social structure: cultivation, cash crops, marketing, economics, consumer-cultivator-dealer networks;

interclass relationships and processes of avoidance or cooperation; parent-child, peer and mate relationships; folk medicine; folk religious doctrines; gossip sanctions; personality and culture; interclass stereotypes; legal and church sanctions; perceived requisites of behavioural changes for social mobility; and adaptive strategies.

Ganja figures strongly in the economic realm. On the lowest rung of the Jamaican socioeconomic ladder, poor families with few marketable skills make their living however they can, often engaging in several diverse economic activities. One of these activities may be the cultivation of cannabis. Vera Rubin and Lambros Comitas point out that vendors of cannabis typically lead a stable family life and are otherwise law abiding and conservative.

Working-class Jamaicans often believe that use of ganja makes work go more pleasantly and allows them to work harder; however, the middle and upper classes, who employ the lower classes and supervise their work, believe the drug is detrimental to work performance. Melanie Dreher has investigated the interplay of these differing views in the setting of a Jamaican sugar estate. Three farms were contrasted, all having different proportions of cannabis smokers to nonsmokers. Dreher examined productivity figures by categories of smokers and nonsmokers and found no significant differences in the work performance of smokers and nonsmokers. The results support neither the views of the workers nor the managers.

Dreher's study illustrates the increasing use of quantitative data in anthropological work on drug use, but it also points to the levels of subtlety and sophistication that can be added to research by the inclusion of qualitative material. This dimension of research is one of anthropology's strongest assets; it allows for deeper and more accurate understanding of complex questions and helps restrain impulses to hasty generalization.

Dreher has also studied the use of cannabis among Jamaican women. She examined the different patterns of ganja use among women in two similar Jamaican villages, where women in one village seemed more inclined to smoke cannabis than in the other. At the time of this study, smoking was contrary to norms for Jamaican women, although they routinely made cannabis teas for medicinal purposes. In the village with the higher rate of smokers, women were found to have more economic opportunity and thus more independence.

Women in the village with fewer smokers found it adaptive to conform to the norms, so that they would not alienate men who were potential husbands and sources of support. By taking a cross-cultural look at what is considered to be a growing problem in the United States, Dreher has undertaken applied research to shed light on an important policy question.

A common conceptual framework for the behaviour of drug users whose conduct is outside society's usual limits is the socialpsychological notion of deviance. Dreher argues that the behaviour of the women she studied is better understood anthropologically in terms of intracultural variation. Again, we see a contextual focus, a perspective that, instead of looking solely at individual action, allows for the connecting of individual action to sociocultural structure. Currently more tolerance for ganja smoking has developed among lower-class Jamaican women.

Use of the substance fits into a constellation of personal characteristics to which the term "roots daughter" has been given. Connected to the ideas generated by the Rastafarian religion about what is "natural" and African, a "roots daughter" is dignified, independent and intelligent. Findings of this later research indicate changing attitudes and thereby remind us of culture's dynamic nature. Anthropological research often probes the role of ritual in human action, as in the case of the Colombian work of William Partridge. He found that smoking of cannabis was associated with work life through the action of ritual. An important part of worker comradeship is sharing cannabis when one can afford it; those who have it share it with whomever may not have it at that time. Workers who use but do not provide cannabis for sharing at work breaks are considered undependable and isolate themselves from the social networks from which work gangs are developed in the agricultural economy. In this fashion, meaning is shifted from one social context to another. In a domain similar to ritual, Rubin has looked at the first experience of smokers as a rite of passage, an experience that strongly influences whether boys become regular smokers. If their first experience is a good one, they will probably become regular smokers; if not, they tend not to become users.

The practice of learning about the cultural intricacies of studied groups makes the anthropological enterprise often more time-consuming than other research approaches. In the Costa Rican work of J. Bryan Page, for example, a command of "proper" Spanish was not adequate to the task of following conversations in the argot of the drug culture on the street. The arcane vernacular was found to be useful to the speakers; one use was the concealment of illegal economic activity, economic activity again motivated by need.

Research in Costa Rica involved life histories, participant observation and ethnographic interviews and subdivided smokers into different categories whose members found their smoking experiences to be shaped considerably by conditioned expectations and socioeconomic factors. Users who enjoyed economic stability and good social support found the smoking of cannabis to enhance activity; those users with less secure lives did not have universally pleasing experiences.

In a keynote address to the Alcohol and Drug Study Group, Dreher documented several theoretical and methodological contributions from anthropology, especially as distinct from the research forthcoming from sociology and social psychology. Not surprisingly, the use of cannabis in modern society is viewed by anthropologists as a "sociocultural phenomenon rather than an individual characteristic". The holistic and comparative approach, the use of society or community as the unit of analysis, the reliance upon ethnohistorical and ethnographic studies and the combination of qualitative and quantitative methods make anthropological research on cannabis unique.

The cannabis question is obviously a loaded issue in American society at the present and therefore it is not surprising that there is resistance to accepting the results of these "controversial" anthropological findings. However, the same skepticism regarding these conclusions about the potential functional role of ganja use in certain cultural contexts is remarkably familiar to some of the resistance alcohol researchers in anthropology have encountered when they call into question certain widespread assumptions about the etiology and diagnosis of alcoholism. The question of where to draw the line between acceptable usage—if usage of a particular drug is thought ever to be acceptable—and dysfunctional-pathological usage is extremely complicated. This message is probably one of the most important ones that anthropology offers, especially on controversial drugs such as cannabis.

CLASSES OF NATURAL DRUGS

The Controlled Substances Act (CSA) regulates five classes of drugs: narcotics, depressants, stimulants, hallucinogens, and anabolic steroids. Each class has distinguishing properties, and drugs within each class often produce similar effects. However, all controlled substances, regardless of class, share a number of common features. It i s the purpose of this introduction to familiarise the reader with some of these shared features and to give definition to terms frequently associated with these drugs. All controlled substances have abuse potential or are immediate precursors to substances with abuse potential. With the exception of anabolic steroids, controlled substances are abused to alter mood, thought, and feeling through their actions on the central nervous system (brain and spinal cord).

Fig. Confiscated Cocaine Packaged in Multi-kilogram Bricks

Some of these drugs alleviate pain, anxiety, or depression. Some induce sleep and others energise. Though therapeutically useful, the "feel good" effects of these drugs contribute to their abuse. The extent to which a substance is reliably capable of producing intensely pleasurable feelings (euphoria) increases the likelihood of that substance being abused. When drugs are used in a manner or amount inconsistent with the medical or social patterns of a culture, it is called drug abuse. In legal terms, the non-sanctioned use of substances controlled in Schedules I through V of the CSA is considered drug abuse. While legal pharmaceuticals placed under control in the CSA are prescribed and used by patients for medical treatment, the use of these same pharmaceuticals outside the scope of sound medical practice is drug abuse. In addition to having abuse potential, most controlled substances are capable of producing dependence, either physical or psychological.

Physical dependence refers to the changes that have occurred in the body after repeated use of a drug that necessitates the continued administration of the drug to prevent a withdrawal syndrome. This withdrawal syndrome can range from mildly unpleasant to life-threatening and is dependent on a number of factors.

The type of withdrawal experienced is related to: the drug being used; the dose and route of administration; concurrent use of other drugs; frequency and duration of drug use; and the age, sex, health, and genetic makeup of the user. Psychological dependence refers to the perceived "need" or "craving" for a drug. Individuals who are psychologically dependent on a particular substance often feel that they cannot function without continued use of that substance. While physical dependence disappears within days or weeks after drug use stops, psychological dependence can last much longer and is one of the primary reasons for relapse (initiation of drug use after a period of abstinence).

Contrary to common belief, physical dependence is not addiction. While addicts are usually physically dependent on the drug they are abusing, physical dependence can exist without addiction. For example, patients who take narcotics for chronic pain management or benzodiazepines to treat anxiety are likely to be physically dependent on that medication. Addiction is defined as compulsive drug-seeking behaviour where acquiring and using a drug becomes the most important activity in the user's life. This definition implies a loss of control regarding drug use, and the addict will continue to use a drug despite serious medical and/or social consequences. The National Institute on Drug Abuse (NIDA) estimates that about five million Americans suffer from drug addiction.

Individuals that abuse drugs often have a preferred drug that they use, but may substitute other drugs that produce similar effects (often found in the same drug class) when they have difficulty obtaining their drug of choice. Drugs within a class are often compared with each other with terms like potency and efficacy. Potency refers to the amount of a drug that must be taken to produce a certain effect, while efficacy refers to whether or not a drug is capable of producing a given effect regardless of dose. Both the strength and the ability of a substance to produce certain effects play a role in whether that drug is selected by the drug abuser.

It is important to keep in mind that the effects produced by any drug can vary significantly and is largely dependent on the dose and route of administration. Concurrent use of other drugs can enhance or block an effect and substance abusers often take more than one drug to boost the desired effects or counter unwanted side effects. The risks associated with drug abuse cannot be accurately predicted because each user has his/her own unique sensitivity to a drug. There are a number of theories that attempt to explain these differences, and it is clear that a genetic component may predispose an individual to certain toxicities or even addictive behaviour.

Youths are especially vulnerable to drug abuse. Young Americans engaged in extraordinary levels of illicit drug use in the last third of the twentieth century. Today, the majority of young people (about 53 percent) have used an illicit drug by the time they leave high school and about 25 percent of all seniors are current (within the past month) users. The behaviours associated with teen and preteen drug use often result in tragic consequences with untold harm to others, themselves, and their families. For example, an analysis of data from the National Household Survey on Drug Abuse indicates that youngsters between the ages of 12 and 17 who have smoked marijuana within the past year are more than twice as likely to cut class, steal, commit assault, and destroy property than are those who did not smoke marijuana. The more frequently a youth smokes marijuana, the more likely he or she is to engage in these antisocial behaviours. Each of the five classes of drugs is reviewed and various drugs within each class are profiled. Although marijuana is classified in the CSA.

There are also a number of substances that are abused but not regulated under the CSA. Alcohol and tobacco, for example, are specifically exempt from control by the CSA. In addition, a whole group of substances called inhalants are commonly available and widely abused by children. Control of these substances under the CSA would not only impede legitimate commerce, but would likely have little effect on the abuse of these substances by youngsters. An energetic campaign aimed at educating both adults and youth about inhalants is more likely to prevent their abuse.

PLANTING SEEDS

Only recently, 10,000-20,000 years ago, nomadic tribes began to settle down in some parts of the world and formed small agricultural communities. This development was one of the most profound changes in the history of the human race. Planting seeds, domesticating animals, and erecting permanent shelters for humans and animals improved nutrition, provided more comfortable, though still primitive, living accommodations, and started the societal bonds that developed into villages, towns, and cities, where the first truly historical times began. At this stage, when people began to live close together, certain rules were established to bring needed order out of unstructured conglomerated living. Strong individuals emerged as the leaders of tribal or village communities.

Ambitious leaders increased their power by shrouding their personalities and decisions in mystery, elevating them to an exalted reputation. Thus emerged the concept of superhuman beings, of gods with powers over natural phenomena such as fertility, family bonds, lightning, and the awesome power of storms. When the demands of multiple deities for sacrifices grew excessive, all these powers and supernatural beliefs and superstitions were united in monotheistic religions, often after catastrophic wars and through a longing for peace. The deities were represented ritualistically by priests and priestesses, and valuable sacrifices offered by the people further enhanced the splendor and power of these leading personalities.

Authority

Part of the influence of these commanding individuals came from the personal help they could give to sick, wounded, and otherwise afflicted members of their groups. The medicine man, witch doctor, or shaman was the first person to turn to in distress. Sometimes authority lay in an influential woman who functioned as priestess, nurse, and midwife. In most cases, power, enforced by superstition, centered in a dominant male recognised as healer, priest, judge, and leader in peace and war. A knowledge of healing herbs formed part of his power base and prerogative; among his duties was applying such healing products when needed.

Botanical specimens taken from trees, shrubs, and other plants formed the mainstay of most of these early medicines. Sometimes the shaman deepened the mystery of the power of his medication by adding parts of animals, and even human hearts. Some mind-induced (psychosomatic) illnesses were probably improved by suggestions, just as they are today. Also, some functional disorders responded to the active chemicals in suitably chosen plants. It must have been hard for a medicine man to decide which plant to give a patient. Trial and error was the order of the day.

Even after a healing effect was found in a certain plant, it was probably used for a variety of illnesses, whether identical in symptoms or only vaguely similar. Modern physicians still face complicated choices in diagnosis and drug therapy in spite of help from analytical tests, X-rays, ultrasound, nuclear magnetic resonance visualisation, blood cell counts, urinalysis, and other laboratory aids. We can only imagine the dilemma of an ancient medicine man, who was guided only by vague symptoms such as a generalised pain, nausea, fever, or convulsions. In such a dilemma, and without any knowledge of anatomy or pathology, some herbal concoction was given in the hope that it would work. The medicine man undoubtedly added prayers or exorcisms to the medication and believed sincerely that his ministrations would aid the afflicted. Cynical healers also appeared. In Roman times, predictions about world affairs as well as the course of a patient's disease were based on the inspection of the intestines of newly slaughtered chickens. There was a saying that described one of these healer-seers (called a haruspex) encountering another member of the guild: Haruspex, haruspicem videns ridet ("When one haruspex sees another, he laughs").

ANORECTIC DRUGS

A number of drugs have been developed and marketed to replace amphetamines as appetite suppressants. These anorectic drugs include benzphetamine, phendimetrazine, and phentermine. These substances are in Schedule III or IV of the CSA and produce some amphetamine-like effects. Of these diet pills, phentermine is the most widely prescribed and most frequently encountered on the illicit market. Two Schedule IV anorectics often used in combination with phentermine, fenfluramine and dexfenfluramine, were removed from the U.S., market because they were associated with heart valve problems.

Khat

For centuries, khat, the fresh young leaves of the Catha edulis shrub, has been consumed where the plant is cultivated, primarily East Africa and the Arabian Peninsula. There, chewing khat predates the use of coffee and is used in a similar social context. Chewed in moderation, khat alleviates fatigue and reduces appetite. Compulsive use may result in manic behaviour with grandiose delusions or in a paranoid type of illness, sometimes accompanied by hallucinations. Khat has been smuggled into the United States and other countries from the source countries for use by emigrants. It contains a number of chemicals, among which are two controlled substances, cathinone and cathine. As the leaves mature or dry, cathinone is converted to cathine, which significantly reduces its stimulatory properties.

DEPRESSANTS

Historically, people of almost every culture have used chemical agents to induce sleep, relieve stress, and allay anxiety. While alcohol is one of the oldest and most universal agents used for these purposes, hundreds of substances have been developed that produce central nervous system depression. These drugs have been referred to as downers, sedatives, hypnotics, minor tranquilisers, anxiolytics, and anti-anxiety medications. Unlike most other classes of drugs of abuse, depressants are rarely produced in clandestine laboratories. Generally, legitimate pharmaceutical products are diverted to the illicit market. A notable exception to this is a relatively recent drug of abuse, gamma hydroxybutyric acid (GHB).

Choral hydrate and paraldehyde are two of the oldest pharmaceutical depressants still in use today. Other depressants, including gluthethimide, methaqualone, and meprobamate have been important players in the milieu of depressant use and abuse. However, two major groups of depressants have dominated the licit and illicit market for nearly a century, first barbiturates and now benzodiazepines.

Barbiturates were very popular in the first half of the 20th century. In moderate amounts, these drugs produce a state of intoxication that is remarkably similar to alcohol intoxication. Symptoms include slurred speech, loss of motor coordination, and impaired judgement. Depending on the dose, frequency, and duration of use, one can rapidly develop tolerance, and physical and psychological dependence on barbiturates. With the development of tolerance, the margin of safety between the effective dose and the lethal dose becomes very narrow. That is, in order to obtain the same level of intoxication, the tolerant abuser may raise his or her dose to a level that may result in coma or death. Although many individuals have taken barbiturates therapeutically without harm, concern about the addiction potential of barbiturates and the ever-increasing number of fatalities associated with them led to the development of alternative medications. Today, less than 10 percent of all depressant prescriptions in the United States are for barbiturates.

Benzodiazepines were first marketed in the 1960s. Touted as much safer depressants with far less addiction potential than barbiturates, today these drugs account for about one out of every five prescriptions for controlled substances. Although benzodiazepines produce significantly less respiratory depression than barbiturates, it is now recognised that benzodiazepines share many of the undesirable side effects of the barbiturates. A number of toxic central nervous system effects are seen with chronic high-dose benzodiazepine therapy, including headaches, irritability, confusion, memory impairment, and depression. The risk of developing over-sedation, dizziness, and confusion increases substantially with higher doses of benzodiazepines.

Prolonged use can lead to physical dependence even at doses recommended for medical treatment. Unlike barbiturates, large doses of benzodiazepines are rarely fatal unless combined with other drugs or alcohol. Although primary abuse of benzodiazepines is well documented, abuse of these drugs usually occurs as part of a pattern of multiple drug abuse. For example, heroin or cocaine abusers will use benzodiazepines and other depressants to augment their "high" or alter the side effects associated with over-stimulation or narcotic withdrawal. There are marked similarities among the withdrawal symptoms seen with most drugs classified as depressants. In the mildest form, the withdrawal syndrome may produce insomnia and anxiety, usually the same symptoms that initiated the drug use. With a greater level of dependence, tremors and weakness are also present, and in its most severe form, the withdrawal syndrome can cause seizures and delirium. Unlike the withdrawal syndrome seen with most other drugs of abuse, withdrawal from depressants can be life threatening.

Barbiturates

Barbiturates were first introduced for medical use in the early 1900s. More than 2,500 barbiturates have been synthesised, and at the height of their popularity, about 50 were marketed for human use. Today, about a dozen are in medical use. Barbiturates produce a wide spectrum of central nervous system depression, from mild sedation to coma, and have been used as sedatives, hypnotics, anesthetics, and anticonvulsants. The primary differences among many of these products are how fast they produce an effect and how long those effects last. Barbiturates are classified as ultrashort, short, intermediate, and long-acting.

The ultrashort-acting barbiturates produce anesthesia within about one minute after intravenous administration. Those in current medical use are the Schedule IV drug methohexital (Brevital®) and the Schedule III drugs thiamyl (Surital®) and thiopental (Pentothal®). Barbiturate abusers prefer the Schedule II short-acting and intermediate-acting barbiturates that include amobarbital (Amytal®), pentobarbital (Nembutal®), secobarbital (Seconal®) and Tuinal (an amobarbital/secobarbital combination product). Other short and intermediate-acting barbiturates are in Schedule III and include butalbital (Fiorina®), butabarbital (Butisol®), talbutal (Lotusate®) and aprobarbital (Alurate®). After oral administration, the onset of action is from 15 to 40 minutes, and the effects last up to six hours. These drugs are primarily used for insomnia and preoperative sedation. Veterinarians use pentobarbital for anesthesia and euthanasia.

Long-acting barbiturates include phenobarbital (Luminal®) and mephobarbital (Mebaral®), both of which are in Schedule IV. Effects of these drugs are realised in about one hour and last for about 12 hours, and are used primarily for daytime sedation and the treatment of seizure disorders.

Paraldehyde

Paraldehyde (Paral®) is a Schedule IV depressant used most frequently in hospital settings to treat delirium tremens associated with alcohol withdrawal. Many individuals who become addicted to paraldehyde have been initially exposed during treatment for alcoholism and, despite the disagreeable odour and taste, come to prefer it to alcohol. This drug is not used by injection because of tissue damage, and taken orally, it can be irritating to the throat and stomach. One of the signs of paraldehyde use is a strong, characteristic smell to the breath.

Chloral Hydrate

The oldest of the hypnotic (sleep inducing, depressants, chloral hydrate was first synthesised in 1832. Marketed as syrups or soft gelatin capsules, chloral hydrate takes effect in a relatively short time (30 minutes) and will induce sleep in about an hour.

A solution of chloral hydrate and alcohol constituted the infamous "knockout drops" or "Mickey Finn." At therapeutic doses, chloral hydrate has little effect on respiration and blood pressure; however, a toxic dose produces severe respiratory depression and very low blood pressure. Chronic use is associated with liver damage and a severe withdrawal syndrome. Although some physicians consider chloral hydrate to be the drug of choice for sedation of children before diagnostic, dental, or medical procedures, its general use as a hypnotic has declined. Chloral hydrate, Noctec® and other compounds, preparations, or mixtures containing choral hydrate are in Schedule IV of the CSA.

Glutethimide and Methaqualone

Glutethimide (Doriden®) was introduced in 1954 and methaqualone (Quaalude®, Sopor®) in 1965 as safe barbiturate substitutes. Experience demonstrated, however, that their addiction liability and the severity of withdrawal symptoms were similar to those of barbiturates. By 1972, "luding out," taking methaqualone with wine, was a popular college pastime. Excessive use leads to tolerance, dependence, and withdrawal symptoms similar to those of barbiturates. In the United States, the marketing of methaqualone pharmaceutical products stopped in 1984, and methaqualone was transferred to Schedule I of the CSA. In 1991, glutethimide was transferred into Schedule II in response to an upsurge in the prevalence of diversion, abuse, and overdose deaths. Today, there is little medical use of glutethimide in the United States.

Benzodiazepines

The benzodiazepine family of depressants is used therapeutically to produce sedation, induce sleep, relieve anxiety and muscle spasms, and to prevent seizures. In general, benzodiazepines act as hypnotics in high doses, anxiolytics in moderate doses, and sedatives in low doses. Of the drugs marketed in the United States that affect central nervous system function, benzodiazepines are among the most widely prescribed medications. Fifteen members of this group are presently marketed in the United States, and about 20 additional benzodiazepines are marketed in other countries. Benzodiazepines are controlled in Schedule IV of the CSA.

Short-acting benzodiazepines are generally used for patients with sleep-onset insomnia (difficulty falling asleep) without daytime anxiety. Shorter-acting benzodiazepines used to manage insomnia include estazolam (ProSom®), flurazepam (Dalmane®), temazepam (Restoril®) and triazolam (Halcion®). Midazolam (Versed®), a short-acting benzodiazepine, is utilised for sedation, or treating anxiety and amnesia in critical care settings and prior to anesthesia. It is available in the United States as an injectable preparation and as a syrup (primarily for pediatric patients).

Benzodiazepines with a longer duration of action are utilised to treat inso mnia in patients with daytime anxiety. These benzodiazepines include alprazolam (Xanax®), chlordiazepoxide (Librium®), clorazepate (Tranxene®), diazepam (Valium®), halazepam (Paxipam®), lorzepam (Ativan®), oxazepam (Serax®), prazepam (Centrax®) and quazepam (Doral®). Clonazepam (Klonopin®), diazepam, and clorazepate are also used as anticonvulsants. Benzodiazepines are classified in the CSA as depressants. Repeated use of large doses or, in some cases, daily use of therapeutic doses of benzodiazepines is associated with amnesia, hostility, irritability, and vivid or disturbing dreams, as well as tolerance and physical dependence. The withdrawal syndrome is similar to that of alcohol and may require hospitalisation. Abrupt cessation of benzodiazepines is not recommended and tapering-down the dose eliminates many of the unpleasant symptoms.

Given the millions of prescriptions written for benzodiazepines, relatively few individuals increase their dose on their own initiative or engage in drug-seeking behaviour. Those individuals who do abuse benzodiazepines often maintain their drug supply by getting prescriptions from several doctors, forging prescriptions, or buying diverted pharmaceutical products on the illicit market. Abuse is frequently associated with adolescents and young adults who take benzodiazepines to obtain a "high."

This intoxicated state results in reduced inhibition and impaired judgement. Concurrent use of alcohol or other depressant with benzodiazepines can be life threatening. Abuse of benzodiazepines is particularly high among heroin and cocaine abusers. A large percentage of people entering treatment for narcotic or cocaine addiction also report abusing benzodiazepines. Alprazolam and diazepam are the two most frequently encountered benzodiazepines on the illicit market.

Flunitrazepam

Flunitrazepam is a benzodiazepine that is not manufactured or legally marketed in the United States, but is smuggled in by traffickers. In the mid-1990s, flunitrazepam was extensively trafficked in Florida and Texas. Known as "rophies," "roofies," and "roach," flunitrazepam gained popularity among younger individuals as a "party" drug. It has also been utilised as a "date rape" drug. In this context, flunitrazepam is placed in the alcoholic drink of an unsuspecting victim to incapacitate them and prevent resistance from sexual assault. The victim is frequently unaware of what has happened to them and often does not report the incident to authorities. A number of actions by the manufacturer of this drug and by government agencies have resulted in reducing the availability and abuse of flunitrazepam in the United States.

Gamma Hydroxybutyric Acid (GHB)

In recent years, gamma hydroxybutyric acid (GHB) has emerged as a significant drug of abuse throughout the United States. Abusers of this drug fall into three major groups:

- Users take GHB for its intoxicant or euphoriant effects;
- Bodybuilders who abuse GHB for its alleged utility as an anabolic agent or as a sleep aid; and
- Individuals who use GHB as a weapon for sexual assault.

These categories are not mutually exclusive and an abuser may use the drug illicitly to produce several effects. GHB is frequently taken with alcohol or other drugs that heighten its effects and is often found at bars, nightclubs, rave parties, and gyms. Teenagers and young adults who frequent these establishments are the primary users. Like flunitrazepam, GHB is often referred to as a "date-rape" drug. GHB invol-vement in rape cases is likely to be unreported or unsub-stantiated because GHB is quickly eliminated from the body making detection in body fluids unlikely. Its fast onset of depressant effects may render the victim with little memory of the details of the attack.

GHB produces a wide range of central nervous system effects, including dose-dependent drowsiness, dizziness, nausea, amnesia, visual hallucinations, hypotension, bradycardia, severe respiratory depression, and coma. The use of alcohol in combination with GHB greatly enhances its depressant effects. Overdose frequently requires emergency room care, and many GHB-related fatalities have been reported. Gamma butyrolactone (GBL) and 1,4-butanediol are GHB analogues that can be used as substitutes for GHB. When ingested, these analogues are converted to GHB and produce identical effects. GBL is also used in the clandestine production of GHB as an immediate precursor. Both GBL and 1,4-butanediol have been sold at health food stores and on various internet sites.

The abuse of GHB began to seriously escalate in the mid-1990s. For example, in 1994, there were 55 emergency department episodes involving GHB reported in the Drug Abuse Warning Network (DAWN) system. By 2002, there were 3,330 emergency room episodes. DAWN data also indicated that most users were male, less than 25 years of age, and taking the drug orally for recreational use. GHB was placed in Schedule I of the CSA in March 2000. Gamma butyrolactone (GBL) was made a List I Chemical in February 2000. GHB has recently been approved as a medication (Xyrem) for the treatment of cataplexy associated with some types of narcolepsy. This approved medication is in Schedule III of the CSA.

NATURAL DRUGS

The notion of the 'status quo' in a living system is a myth. Stasis does not occur in any living entity or network. Medicine deals with life and hence, is alive itself. Naturally, then, it is no surprise that the philosophical model of medicine under which the western world has operated for the better part of the twentieth century is changing fundamentally. While change in life is always confronting it is equally inevitable. The true question lies not in whether the current dominant medical paradigm will change but rather how it will change.

The particular phase of evolution facing twenty-first century medicine is unique but not entirely new. Global mythology reveals that the quest for a world-view that yields health and wellbeing is timeless. In Central African mythology there is a dangerous one-legged, one-armed half man known as the 'Chiruwi', or 'mysterious thing'. This creature challenges anyone he encounters to a fight with the promise that if vanquished he will reveal many medicines and the lucky victor will become skilled in the healing arts. Perhaps western medicine is currently in the process of dealing with a cultural Chiruwi. Another mythical figure relevant to the medical situation of the third millennium comes in the form of the Greek God of the mountainside, Pan. This playful and energetic god could be fearsome and he instilled in humans who ventured carelessly into his domain a sudden and groundless fright, or 'pan-ic'. On the other hand Pan offered bounty to those who honoured him. To these worshippers he would offer health and the wisdom to reach the universal source.

In an unconscious homage to Pan and the Chiruwi there is a distinct global movement to embrace something beyond the biomedical model of medicine that has been the beloved child of science. Global organisations, governments, science and individuals are turning in ever greater numbers to the complementary paradigm of medicine. In a sense, there is a metaphoric move towards honouring the nature god Pan in the hope that perhaps he will offer the wisdom that will lead to truly fundamental healing. This is not to say that there is a mood to discard totally the advances of biomedical medicine. A Luddite yearning for simpler days past will not move medicine forward. Rather what is occurring is the shaping of a post-modern medical paradigm.

TERMS OF ENGAGEMENT

To begin a discussion of the changes that are taking place in medicine it is necessary to define terms. For much of the twentieth century the biomedical model was the dominant medical paradigm and as suggested, refer to it as 'orthodox medicine'. It features a drug and surgical approach to disease management. Underpinning these curative tools is a scientific and reductionist attitude towards the human body that tracks disease to its biochemical roots. Essentially the orthodox philosophy sees medical science as applied biology and its practitioners as diagnostic biochemists. It is based on the principle of Cartesian Dualism which implies that, even if there were no mind in it, the nerves, muscles and blood vessels of the body would have the same functions. This separation of mind and body was espoused by Rene Descartes in the seventeenth century and has served as the underpinning of the orthodox approach since, that time.

The separation of mind and body paved the way for the development of a medical approach based on measurable quantities. This was what allowed medicine to become a science and possess directly predictive properties. The dualist notion was born just as the scientific revolution of the seventeenth century was under way. Isaac Newton's mechanistic theories were emerging and it was only natural that the orthodox science of medicine came to view the human body as a machine. Orthodox medicine has been incredibly successful. It has prolonged lives and, along with improvements in hygiene, been part of a social and medical milieu that has seen many diseases eradicated. Yet the latter part of the twentieth century and now the twentyfirst century has seen a widespread movement towards another medical model.

'Complementary medicine' is a term that encompasses a range of practices and philosophies. Some are ancient and some are relatively new but what they have in common is that they are based on theories or explanatory mechanisms that are not in keeping with the orthodox biomedical model. Some of the leading medical modalities that come under the heading of 'complementary medicine' including among their number herbal medicine, homoeopathy, acupuncture, nutrition, massage and counselling. These modalities are diverse but they hold common central attitudes. Complementary medicine rejects the dualism of orthodox philosophy while accepting the science of medicine. Fundamentally, complementary modalities embrace 'a concept of medicine as a science of the human person and insist on an understanding of disease as something that involves a systemic dislocation of the whole person, not just of the body'. Within complementary medical modalities there are variations on this philosophy yet the holistic view that health is a dynamic interplay between mind, body and spirit is common to all. Certainly the premises of complementary medicine are finding increasing support with scientists, governments and people the world over.

The People

On a global scale there is substantial and justifiable acceptance of the many benefits that orthodox medicine brings. Simultaneously, however, the western world, birthplace of orthodox medicine, is embracing complementary medicine in an emphatic way. The World Health Organisation (WHO) estimates that the global market for complementary therapies stands at 60 billion US dollars a year and is growing steadily. This represents a worldwide trend that is reflected in individual countries in significant ways.

Data from the British parliament support the idea that complementary medicine use in Britain is high and is increasing. In 1999 in Britain 93 million pounds was spent on complementary medicine and five million Britons visited the 50 000 complementary practitioners who operate there. By 2002 the amount that Britons spent on complementary medicines had risen to 126 million pounds. Australia is also experiencing an exponential increase in the use of complementary medicine. It is estimated that Australians spent 2.3 billion Australian dollars on complementary therapies and medicine in the year 2000, which is a 120 per cent increase on what they spent in 1993. Overall in 2000 Australian people spent four times more money from their pockets on complementary medicine than they spent on prescription drugs.

The same report indicated that in the USA expenditure on complementary medicines in 2000 was 34 billion US dollars. The popularity of complementary medicine with the American people is widely attested. Nearly 68 per cent of all American adults have used at least one complementary therapy at some time in their lives and of these 50 per cent continue to use complementary medicine twenty years later. High retention rates such as this reflect the fact that complementary medicine and its philosophy can become a way of life for those who use it. A more recent study based in El Paso, Texas, suggests that the number of people using complementary medicine may be still increasing. This study found that 77 per cent of those surveyed were using complementary medicine.

In France 75 per cent of the population has used complementary medicine at least once and in Germany 77 per cent of pain clinics provide acupuncture. In Germany as well, a study of adults with hayfever, asthma, eczema and food allergies found that more than one in four of these people used complementary medicine to treat their allergic condition. Of these people 78 per cent used the complementary treatments because they assumed there would be fewer side effects. Although the assumption that 'natural is safe' is a fallacious one, the perceived relative safety of complementary medicine is proving a significant factor in driving people to embrace it as part of their health regime. Equally, however, people clearly perceive that complementary medicine is efficacious. A study from a New York City pain clinic found that 85 per cent of patients had used complementary medicine to relieve their pain. 60 per cent believed that their complementary medicines had worked.

This points to the fact that the ground swell of support for complementary medicine is not arising out of some misguided search for absent nurturing or 'feel-good' treatments. When it comes to choosing health options individuals are pragmatic enough to insist that their treatments are efficacious. Often, people will turn to complementary medicine if they perceive that orthodox medicine is not delivering the health outcomes they desire. The surging tide of pragmatism that is seeing people across the world turn to a combination of orthodox and complementary medicine is leading science to address the wisdom that lies in these complementary therapies. In the face of immense public interest science is turning its attention to the question of health and health management. As this process occurs what is unfolding is that, both on a philosophical and fundamental level, science can embrace and learn from a 'complementary' approach to healing.

BODIES OF INTELLIGENCE

Complementary medicine treats the body as a holistic organism with dynamic processes that operate to keep that body in balance. Implicit to this paradigm is the acceptance that all parts of a human being are connected and that disease in one area can result from imbalance in another area. This conflicts with the traditional approach of orthodox medicine that targets symptoms in discrete body areas. Evolving scientific theory, however, is finding ways to encompass the holistic philosophy. One of the newer scientific world-views is complexity theory. This theory is applied to a variety of fields including ecology, physics, economics and computing where complex structures exist. In essence, complexity theory holds that once the rules governing these systems are found, it is possible to make effective predictions and even to control apparently complex entities. One salient point that is also emerging from complexity theory is that the properties of the system cannot necessarily be tracked to individual components but arise from the whole system. The implications of this notion for medicine are clear.

If complexity theory holds true then disease arises from the whole body, not its discrete parts. Effectively, pathology arises 'when the body self-organises in response to some disturbance but becomes confused and ends up worse at self-regulation than before'. In this light the mechanistic notions of orthodox medicine cannot hope to achieve cure but can only patch holes in the network that is the human body. Under complexity theory, genes, lifestyle and environment all interact to impact on the network that is the human body and the result is either disease or health. This paradigm is coincident with theories of holistic healing and reflects that at the true sharp end of science, the underpinnings of holistic medicine are being reinforced.

The complete coalescence of science and complementary theory will not happen immediately. Nevertheless the signs and symptoms that science is actively considering complementary medicine are manifold. Deliberate investigation of complementary medicines is taking place within the orthodox scientific establishment in many ways. There is a plethora of studies in the medical and scientific journals examining the components of complementary medicine from acupuncture to herbal medicine to yoga. A striking illustration of modern science meeting traditional methods is research in Wales that may result in a 600-year-old yielding new drugs.

In a village called Myddfai in Wales in the early thirteenth century a physician called Rhiwallon founded a line of healers that spread throughout the Welsh land. Over 500 remedies using more than 200 different plants. Modern pharmacists are engaged in examining the Myddfai cures to see what chemicals they may yield that can be turned into pharmaceutical drugs. Similar pursuits are occurring in the southern hemisphere. In Australia the Commonwealth Scientific and Industrial Research Organisation (CSIRO) has announced that it will be turning considerable attention to the investigation of the healing properties of herbs. In addition to examining traditional western herbs such as garlic, the CSIRO also intends to examine native plants for their economic and medicinal potential. The dilemma facing scientific enquiry into herbs is that isolating so-called 'active ingredients' from herbs violates holistic principles and is likely to have a negative effect on efficacy. Still, interest is the first step towards understanding. Herbs, however, are not the only ancient medicines that are experiencing a rising tide of scientific interest. If ever there was a symbol through which the orthodox medical establishment has expressed its disdain for the healing wisdom that had passed before, it was the leech.

The image of patients festooned with leeches has been emblematic of the supposed naivety of historical healing. Indeed, the leech was a popular medical tool throughout much of history. Babylonian writings suggest that the leech was being used medicinally as long as 3500 years ago. The Egyptians also embraced leeches as healing tools and the Europeans took up the annelids with equal verve. So popular was leech therapy with medieval European healers that physicians of this time were actually known as 'leeches'. It is estimated that by 1850 French physicians were using 100 million leeches per year so it is not surprising that in the nineteenth century the European leech was dancing along the edge of extinction.

Medicinal leeches were used over the centuries for bloodletting and were applied to congested or inflamed parts of the human body. Many ailments were subjected to the application of leeches. In America in the nineteenth century, leeches were used as a common home remedy to treat gum disorders and haemorrhoids, and to relieve the pain of large bruises. The advent of improved understanding of bacterio-logy and infection in the latter nineteenth century led to a significant focus on hygiene and a drop in popularity for the leech. Thus in the twentieth century, orthodox medicine scoffed at the apparently farcical adherence to the leech therapy of earlier centuries. In the last decade of the second millennium CE, however, leeches have made a medical comeback.

It has been found that in cases of venous congestion, where re-establishing the flow of blood is essential, leeches have great therapeutic value. As they consume their meal of blood, leeches promote blood flow through the tissue. Even after a leech is full of blood and detaches from the body, the anticoagulants it secretes into the tissue allow the wound to ooze blood for hours afterwards. These effects cannot be replicated by application to the site of anticoagulants such as heparin. Evidence is also growing to support the pain-relieving effects of leeches for conditions such as arthritis.

One study has evaluated the effectiveness of leeches in relieving arthritis of the knee. This study took place at the University of Essen, Germany, and involved sixteen patients with confirmed osteoarthritis of the knee. Ten of the patients were treated with leeches and six were controls. For those in the leech group, four leeches were applied around the knee joint for a period of 80 minutes each day. Both the control and the leech group were given conventional treatment for pain excluding non-steroidal anti-inflammatories. Those in the leech group experienced a rapid relief of pain which was faster than those in the control group, with sustained improvement after four weeks and no major complications.

This is just one of many studies reporting that leech therapy can reduce pain and improve joint mobility in arthritis. To testify that orthodox medicine is genuinely interested in leeches, there are moves afoot to develop mechanical leeches that secrete chemicals similar to real leeches but which are also insatiable. Time will tell whether a mechanical leech can provide all of the subtle benefits that the living creature can yield. This is not to particularly advocate leech therapy, but to illustrate that science is increasingly validating ancient medical wisdom.

Nutrients or Nutraceuticals

At the most basic level food is the root of health and, perforce, disease. The axiom that food should be synonymous with medicine has now passed into popular consciousness and parlance. Perhaps the most tangible manifestation of the acceptance by orthodox science and medicine of the curative power of nutrition is the development of what are known as 'functional foods' or 'nutraceuticals'.

Since, the British navy realised that limes could ward off scurvy, three centuries of investigation has revealed that food contains substances that are vital to health. In that time many vitamins and minerals have been discovered. In the early part of the twenty-first century, however, another revolution is taking place in the foods available to human beings. The modern diet is based on a few central plants that have been domesticated and bred to be increasingly bland. This narrowing and blanching of the human food chain has resulted in a relative deficiency of healthpromoting substances in the average western diet. The missing substances are neither nutrients nor toxins but lie somewhere in between and their lack is resulting in modern humans being 'pharmacologically impoverished'. To make up for this deficit, modern science has turned its attention to the development of 'nutraceuticals'.

These nutraceuticals are foods that richly supply the consumer with substances that are generally missing from modern food but which are health promoting. Leading among these are yoghurts containing Lactobadllus bacteria and margarines fortified with plant sterols. It is estimated that by the year 2006 the market in nutraceuticals will be worth 35.4 billion US dollars annually. This is not to advocate nutraceuticals, as the option of consuming a greater diversity of fruit and vegetables would be a premium, and far preferable, strategy in establishing health. Nevertheless this immensely popular sector of health management reflects clearly that orthodox science and the medicine that it inspires has embraced one of the premises of complementary medicine, which is the simple notion that lifestyle, including diet, has real and significant impacts on health. Thus on many levels orthodox medicine and complementary medicine are already moving together. At government and bureaucratic levels, this move is happening at an even more rapid pace.

MECHANISM OF DRUG ACTION

AGONISM AND ANTAGONISM

To appreciate fully the usefulness of receptor proteins as drug targets one needs to realize another important concept in drug action. Receptor proteins are primarily made for the recognition of endogenous ligands, which upon binding to the receptor proteins give rise to a cellular effect.

These ligands are called agonists. In many instances one can foresee the usefulness of exogenously administered (synthetic) agonists. For example, in patients with diabetes one can control the energy metabolism by injection of recombinant insulin to supplement the reduced endogenous levels. In asthmatic patients one can relieve airway problems by the inhalation of so-called β2-agonists that will relax the smooth muscle of the airways.

?The concept of physiological antagonism is important in drug action, but not as important as a more direct approach to counteract the action of endogenous signalling molecules, the use of receptor antagonists. It had already been recognized in the early days of medicinal chemistry and pharmacology that certain ligands do not produce an effect on their own upon interaction with a receptor protein (i.e. act as agonist). In contrast, these ligands can inhibit the action of endogenous signalling molecules by the simple binding to the appropriate receptor protein. This concept can be easily understood if one considers the 'lock-and-key' principle for ligand– receptor interaction. Only keys (ligands) that fulfil all criteria for fitting into the lock (receptor) can open the door (produce an effect). Some keys can fit in the lock, but not perfectly. Consequently, they cannot open the door yet, by fitting into the lock, these keys prevent the original key from fitting into the lock and opening the door.

The concept of receptor antagonism is extremely important in medicinal chemistry and is very often the underlying mechanism of drug action. To prevent e.g. the constriction of airway smooth muscle in asthmatic conditions one can e.g. administer receptor antagonists that prevent the actions of the signalling molecules causing muscle contraction (e.g. histamine and leukotriene antagonists).

RECEPTOR SUBTYPES

It is now well known that endogenous chemical messengers often use more than one receptor protein to elicit their effects. The fact that endogenous signals can act as ligand for more than one receptor protein substantially increases the versatility of signalling molecules.

The same signalling molecule can produce different effects in different tissues via interaction with different receptor proteins. Initially, receptor subtypes were discovered on the basis of differential sensitivity of specific physiological responses to drugs.

However, in the genomics era the discovery of new genes has revolutionised this area and has brought us a large number of previously unknown receptor proteins. This shift in paradigm is nicely illustrated by the developments in the field of histamine receptors.

The first evidence for the existence of a specific histamine receptor came from the work of Bovet and Staub (1937), who reported the weak competitive inhibitory effects of piperoxan against the effects of histamine. This discovery led to the development of several potent socalled 'antihistamines' which were able to inhibit the contraction of histamine of airway smooth muscle and to relieve certain symptoms of allergic disease. However, none of these compounds could block the stimulatory actions of histamine on e.g. gastric acid secretion. This led Ash and Schild to the hypothesis that there are probably two subtypes of histamine receptors referred to as the histamine H_1 and H_2 receptors. Both subtypes can be stimulated by histamine, but the actions of histamine can be antagonised by a subtype-specific antagonist.

This hypothesis was definitively established in 1972 when Nobel-prize winning Sir James Black and his co-workers identified burimamide a compound that competitively antagonised the actions of histamine on gastric acid secretion. Various burimamide derivatives have been developed as H_2 antagonists for the treatment of gastric ulcers and have been major block-busters for many years. The observation in 1983 that histamine can inhibit its own synthesis and release from brain slices suggested the existence of a histamine H_3 receptor; this hypothesis was confirmed by the discovery of unique agonists and antagonists for this receptor as well. From these developments one can learn that for many years receptor classification has relied on the development of receptor subtype selective agonists and antagonists. However, with the introduction of molecular biology in the area of receptor research, this paradigm has completely changed. In the last 15 years most new receptors have been suggested on the basis of the discovery of new genes.

Again, the field of histamine receptors nicely illustrates this development. Until Autumn 2000 three genes encoding the previously known H_1, H_2 and H_3 receptor were cloned. A new gene, encoding an H_4 receptor subtype, was recently identified in a genome database (so-called 'in silico cloning'), as a gene that was very similar to the H_3 receptor gene.

The encoding receptor is expressed in the intestine and immune cells and binds several histamine receptor ligands with a distinct profile. The in vivo function of this new receptor is so far unknown, but potentially offers new options for therapeutic intervention. Often it is seen that endogenous ligands mediate their actions through one or more receptors of just one superfamily. For example, all of the known histamine receptors (H_1–H_4) belong to the family of G-protein coupled receptors. However, there are some important exceptions to this rule. Within the family of (simple) biogenic amine neurotransmitters important differences can already be noticed. Whereas histamine, dopamine and noradrenaline act only via (various) G-protein coupled receptors, acetylcholine and serotonin act via both G-protein coupled receptors and ligand-gated ion channels.

DRUG SELECTIVITY

To understand drug action one should be aware of the important issue of drug selectivity. In general, a drug is known for its activity at one target. However, some drugs can display activity at a variety of targets. The specificity of the ligand–receptor interaction can be compared with the combination of a lock with specific keys. Some locks can accommodate a variety of keys, whereas the opposite is also true. Some keys can fit more than one lock. Often such drugs show preference for one type of receptor, but at higher concentrations they can also interact with other targets.

The three structurally closely related H_3 agonists imetit, immepip and (R)-α-methylhistamine are e.g. all equipotent at the H_3 receptor (a Gprotein coupled receptor), but in vivo two of them also show agonistic activity at either the serotonin $5HT_3$ receptor (a ligand-gated ion channel; imetit) or the α_2-adrenergic receptor (a G-protein coupled receptor; (R)-α-methylhistamine). This example illustrates that for newly developed drugs one needs to know the full spectrum of its biological activities at a variety of known targets in order to be able to explain its action in complex physiological systems. Of course the in vivo application is the most challenging one in this respect. Often drug side-effects can be explained on the basis of a known interaction with another receptor site.

DRUG DISCOVERY

It is fitting to include in *Neuropsychopharmacology: The Fifth Generation of Progress* a group concerned with many of the important issues related to:

- The development of new drugs (*e.g.*, the role of preclinical models to develop and test targets for new drugs;
- The role of biological markers and imaging studies to provide indices of drug action in humans or predict clinical response;
- Advances in clinical trial design;
- The optimal utilisation of existing drugs (*e.g.*, pharmacokinetics);
- Pharmacogenomics in relation to drug metabolism, discovery and development;
- The ethical issues concerning clinical trials and neuropsychiatric research in general;
- Governmental (at least United States) regulation of the process of developing and utilising new drugs; and
- The evaluation of the impact of drug treatment on outcome of neuropsychiatric disorders from the perspectives of economics, clinical endpoints and humanistic considerations.

Although the majority of these issues are the same in many parts of the world where this book will be read, there are clear differences in ethical attitudes towards clinical research and governmental regulation of drug development and utilisation around the world (despite the Declaration of Helsinki). It is beyond the scope of this volume to consider the worldwide variations in these issues. New drug development depends so heavily on our understanding of the function of the brain in general, neurotransmitter and modulator receptors and elimination mechanisms, theories of the etiology of the major neuropsychiatric disorders, efficacy and side effects of existing treatments and an understanding of the mechanism of action of existing drugs.

The process of new drug development has changed greatly in the few years since, the last volume in this series and is likely to change even more rapidly in the immediate future. A great advance was the development of combinatorial chemistry and rapid, robotic characterisation of the pharmacologic profile of the vast libraries of compounds produced by this shotgun approach, which, when successful, leads to the elegant and expensive custom syntheses of candidate compounds by sophisticated organic chemical procedures that are now often computer-derived or by methods involving cell and molecular biology to produce peptides or other organic substances.

Together with greatly improved methods for analysis of structure–activity relationships, it has been possible to develop putative pharmaceuticals with the desired pharmacologic profile. The old cliche′, beware what you desire because you may get it, is relevant here, because it much easier now to come up with the desired pharmacologic profile than it is to be certain that what is sought is what should be sought. There is not yet sufficient understanding of what is needed in the way of an optimal antidepressant, anxiolytic, antipsychotic, mood stabiliser, antidementia, or other type of drug, especially when a truly novel compound is sought. Geyer and Markou on the role of preclinical models in the development of psychotropic drugs mainly focuses on animal models for the major psychiatric disorders.

These authors point out that this approach may seem somewhat old-fashioned compared to approaches such as high throughput screening and utilisation of molecular biological techniques to develop targets based on gene expression and identification methods; however, they correctly state that preclinical models are *required* (emphasis added) to provide initial assessment of the functional effects of novel compounds in the integrated organism.

We are not yet to the point where new chemical entities go directly into patients or even normal volunteers without some evidence that clinically relevant effects might be present. We can expect major advances in the development of preclinical models as our knowledge of disease processes and our ability to alter the genome in laboratory animals increase. Knockout and knockin mouse models will increasingly guide drug discovery and testing.

The importance of research designed to identify new drug targets based on the Human Genome Project and the ensuing effort to characterise the genes involved in neuropsychiatric disorders and the action of drugs used to treat neuropsychiatric. The use of biomarkers (*i.e.*, natural history markers), biological activity marker and surrogate markers is thoroughly explored by Wong and colleagues, the importance of biomarkers as a means to reduce the cost of drug development, improve the ability to predict outcome and expedite the identification of desired endpoints (*e.g.*, no more than 80 per cent occupancy of striatal D2 receptors in order to minimize the development of extrapyramidal side effects), is increasing all the time. The extraordinary development of a variety of brain imaging methods, including magnetic resonance imaging, functional magnetic resonance imaging, single photon emission computed tomography and positron emission tomography, appears to be particularly suited for this purpose. Given the cost associated with a failed clinical trial, someday it may be possible to bring brain imaging into routine clinical practice to guide drug dosage and choice.

However, more classical methods such as neuroendocrine testing or examining the effect of treatments on peripheral processes such as changes in saliva, serum and blood cells still can be valuable at various stages of drug development. Clinical trials abound in psychiatry. Good clinical trials are much more rare. Problems in trial design, identification of appropriate patients, recruitment, retention and ethical issues surrounding the use of placebos are very much with us and show signs of becoming more rather than less intractable in the near future.

The cost of clinical trials in Western countries has grown enormously, leading to fewer and smaller trials that are often market-driven rather than designed to answer the most important research questions. Kane describes a number of efforts that have been made to improve clinical trial design and to cope with the increasing limitations that current ethical viewpoints have placed on this process.

It is a sign of progress that broader outcome measures other than global psychopathology are increasingly the focus of clinical trials. For example, the recognition that cognition may be a more important endpoint than the reduction of positive or negative symptoms in the evaluation of a new drug for schizophrenia is an enormous advance because it refocuses the goal of new drug development and allows for distinguishing between new and existing drugs on a much more meaningful basis. Advances in this area yield the information needed to use drugs wisely.

This area of research has matured to the point where the fundamental principles are well understood and can be readily incorporated into the processes of drug development and utilisation. Information about drug interactions that affect efficacy, elimination and toxicity are ever more essential in the current area of polypharmacy. O¨ zdemir and colleagues pharmacogenetics, the field that explores individual differences in drug responses that depend on genetic factors and genetic–environmental interactions. A key part of this refers to genetic variations in the liver enzymes that metabolise drugs. Greenblatt and colleagues consider this as well as the genetic factors that determine pharmacodynamics and thus directly impact on efficacy and side effects.

It is clear that this is critically important to psychopharmacology and will become even more so with the completion of the Human Genome Project. They also consider how genomic research will play an increasingly important role in drug discovery and development, including the design of safer and more efficient clinical trials. Their term "personalised therapeutics" is provocative. Have doctors not tried to do this since, time immemorial?

Genetic information will aid the process to be more science than art (rather than the reverse). Is medicine in need of new ethical compasses in clinical research? This would seem to be the view of the authors of the United States-based National Bioethics Advisory Commission (NBAC) or the newest version of the Declaration of Helsinki. NBAC, in particular, has singled out research on the mentally ill for more stringent regulation and unique standards. The ethical aspects of neuropsychiatric research by Pinals and Appelbaum thoughtfully analyses these recommendations as well as the fundamental principles that should guide policy in this area.

Professional societies such as the ACNP have developed guidelines for investigators in an effort to show that there is an awareness of the obligation to protect subjects who agree to research to the greatest possible extent with minimal diminution of the information to be gained from the study.

A "Safe" study from which we can learn little or nothing of use because of design flaws may be less ethical than others where the absolute risk may be greater, but still within acceptable limits, whereas the potential gain in knowledge is far higher. Institutional review boards (IRB) are becoming increasingly restrictive around clinical research with the mentally ill, based in part on a poor understanding of the intactness of their decisional capacities. It is imperative that methods to assure competence to give consent that meets the legitimate concerns of IRBs are employed in all trials. Paul Leber, formerly head of the Division of Psychopharmacology of the United States Food and Drug Administration (FDA), explicates the policies of his former employer with the oratorical flourish he is renowned for now transmuted into the written word with equal elegance.

Mahmoud and colleagues succinctly and clearly describe the process of evaluating treatment outcomes utilising the Economic, Clinical, Humanistic outcomes (ECHO) model. These three perspectives on outcome are intimately intertwined. Trouble arises when any one of the aspects is overemphasised to the neglect of the others.

There has been increasing awareness of the need for societal consensus on the importance of outcomes as the cost of achieving the best outcomes now possible has risen greatly as a percentage of gross national product in both developing and developed countries. Advances in medical research will surely suffer if there is insufficient attention to demonstrating that new therapies are more cost-effective, not just more effective, than previous methods.

The critical issue of distinguishing between efficacy and effectiveness research and the need for both are thoroughly. In conclusion, this section on new drug development and clinical research issues should be of great importance to any reader who is interested in the broader picture of alleviating neuropsychiatric disease burden through psychophar-macology.

HISTORICAL EVOLUTION

With the advent of computers and the ability to store and retrieve chemical information, serious efforts to compile relevant databases and construct information retrieval systems began. One of the first efforts to have a substantial long-term impact was to collect the crystal structure information for small molecules by Olga Kennard.

The Cambridge Structural Database (CSD) stores crystal structures of small molecules and provides a fertile resource for geometrical data on molecular fragments for calibration of force fields and validation of results from computational chemistry.

As protein crystallography gained momentum, the need for a common repository of macromolecular structural data led to the Protein Data Base (PDB) originally located at Brookhaven National Laboratories. These efforts focused on the accumulation and organisation of experimental results on the three-dimensional structure of molecules, both large and small. Todd Wipke recognised the need for a chemical information system to handle the increasing numbers of smallmolecules generated in industry and thus MDL and MACCS were born.

With the advent of computers and the availability of oscilloscopes, the idea of displaying a three-dimensional structure of the screen was obvious with rotation providing depth cueing. Cyrus Levinthal and colleagues utilised the primitive computer graphics facilities at MIT to generate rotating images of proteins and nucleic acids to provide insight into the three-dimensional aspects of these structures without having to build physical models.

Scientific American in 1965 was sensational and inspired others (including myself) to explore computer graphics (1966/1967) as a means of coping with the 3D nature of chemistry. Physical models were useful accepted tools for medicinal chemists, but physical overlap of two or more compounds was difficult and exploration of the potential energy surface hard to correlate with a given conformation of a physical model. As more and more chemical data accumulated with its implicit information content, a multitude of approaches began to extract useful information. Certainly, the shape and variability in geometry of molecular fragments from CSD was mined to provide fragments of functional groups for a variety of purposes. As series of compounds were tested for biological activity in a given assay, the desire to distill the essence of the chemical requirements for such activity to guide optimisation was generated. Initially, the efforts focused on congeneric series as the common scaffold presumably eliminated the molecular alignment problem with the assumption that all molecules bound with a common orientation of the scaffold. This was the intellectual basis of the Hansch approach (quantitative structure-activity relationships, QSAR), in which substituent parameters from physical chemistry were used to correlate chemical properties with biological activity for a series of compounds with the same substitution pattern on the congeneric scaffold.

KNOWN VERSUS UNKNOWN TARGETS

Intellectually, the application of molecular modeling has dichotomised into those methods dealing with biological systems where no structural information at the atomic level is known, the unknown receptor and those systems that have become relatively common, where a three-dimensional structure is know from crystallography or NMR spectroscopy. The Washington University group has spent most of its efforts over the last three decades focused on the common problem encountered where one has little structural information. Others, such as Peter Goodford and Tak Kuntz, have taken the lead in developing approaches to therapeutic targets where the structure of the target was available at atomic resolution. The seminal work of Goodford and colleagues on designing inhibitors of the 2,3-diphosphorylglycerate (DPG) binding site on hemoglobin for the treatment of sickle-cell disease certainly stimulated many others to obtain crystal structures of their therapeutic target. The most dramatic example of computer-aided drug design of which I am aware is the development of superoxide dismutase mimetics of below 500 molecular weight by Dennis Riley of Metaphore Pharmaceuticals. By understanding the redox chemistry of manganese superoxide reductase, Riley was able to design a totally novel pentaazacrown scaffold complexed with manganese that catalyses the conversion of superoxide to hydrogen peroxide at diffusion-controlled rates.

This is the first example of a synthetic enzyme with a catalytic rate equal to or better than nature's best. The advances in molecular biology provided the means of cloning and expressing proteins in sufficient quantities to screen a variety of conditions for crystallisation. Thus, it is almost expected that a crystal structure is available for any therapeutic target of interest. Unfortunately, many therapeutic targets such as G-protein-coupled receptors are still significant challenges to structural biology.

Graph Theory and Molecular Numerology

Considerable litreature developed around the ability of numerical indices derived from graph theoretical considerations to correlate with SAR data. This was a source of mystery to me for some time. A colleague, Ioan Motoc, from Romania, with experience in this arena and a very strong intellect, helped me understand the ability of various indices to be useful parameters in QSAR equations. Ioan correlated various indices with more physically relevant (at least to me) variables such as surface area and molecular volume. Since, computational time was at a premium during the early days of QSAR and such indices could be calculated withminimal computations, they played a useful role and continue to be used. As a chemist, however, I am much more comfortable with parameters such as surface area or volume.

Pharmacophore

The success of QSAR led to efforts to extend the domain to non-congeneric series, where the structural similarity between molecules active in the same bioassay was not obvious. Certainly, the work of Beckett and Casey on opioids to define those parts of the active molecules (pharmacophoric groups) essential for activity was seminal. Kier further developed the concept of pharmacophore and applied it to rationalise the SAR of several systems. Peter Gund and Todd Wipke implemented the first *in silico* screening methodology with a programme to screen a molecular database for pharmacophoric patterns in 1974. LeslieHumber of Ayerst inMontreal exposed me to the wide variety of structures active in the dopaminergic system. Overlaps of apomorphine, chlorpromazine and butaclamol to align the amines while maintaining the coplanarity of an aromatic ring led to a plausible hypothesis of a receptor-bound conformation.

The least-squares fitting of atomic centers did not allow such an overlap, but the use of a centroid of the aromatic ring with normals to the plane for leastsquares fitting accomplished the overlap. There still continues to be developments of methods to generate overlaps of hydrogen-bond donors and acceptors, aromatic rings and so on, to generate a pharmacophore hypothesis from a set of compounds active at a given receptor/enzyme. One method developed early at Washington University was minimization of distances between groups in different molecules assigned by the investigator with no intermolecular interactions. In effect, adding springs to cause the groups to overlap as the energy of the entire set of molecules was minimized excluding any interatomic interactions except those imposed by the springs. As a minimization procedure, the results were dependent on the starting conformation of the set of molecules being minimized and multiple starting conformations were used to generate multiple pharmacophoric hypotheses.

Index